This book gives a broad synthesis of conceptual developments of 20th century field theories, from the general theory of relativity to quantum field theory and gauge theory. The author gives a historico-critical exposition of the conceptual foundations of the theories, revealing a pattern in the evolution of these conceptions of nature.

Theoretical physicists and students of theoretical physics will find in this book an account of the foundational problems of their discipline that will help them understand the internal logic and dynamics of their subject. In addition, the book will provide professional historians and sociologists of science, and especially those of physics, with a conceptual basis for further historical, cultural, and sociological analyses of the theories discussed. The book also contains much material for philosophical (metaphysical, methodological, and semantical) reflection. Finally, the scientifically qualified general reader will find in this book a deeper analysis of contemporary conceptions of the physical world than can be found in popular accounts of the subject.

This fascinating account of a key part of contemporary mathematical physics will be of great interest to physicists, historians, philosophers, and sociologists of science at a range of levels from the professional researcher at one end of the scale to the educated general reader at the other.

In this profound but readable treatise, Professor Cao observes that a consistent description of all observed phenomena of the microworld is at hand. This so-called standard model emerged from a complex interplay between experiment and theory. In recent decades, experimentalists, with immense fortitude and generous funding, identified and studied what appear to be the basic building blocks of matter and the forces they experience. Meanwhile, theorists created and refined a mathematical framework – quantum field theory – in terms of which the standard model is expressed. In all confrontations with experiment, the standard model emerges triumphant. And yet, too many questions remain unaddressed for it to be the last word. Many theoretical physicists believe an entirely new theoretical system to be needed – superstring theory, or something like it.

In the course of these developments, the conceptual basis of the present theory has become obscure. Most physicists are too much concerned with the phenomena they explore or the theories they build to worry about the logical basis and historical origin of their discipline. Most philosophers who explore the relationship between scientific knowledge and objective reality are ill-equipped to deal with the intricacies of quantum field theory, let alone its successor. Cao argues persuasively that we must first understand where we are and how we got here before we can build a better theory or even comprehend the true meaning of the standard model. His lucid account of the development and interpretation of both classical and quantum field theories, culminating in the creation of a gauge field theory of all the forces of nature, will enable both physicists and philosophers to address the questions of what science is and how it evolves.

Sheldon Lee Glashow, Harvard University

Tian Yu Cao's book confronts an essential problem of physics today: field theory is no longer advancing our understanding of Nature at its fundamental workings. In principle, present-day field theory explains all observed phenomena in terms of the quantal 'standard model' for particle physics and the classical Newton–Einstein model for gravity. Yet this structure is manifestly imperfect: the success in particle physics is built on ad hoc, unexplained hypotheses, while classical gravity has not been integrated into the quantum description of non-gravitational phenomena. It has not been possible to make progress on these questions within field theory, which for over a century has provided the framework for presenting fundamental physical laws. In the absence of new experiments to channel theoretical speculation, some theorists have adopted a concept of mathematical elegance and necessity to guide their invention of a new physical idea: string theory. But the inability to test experimentally whether string theory can be a viable replacement for quantum field theory has dismayed other physicists, and prompted science journalists to question our commitment to scientific realism. At this critical point in the development of our discipline, the book provides a welcome overview of field theory, clearly recording the path taken to the present, conflicting positions. Cao belongs to the new breed of historians/philosophers of physics who are thoroughly familiar with the technical complexities of the modern material. His discussion is accurate and rich in detail and insight. The scientifically literate and interested reader will acquire an excellent description of present-day fundamental physical theory. The historically/philosophically minded will find penetrating discussion of the logic behind the ideas. Practicing researchers and students will enjoy a scrupulously researched retelling of the history of our subject.

Roman Jackiw, Massachusetts Institute of Technology

The work is an impressive tour de force, combining masterly understanding of the many technical details of modern field theories, including general relativity, quantum field theory and gauge theory, together with a spirited philosophical defence of the rationality and objectivity of theoretical physics as captured in Cao's espousal of a structural realist position. The book provides a powerful antidote to the postmodern fashion in contemporary history and philosophy of science. It is equally suitable for physicists who want to make some coherent sense of the complicated history of modern developments in field theory, and for historians and philosophers of science who want to base their judgements on technically accurate presentations of the issues.

Michael Redhead, Cambridge University

Following in the footsteps of Ernst Mach and John Merz, Tian Yu Cao has given a masterful, incisive exposition of the conceptual developments of twentieth century foundational theoretical physics, of their historical roots and of their limits. It is a profound critical inquiry into the metaphysical, philosophical and technical assumptions that underlie the physical theories that so impressively and accurately describe nature in the domains thus far accessible to experimental probing: the special and general theory of relativity and quantum field theory, and, in particular, the relativistic quantum gauge field theories that are the essential components of the standard model. As with Mach, Cao's exposition demonstrates how a philosophical inquiry that is sensitive to history can illuminate physical theories, and his book may well prove to be a valuable guidepost in helping chart the future path of theorizing in fundamental physics. It will be eagerly read by physicists, by historians and by philosophers of science.

Silvan S. Schweber, Brandeis University

For Rosa

Conceptual developments
of 20th century field theories

Conceptual developments
of 20th century field theories

TIAN YU CAO

Department of Philosophy,
Boston University

CAMBRIDGE
UNIVERSITY PRESS

PUBLISHED BY THE PRESS SYNDICATE OF THE UNIVERSITY OF CAMBRIDGE
The Pitt Building, Trumpington Street, Cambridge CB2 1RP, United Kingdom

CAMBRIDGE UNIVERSITY PRESS
The Edinburgh Building, Cambridge CB2 2RU, UK http://www.cup.cam.ac.uk
40 West 20th Street, New York, NY 10011-4211, USA http://www.cup.org
10 Stamford Road, Oakleigh, Melbourne 3166, Australia

First published 1997
First paperback edition (with corrections) 1998

Printed in the United Kingdom at the University Press, Cambridge

Typeset in Linotron Times 11/14pt

A catalogue record for this book is available from the British Library

Library of Congress Cataloguing in Publication data

Cao, Tian Yu, 1941–
    Conceptual developments of 20th century field theories / Tian Yu
Cao.
        p. cm.
    Includes bibliographical references and index.
    ISBN 0 521 43178 6
    1. Unified field theories.   I. Title.
QC794.6.G7C36   1996
530.1′4′0904 – dc20   96-25090 CIP

ISBN 0 521 43178 6 hardback
ISBN 0 521 63420 2 paperback

Contents

Preface

The aim of this volume is to give a broad synthetic overview of 20th century field theories, from the general theory of relativity to quantum field theory and gauge theory. These theories are treated primarily as conceptual schemes, in terms of which our conceptions of the physical world are formed. The intent of the book is to give a historico-critical exposition of the conceptual foundations of the theories, and thereby detect a pattern in the evolution of these conceptions.

As an important component of culture, a conception of the physical world involves a model of the constitution and workings of nature, and includes assumptions about the mechanisms for fundamental interactions among the ultimate constituents of matter, and an interpretation of the nature of space and time. That is, the conception involves what philosophers usually call metaphysical assumptions. Talking about metaphysics is out of fashion these days. This is particularly so in the profession of science studies, where the primary concern now is with local and empirical successes, social interests, and power relations. Who would care for the ontological status of curved spacetime or virtual quanta when even the objective status of observed facts is challenged by the social constructivists? However, as we shall see in the text, metaphysical considerations are of crucial importance for path-breaking physicists in their investigations. One reason for this is that these considerations constitute essential ingredients of their conceptual frameworks. Yet the cultural importance of metaphysics goes much deeper and wider than its contribution to professional research. My own experience might be illuminating.

When I began to study theoretical physics after reading the philosophical writings of Descartes, Kant, Hegel, Russell, Einstein, Heisenberg, and David Bohm, I was attracted to physics purely by cultural curiosity, trying to obtain a picture of the physical world endorsed by the most recent developments in physics. I was told that the Newtonian picture was not a proper one, that in the 19th century the mechanical worldview was replaced by the electromagnetic one, essentially a field-theoretical picture of the world. I also learnt that the 20th century had witnessed two profound conceptual and ontological revolutions in the physical sciences, which were brought about by relativity theories and quantum theories. As a result, we are equipped with new conceptual frameworks for probing the foundations of the physical world. But what about an integrated picture of the world suggested by these revolutionary theories? When I began doing research in the history and philosophy of science twelve years ago in Cambridge, England, I tried in vain to find such a picture in works by physicists, or by the philosophers and historians of 20th century physics.

Of course I have learnt a lot from Ernst Cassirer, Moritz Schlick, Hans Reichenbach, Karl Popper, Gerald Holton, Adolf Grünbaum, Howard Stein, John Earman, John Stachel, Martin Klein, Thomas Kuhn, John Bell, Abner Shimony, Arthur Fine, Michael Redhead, and many other scholars. For example, I have learnt that some metaphysical presuppositions, such as the principles of universality and correspondence, played an important heuristic role in theory construction by the founders of the revolutionary theories. I have also learnt that for most educated people, some of the metaphysical implications of these theories, such as the absence of the mechanical ether, the removal of the flatness of spacetime, and the impossibility of causal and spatio-temporal descriptions of individual events in the microscopic world have been accepted as important parts of our picture of the world. Yet nowhere could I find an integrated picture, let alone a cogent exposition of its evolution and of the pattern and direction of the evolution. I decided to fill this gap, and the result is this volume.

The book is written primarily for students of theoretical physics who are interested in the foundational problems of their discipline and are struggling to grasp the internal logic and dynamics of their subject from a historical perspective. But I have also done my best to make the text accessible to general readers with a basic scientific education who feel that their cultural curiosity concerning the contemporary conception of nature cannot be satisfied by popular writings. The last audience I have in mind are mainstream historians and philosophers of science. Although the book has provided a basis for further cultural and sociological analyses of these theories, and contains much material for philosophical reflection, the project pursued in this volume under present circumstances is unlikely to be interesting or even acceptable to these scholars. The disagreement comes from different conceptions of science. Detailed arguments against current positions will be given in the introductory and concluding chapters. Here I just want to highlight a few points at issue.

For many science studies scholars, any discussion of a world picture in terms of the ultimate constituents and hidden mechanisms posited by a scientific theory as underlying empirical laws seems equal to presupposing a naive realist position on the unobservable entities and structures of the theory, and this is simply unacceptable. This anti-realist position has a long tradition. For the classical positivists, any statements about unobservables, such as atoms or fields, that go beyond the scope of empirical evidence or logical inference are meaningless and have to be expelled from scientific discourse; thus the world picture problem is a pseudoproblem. For constructive empiricists or sophisticated instrumentalists living in the post-empiricist period, theoretical terms for describing hypothetical unobservables are permitted but accorded no existent status, because these terms are merely tools for saving phenomena and making predictions, or a kind of shorthand for observables. Then a question facing them is what the sources are of the effectiveness of these tools.

To answer this question requires a clarification of the relationship between the tools and the external world. No such clarification has ever been provided by the instrumentalists. Nevertheless, they have tried to discredit the realist interpretation of theoretical terms by appealing to the so-called Duhem–Quine thesis of underdetermination, according to which no theoretical terms can be uniquely determined by empirical data. The convincing power of the thesis, however, entirely rests on taking empirical data as the single criterion for determining the acceptability of the ontology posited by a theory. Once empirical data are deprived of such a privileged status, the simplistic view of scientific theory as consisting only of the empirical, logico-mathematical, and conventional components is replaced by a more tenable one, in which a metaphysical component (e.g. the intelligibility and plausibility of a conceptual framework) is also included and taken to be a criterion for the acceptability of a theory, thus putting scientific theories in a wider network of entrenched presuppositions of the times and in a pervasive cultural climate, then the Duhem–Quine thesis alone is not powerful enough to discredit the realist interpretation of theoretical terms.

More radical is Kuhn's position. If the Duhem–Quine thesis accepts the existence of a multiplicity of conflicting theoretical ontologies, all of which are compatible with a given set of data, and thus nullifies the debate on which ontology should be taken as the real one, Kuhn (1970) rejects the reality of any theoretical ontology. He asks, since whatever ontology posited by a scientific theory is always replaced by another different and often incompatible ontology posited by a later theory, as the history of science seems to have shown us, and there is no coherent direction of ontological development, how can we take any theoretical ontology as the real ontology of the world? Yet a historical fact is that explicitly or implicitly some hypothetical ontologies are always posited in the theoretical sciences. Thus a problem facing Kuhn is why a theoretical ontology is so indispensable in the theoretical structure of science.

Kuhn's work has produced some resonances. Unlike the logical empiricists, who are preoccupied exclusively by abstract logical and semantical analyses of scientific theories, Kuhn has tried to develop his view of science on the basis of the historical examination of actually existing theories. However, some of his followers are not so happy with his limiting scientific practice solely to conceptual aspects. They passionately cry out for the importance of experiments and institutions, social interests and power relations, and so on. In my opinion, however, Kuhn is essentially right in this regard: the core of scientific practice lies in theory construction and theoretical debates. Experiments are important, but their importance would be incomprehensible without being put in a theoretical context. All external factors would be interesting and valuable for our understanding of science, but only if they had some bearing on a theory, on the genesis, construction, acceptance, use, and consequences of a theory. Otherwise they would be irrelevant to our understanding of science. In this regard, Paul Forman's work (1971) is important because it describes the cultural climate in Germany that helped the acceptance of the notion of acausality developed in quantum mechanics, although it does not touch upon the question whether the cultural climate had played any constitutive role in the formation of the notion of acausality.

Questions similar to this are taken up and answered affirmatively by the advocates of the strong programme in the sociology of science, who uphold that science is a social construction (see D. Bloor, 1976; B. Barnes, 1977; A. Pickering, 1984). In a trivial sense few people at present would dispute with them the socio-constructive character of science. Yet the really interesting point at issue is their special position concerning nature. If nature is assumed to play no role in the construction of science, then the social constructivists would have no theoretical resources to address questions concerning the truth status and objectivity of scientific theories, and relativism and skepticism would be inescapable. Yet if nature is allowed to have a role to play in the construction of science, then science is more than a social construction, and the social constructivists would have achieved little in illuminating the nature of science as knowledge of nature.

The latest fashion in science studies follows a radical version of social constructivism that has absorbed much of its rhetoric from the cultural fad of postmodernism. The faddists take science only as an art of rhetoric for persuasion, manipulation, and manufacture of facts and knowledge; knowledge only as a power move, having nothing to do with truth or objectivity; and objectivity only as an ideology, having nothing to do with how scientific knowledge is actually made. They argue that the important tasks for science studies scholars are not to find out who discovered facts and who constructed concepts and theories, but rather who controlled laboratories; not to explain why science works in the sense that the redshift predicted by the general theory of relativity really can be observed, but rather to question who benefitted from science. A problem for the faddists is that they can only talk to each other, and will never be able to talk to scientists seriously about their major practice, that is, about their theoretical activities.

Another influential position is Putnam's (1981) internal realism. This allows us to talk about abstract entities, truths, and reality, but only within a theoretical framework. Since any talking is always conducted within a certain framework, it seems impossible to escape from this position. It should be noticed that this position has a close kinship with Carnap's position on linguistic frameworks (1956). Both reject the external question concerning the objective reality of theoretical entities independent of our linguistic frameworks. The justification for the position, as Putnam puts it, lies in the claim that we have no access to metaphysical reality if it exists at all. Assuming that the ontologies posited by successive theories can be shown to have no connection to each other, then this position is indistinguishable from Kuhn's position. But what if there is a coherent direction of the evolution of ontological commitments in successive theories? Then Putnam will have to face the old question raised by the realists: what is the noumenal basis for the coherent direction?

Thus to justify a conceptual history of physics focusing on the underlying assumptions about the ultimate constitution and workings of nature, we have to answer two questions. First, why are these metaphysical assumptions indispensable for physics? Second, do we have any access to metaphysical reality? An affirmative answer to the second question will be given in the text. Here is a brief outline of my position on the first question.

As is well known (see E. A. Burtt, 1925; A. Koyré, 1965), at the end of the medieval period there was a decline of Aristotelian philosophy and a revival of Neoplatonism with a Pythagorean cast. The latter took mathematics as the foundation of reality and the universe as fundamentally mathematical in its structure. It was assumed that observable phenomena must conform to the mathematical structures, and that the mathematical structures should have implications for further observations and for counterfactual inferences which went beyond what were given. Since then there has been a strong tendency, particularly among mathematical physicists, to take mathematical structures as conceptual frameworks for describing the elementary entities of the physical world and their behaviors.

Another aspect of the changing metaphysics in the same period was a replacement of the final cause by the efficient cause in the conception of causality, which was concomitant with the replacement of the power of authority by the power of rational, i.e. causal, reasoning, that is, with the rise of scientific rationality itself. Thus forces, instead of the Aristotelian telos, as agents of causality were taken to be the metaphysical foundations of natural phenomena. In some sense, all subsequent developments in physics can be regarded as being driven by searching for a model, mechanical or otherwise, for describing forces, understood as causal agents.

The concurrence of these changes led to a rise, in the 17th century, of a hypothetico-deductive method in physics, as developed by Descartes, Boyle, and, to some extent, Newton, for explanation and prediction. It is in this particular structure of physical theory that we can find a deep root for the indispensability of ontological assumptions. Forces, fields, the ether, rigid or dynamical spacetime, virtual quanta, confined quarks, gauge potentials, all these hypothetical (at certain stages of development, they were called metaphysical) entities are indispensable for theoretical physics because they are required by the historically emergent hypothetico-deductive method that is inherent in the discipline. The assumption of some ultimate ontology in a theory provides the basis for reducing some set of entities to another simpler set, thus endowing the theory with a unifying power. No proper understanding of theoretical physics and its power would be possible without paying enough attention to this characteristic of its theoretical structure. In this regard, I think Meyerson (1908) is right when he holds that modern science as an institution which emerged since the time of Copernicus is only a further stage of a natural metaphysics, whereby commonsense assumes the existence of permanent substance underlying observable phenomena.

The treatment of field theories in this volume is highly selective. Considering the rich content of the subject it could not be otherwise. The selection is guided by my view of scientific theory in general and my understanding of field theories in particular. These have provided perspectives from which various topics are examined and interpreted, and thus have determined, to a large extent, the significance of the topics in the evolution of the subject. The general framework in which material is selected and interpreted relies heavily on some organizing concepts, such as those of metaphysics, ontology, substance, reality, causality, explanation, progress, and so on. These concepts, however, are often vague and ambiguous in the literature. To clear the air, I devote chapter 1 to spelling out my usage of these concepts and to addressing some topics of methodological importance. The departure point of the story, namely the rise and crisis of classical field theories up to Lorentz's work, is outlined in chapter 2. The main body of the text, in accord with my understanding of the structure of the developments that I wish to elaborate, is divided into three parts: the geometrical programme, the quantum field programme, and the gauge field programme. Each part consists of three chapters: prehistory, the formation of conceptual foundations, and further developments and assessment. The philosophical implications of the developments, especially those for realism and rationality, are explored in the concluding chapter.

A remark about the Bibliography. Only those works that have actually been used in the preparation of this volume are listed in the Bibliography. In addition to original contributions of crucial importance, recent scholarly works that provide interpretations of the original works are also listed in the Bibliography. Yet no effort has been made to offer an exhaustive bibliography of the secondary literature; only those works having direct bearing on my interpretation of the subject are included. As to the general background of the intellectual history of modern times, to which the first two chapters frequently refer, I simply invite readers to consult a few of the outstanding historiographical works instead of giving out detailed references to the original texts, which in fact can be found in the books suggested.

My work on this project proceeded in two phases. During the first phase (from 1983 to 1988 at the University of Cambridge, England) I benefitted importantly from many discussions with Mary Hesse and Michael Redhead, my supervisors, and with Jeremy Butterfield, my closest friend in Cambridge. Each of them read several earlier versions of the manuscript and made numerous remarks and suggestions for revision. I express my deep gratitude for their invaluable criticisms, help, and, most importantly, encouragement. I am also grateful to Henry K. Moffatt for his concern, encouragement, and help, and to David Wood for friendship and help.

The second phase began with my moving from Cambridge, England, to Cambridge, Massachusetts, USA, in 1988. During the last seven years, I have been fortunate to have numerous opportunities to discuss matters with Silvan S. Schweber and Robert S. Cohen, to both of whom I owe a considerable debt; I have also had some detailed discussions with John Stachel and Abner Shimony. I have been deeply impressed by their knowledge and understanding of contemporary physics and philosophy, and greatly appreciate their significant criticisms and suggestions about all or part of the manuscript. Since the mid-1980s I have benefitted from a long-term friendship with Laurie Brown and James T. Cushing, and I am indebted to them. I am grateful to Peter

Harman for stimulation and encouragement. I am also grateful to many physicists for clarifying conversations, among them Stephen Adler, William Bardeen, Sidney Coleman, Michael Fisher, Howard Georgi, Sheldon Glashow, David Gross, Roman Jackiw, Kenneth Johnson, Leo Kadanoff, Francis Low, Yoichiro Nambu, Joseph Polchinski, Gerardus 't Hooft, Martinus Veltman, Steven Weinberg, Arthur Wightman, Kenneth Wilson, Tai Tsun Wu, and Chen Ning Yang.

The research for this work was supported first by a combination of an Overseas Research Studentship from the ORS Committee of the United Kingdom, a Chancellor's Bursary from the University of Cambridge, and an Overseas Research Student Fees Bursary from Trinity College of Cambridge University (1983–85); and then by a Research Fellowship from Trinity College (1985–90), by a grant from the National Science Foundation [Grant DIR-No. 9014412 (4-59070)] (1990–91), and a grant from Brandeis University (1991–92). Without such generous support it would have been impossible to complete this work. In addition, the effectiveness of my work in the second phase has been much enhanced by affiliations with Harvard University, Brandeis University, Boston University, and Massachusetts Institute of Technology. To all of these institutions I am very deeply indebted. I am especially grateful to Erwin N. Hiebert and Gerald Holton of Harvard, Silvan S. Schweber of Brandeis, Robert S. Cohen of Boston University, and Jed Buchwald and Evelyn Simha of the Dibner Institute for the History of Science and Technology, MIT for their hospitality.

My most extended debt is to the members of my family, who, under difficult conditions, have given emotional and practical support to my pursuit of academic excellence without any reservation, and have tolerated my 'singlemindedness'. For all these and many things else, I thank them.

Boston University   *T. Y. C.*

1

Introduction

The treatment of the subject in this monograph is selective and interpretive, motivated and guided by some philosophical and methodological considerations, such as those centered around the notions of metaphysics, causality, and ontology, as well as those of progress and research programme. In the literature, however, these notions are often expressed in a vague and ambiguous way, and this has resulted in misconceptions and disputes. The debates over these motivations, concerning their implications for realism, relativism, rationality, and reductionism, have become ever more vehement in recent years, because of a radical reorientation in theoretical discourses. Thus it is obligatory to elaborate as clearly as possible these components of the framework within which I have selected and interpreted the relevant material. I shall begin this endeavor by recounting in section 1.1 my general view on science. After expounding topics concerning the conceptual foundations of physics in sections 1.2–1.4, I shall turn to my understanding of history and the history of science in section 1.5. The introduction ends with an outline of the main story in section 1.6.

1.1 Science

Modern science as a social institution emerged in the 16th and 17th centuries as a cluster of human practices by which natural phenomena could be systematically comprehended, described, explained, and manipulated. Among important factors that contributed to its genesis we find crafts (instruments, skills, and guilds or professional societies), social needs (technological innovations demanded by emerging capitalism), magic, and religion. As an extension of everyday activities, science on the practical level aims at solving puzzles, predicting phenomena, and controlling the environment. In this regard, the relevance of crafts and social needs to science is beyond dispute.[1]

Yet as a way of meeting human beings' curiosity about the nature of the cosmos in which they live, of satisfying their desire to have a coherent conception of the physical world (an understanding of the construction, structures, laws, and workings of the world, not in terms of its appearances, but in terms of its reality, that is, in terms of its true picture, its ultimate cause, and its unification), more pertinent to the genesis of modern science, however, were certain traditions in magic and religion, namely the Renaissance Hermetism and the Protestant Reformation, as pointed out by Frances Yates (1964) and Robert Merton (1938) respectively. In these traditions, the possibility and ways of understanding, manipulating, and transforming the physical world were rationally argued and justified by appealing to certain preconceptions of the physical world, which were deeply rooted in the human mind but became dominant in modern thoughts only through the religious Reformation and the rise of modern science.

The most important among these preconceptions presumes that the physical world has a transcendental character. In Hermetic tradition a pagan cosmology of universal harmony was assumed, in which idols possessed occult power, yet humans shared with these transcendental entities similar properties and capacities, and could have exchange and reciprocity with them. In religious traditions the transcendence of the world lay in God's consciousness, because the very existence of the world was a result of God's will and the working of nature was designed by him. This transcendence assumption has underlain the basic ambiguity of modern science, which is both mystical and rational at the same time. It is mystical because it aims at revealing the secrets of nature, which are related either to the mysteriously preestablished universal harmony or to divine Providence. It is rational because it assumes that the secrets of nature are approachable by reason and accessible to human beings. The rationalist implication of the transcendence assumption was deliberately elaborated by Protestant theologians in their formulation of cosmology. Not only was the formulation, in addition to other factors, crucial to the genesis of modern science, but it has also bequeathed some essential features to modern science.

According to Protestant cosmology, God works through nature and acts according to regular laws of nature, which are consciously designed by him and thus are certain, immutable, inevitable, and in harmony with each other. Since God's sovereignty was thought to be executed through regular channels and reflected in the daily happenings of the world, the orderly world was believed to be fully susceptible to study by scientists who tried to find out causes and regularities of natural phenomena with the help of their empirical experience. The cosmological principles of Protestant theology (which related

God with natural phenomena and their laws) provided religious motivations and justifications for the study of nature. For the followers of Calvinism, the systematic, rational, and empirical investigations of nature were vehicles to God, or even the most effective means of begetting in man a veneration of God. The reason for this was that the incessant investigations of, and operations upon, nature would gradually unfold reason, increasingly approximate perfection, and finally discover the true nature of the works of God and glorify God. This transcendental motivation has guided modern theoretical sciences ever since their emergence. And a secularized version of it is still prevalent in contemporary scientific literature.[2]

Although its ethical and emotional motivations for rationally and systematically understanding and transforming the world originated from Puritan values, which were in concordance with the ethics of capitalism (systematically calculated conduct in the domains of economy, administration, and politics), modern science as an intellectual pursuit was mainly shaped by the revived ancient Greek atomism and rediscovered Archimedes, and particularly by the Renaissance Neoplatonism. The latter was tied to Platonic metaphysics, or to a mystical cosmology of universal harmony and system of correspondence, yet aimed at a rational synthesis of human experience of natural phenomena with the help of mathematical mysticism and mathematical symbolism, and thus attracted the curiosity and fired the imagination of Copernicus and Kepler, as well as Einstein, Dirac, Penrose, Hawking, and many contemporary superstring theorists.

Another preconception prevalent in the Renaissance magic was about the 'uniformity of nature'. Its magicians believed that same causes always led to same effects, and as long as they performed the ritual acts in accordance with the rules laid down, the desired results would inevitably follow. Although the belief in associative relations between natural events had only an analogical basis, it was surely the precursor of the mechanical idea that the succession of natural events is regular and certain, and is determined by immutable laws; that the operation of laws can be foreseen and calculated precisely; and that the element of chance and accident is thus banished from the course of nature.

The carving out of a domain of nature with its regular laws helped to move God further and further away from the ideas of causality in empirical science, to demarcate nature as separated from the domain of supernatural phenomena, and to take naturalistic causes as a ground for the explanation of natural phenomena. Related with this was the belief that natural forces were manipulable and controllable. Without this belief there would be no practice in astrology, alchemy, and magic operated in symbolic languages. The mathematical symbolism was respected just because it was believed to be the key to the operation by which natural forces could be manipulated and nature conquered.

It is interesting to notice that the occult and scientific perspectives coexisted and overlapped in the 16th and 17th centuries when science was in formation, and that the magical and religious preconceptions helped to shape the characteristics of science, such as (i) rationalism and empiricism as well as objectivity, which were related with the transcendental character of nature as conceived by Protestant cosmology; (ii) causal reasoning that was based on the idea of the 'uniformity of nature'; (iii) mathematical symbolism in its theoretical formulations; and (iv) the will to operate as manifested in its experimental spirit.

Yet, unlike magic and religion, science has its distinctive tools for its undertaking: the understanding and manipulation of the world. Important among them are (i) professional societies and publications; (ii) rational criticism and debates based on a skeptical spirit and a tolerance of differences; (iii) empirical observation and experiments, logic, and mathematics, the systematic use of which leads to distinctive modes of demonstration; and, most importantly, (iv) fruitful metaphors, conceptual schemes, and models, with which theoretical structures the structures and workings of the world can be approached, described, and understood.

A scientific theory must have some empirical statements, in the sense of having falsifiable consequences, and hypothetic statements that are not individually falsifiable but are crucial for understanding and explaining phenomena. The hypothetic statements are expressed in theoretical terms: unobservable entities and mechanisms as well as abstract principles. There is no dispute among philosophers of science about the function of theoretical terms in a theory as heuristic devices for organizing experiences. What is at issue is their ontological status: should we take them realistically? For sense data empiricism, the answer is simply negative. But Comte's anathema against the undulatory theory and Mach's opposition to the atom turned out to be grave mistakes. For conventionalism, the foundation of science is much wider than sense data and includes, in addition to logic and mathematics, conventions. In particular, it takes conventional mathematical expressions as foundations of reason to which observables and unobservables are subordinated. But 'how real are these constitutive conventions?' is a question to which the conventionalists are reluctant and unable to answer. For internal realism, the reality of theoretical terms is accepted, but only within the theory in which these terms appeared, and cannot be divorced from the theory. The reason is that we have no access to a metaphysical reality if it exists at all. My disagreement with internal realism will be given here and there in the text.

Concerning the ontological status of theoretical terms, the position that I am going to defend in the last chapter is structural realism. Briefly stated, the position holds that the structural relations (often expressed directly by mathematical structures, but also by models and analogy indirectly) in a successful theory should be taken as real, and the reality of unobservable entities are gradually constituted and, in an ideal situation, finally determined in a unique way by these structural relations.

An immediate objection to this position is that this is a disguised phenomenalism, in which empirical truths of observables are replaced by mathematical truths of observables, and in which there is no room for the reality of unobservables. Critics would argue that the problem of giving an interpretation proves to be much more difficult than just writing out equations that summarize the observed regularities.

To anticipate my arguments in the last chapter, suffice it to point out that in addition to the structural relations for observables, there are structural relations for unobservables, which are more important for understanding and explanation. To the objection that any such structural relation must be ontologically supported by unobservable entities, my answer is that in any interpretation, while structural relations are real in the sense that they are testable, the concept of unobservable entities that are involved in the structural relations always has some conventional elements, and the reality of the entities is constituted by, or derived from, more and more relations in which they are involved. Once we have accepted the ontological primacy of structures over entities, we have more flexibility in accommodating changing interpretations of entities.

1.2 Metaphysics

Metaphysics, as I understand it, consists of presuppositions about the ultimate structure of the universe. First, it is concerned with the question about what the world is really made of, or what the basic ontology of the world really is. Is the world made of objects, properties, relations, or processes? If we take objects as the basic ontology, then further questions follow: What categories of objects are there? Are there mental objects as well as physical objects? What are the basic forms of the physical objects – particle, field, or some other form? In addition, what is the nature of space and time? A difficult question central to ontological discussion is about the criteria of reality, because metaphysicians are always concerned with real or fundamental or primary entities rather than epiphenomena or derivatives. A classical answer to this question by modern philosophers, such as Descartes and Leibniz, is that only a substance is real. A substance exists permanently by itself without the aid of any other substance, and is capable of action without any external cause. Yet, as we shall see, there can be other conceptions of reality, based on the concepts of potentials, structures, or processes instead of substance.

Second, metaphysics is also concerned with principles that govern the fundamental entities of the world. For example, there is a principle of identity, which says that individuals should be able to change over time and remain one and the same. Similarly, the principle of continuity says that no discontinuous changes are possible. There are many other metaphysical principles that have played important regulative or heuristic roles in the construction of scientific theories, such as the principles of simplicity, unity, and spatio-temporal visualizability. But the most important among these principles is the principle of causality, which is supposed to dictate the working of nature and helps to make the action of entities intelligible.

Thus, by appealing to ontological assumptions and regulative principles, metaphysics supplies premises and bolsters the plausibility of scientific arguments, and is quite different from empirical, practical, and local statements about observed phenomena and their regularities. Traditionally, metaphysics is highly speculative. That is, its assertions are unexamined presuppositions and are not required to be empirically testable. However, these presuppositions of epistemic and ontic significance are so entrenched in a culture that they appear to scientists as commonsense intuitions. Since these entrenched assumptions give a seemingly plausible picture of the world, and have virtually determined the deep structure of the thinking modes of people who share these assumptions, metaphysics comprises an important part of a culture.

William Whewell once said that

physical discoverers have differed from barren speculators, not by having no metaphysics in their heads, but by having good metaphysics while their adversaries had bad; and by binding their metaphysics to their physics, instead of keeping the two asunder.
(Whewell, 1847)

As we shall see in the text, metaphysical assumptions can be fleshed out with physical parameters. More important than this, however, is that metaphysics provides a comprehensive framework of concepts within which specific theories can be proposed and tested. As is well known, ancient science originally developed from metaphysical speculations. But even now science is still associated with this or that world picture provided by metaphysical ideas. An explanation of a phenomenon is always given in terms of a specific world picture. The question of which ontology, matter, field, energy, or spacetime best explains phenomena is extremely important for a physical theory, far more important than the details of its empirical laws. For example, the empirical content of Newtonian mechanics has been modified only slightly by Einstein's theory of relativity, yet no one would deny that this is a great step in the development of physics, since the old ideas about Euclidean space, absolute time and absolute simultaneity, and action at a distance were swept aside, and the world picture thereby changed.

Examples of the interactions between physics and metaphysics are many. The guidance of the metaphysical assumption, concerning the universality of the principle of relativity, in the evolution from the electromagnetic theory of the 19th century to the special and general theories of relativity is widely acknowledged. On the other hand, developments in physics, particularly in quantum mechanics, quantum field theory, and gauge theory, also have profound metaphysical implications and have radically changed our conception of the physical world, as we shall discuss in the main text. A corollary of mutual penetration between physics and metaphysics is this. Not only is metaphysics indispensable for physical research, but physics has also provided us with a direct access to metaphysical reality. For example, experimental investigations of the Aharonov–Bohm effect and Bell inequality have greatly clarified the ontological status of quantum potentials and the nature of quantum states respectively, both of which were thought to be inaccessible metaphysical questions. For this reason, Abner Shimony (1978) calls this kind of research experimental metaphysics, emphasizing the important role of physics in testing metaphysical assumptions.

Thus it is inappropriate to take metaphysics only as an acceptable remainder of scientific theory from which empirical content and logical structures have been removed. Rather, it has special scientific content, in the sense that it provides a basic model of physical reality so that the theory is intelligible, and also in the sense that it prefers certain types of explanation on the basis of certain conceptions of causality. In the case that the mechanical conception of efficient cause is taken as a basis for explanation, this metaphysical assumption has not only determined the hypothetico-deductive structure of physical theory, but also entails a built-in reductionist methodology. Moreover, since the positivists are agnostic with respect to causes, and only the realists take causes seriously, the built-in implication of the metaphysical assumption for realism should also not be ignored.

1.3 Causality

The rise of modern science was accompanied with the replacement of authorities or traditions by causes in explaining phenomena. One of the ultimate goals of science is to understand the world, and this is approached by scientific explanation, that is, by finding out causes for various phenomena. According to Aristotle, however, there are different kinds of cause: material, formal, efficient, and final causes. Before the rise of modern science, teleological explanation based on the notion of final cause was a dominant mode of explanation. With the revival of Neoplatonism, Archimedeanism, and atomism in the Renaissance, there began a transformation in basic assumptions of scientific explanation. Copernicus, Kepler, Galileo, and Descartes, for example, believed that the underlying truth and universal harmony of the world can be perfectly represented by simple and exact mathematical expressions. The mathematization of nature led to a certain degree of popularity of formal cause. But the most popular and most powerful conception of causality, in fighting against the teleological explanation, was a mechanical one based on the notion of efficient cause. Different from final and formal causes, the idea of efficient cause focuses on how the cause is transmitted to the effect, that is, on the mode of this transmission. According to the mechanical view, causality can be reduced to the laws of motion of bodies in space and time, and observable qualitative changes can be explained by purely quantitative changes of unobservable constituting corpuscles.

Mechanical explanation had different variations. According to Descartes, the universe is an extended plenum and no vacuum can exist, any given body is continuously in contact with other bodies, and thus the motion of several parts of the universe can only be communicated to each other by immediate impact or pressure, and no action at a distance would be possible. There is no need to call in the force or attraction of Galileo to account for specific kinds of motion, still less the 'active power' of Kepler. All happens in accordance with the regularity, precision, and inevitability of a smoothly running machine. According to Newton, however, force is the causal principle of motion, although force itself has to be defined by the laws of motion. For Newton, as well as for Huygens and Leibniz, the intelligibility of causality was principally lodged in the concept of force. Then a serious question is about the concrete mechanism for transmitting force. This question is so central to the subsequent development of physics that it actually defines the internal logic of the development. The search for a solution to this question has led to the advent of field theory, quantum field theory, and, finally, gauge theory.

There are different forms of mechanical explanation. First, natural phenomena can be explained in terms of the arrangement of particles of matter that are actually involved in the phenomena, and in terms of the forces acting among them. In the second form, some mechanical models are adopted to represent phenomena. These models are not necessarily taken as representations of reality but are seen as demonstrating that phenomena can in principle be represented by mechanisms. That is, these mechanical constructions rendered phenomena intelligible. Third, mechanical explanation can also be formulated in the abstract formalism of Lagrangian analytic dynamics. The equations of motion obtained thereby are independent of the details of mechanical systems, but phenomena are nevertheless explained in mechanical terms of mass, energy, and motion, and thus are subsumed under the principles of mechanical explanation involved in the formalism, although they are not represented by a specific visualizable mechanical model.

Among the three forms, the use of models is of special importance. Even the abstract formalism of analytic dynamics needs to be illustrated by models. Moreover, since one of the major motivations in physical investigations is to find out the agent of force at the foundational level when direct causes at the phenomenal level fail to explain, the postulation of models involving hypothetic and unobservable entities and mechanisms is unavoidable. Thus the necessity of hypothesis is inherent in the very idea of mechanical explanation, or in the search for efficient causes.

Any hypothesis must be consistent with the fundamental laws of nature and with all the generally accepted assumptions about the phenomena in question. But a hypothesis is only justified by its ability, in conjunction with the fundamental laws and general assumptions, to explain phenomena. Thus its specific content is to be adjusted to permit deduction of statements about the phenomena under investigation. But how can it be possible for a hypothesis about unobservables to be able to explain phenomena? And how can the hypothesis be adjusted so that this goal can be achieved? A tentative answer to these questions, based on the structural realism that I shall argue for in the last chapter, is that only when the structure of a model (any hypothesis is a model), based on analogy drawn from everyday experiences or other known phenomena, is similar to the structure of the phenomena can a hypothesis fulfill its explanatory function.

The hypothetico-deductive structure of physical theory has immediate metaphysical implications: if a set of mutually consistent hypotheses with a set of unobservable entities serves as causes of the phenomenal world, then it seems undeniable that the hypothetic world gives a true picture of the real world, and the phenomenal world can be reduced to this real world. For example, most mechanical explanations suggest a real world with the hidden ontology of unobservable atoms or elementary particles in motion as the substratum underlying the physical reality. There are other possibilities. For example, Leibniz took the intensive continuum of forces as the metaphysical foundation of phenomena. Other physicists in the 18th and 19th centuries went beyond mechanical explanation but still worked within the general framework of the hypothetico-deductive framework, suggesting different non-mechanical ontologies, such as active principles, fire, energy, and force fields.[3] With each different ontology, physicists offered not only a different physical theory or research programme, but also a different conception of a real world that underlies the phenomenal world.

1.4 Ontology

In contrast with appearances or epiphenomena, and also opposed to mere heuristic and conventional devices, ontology as an irreducible conceptual element in the logical construction of reality is concerned with a real existence, that is, with an autonomous existence without reference to anything external. Since an ontology gives a picture of the world, it serves as a foundation on which a theory can be based. This helps to explain its reductive and constitutive roles in the theoretical structure of science.

Although the term ontology often refers to substance, as in the case of the mechanical world view, in which the basic ontology is particles in motion, this is not necessarily so. The concept of ontology, even in the sense of an ultimately true reality, is wider than that of substance, which in turn is wider than entities and individuals. For example, it can be argued, as the Neoplatonists like Kepler would do, that mathematical relations, as they represent the structure of the universe, are the foundations of reality; even forces, as the causal principle, have to be defined in terms of mathematical relations. While it can be argued that any mathematical structure has to be supported by physical relations between entities, from a constitutive perspective a physical entity, if it is not merely an empty name, can only be defined by the relations in which it is involved. This is only one example of what Cassirer calls the 'functional mode of representing reality'. Another example can be found in Whitehead's philosophy of process. According to Whitehead, activity-functioning is not a function of a changeless underlying stuff; rather, a physical object is a connection, a more or less permanent pattern of the basic functioning. He argues that nature is a structure of evolving processes, the reality is the processes, and the substantial things issue out of the process of activity and becoming, which is more fundamental than the things.

This is, of course, a very controversial topic. According to Julius Mayer, who follows Leibniz in taking forces as the primary agency of nature, forces, as the embodiment of nature's activity, should be viewed as non-mechanical yet substantial entities. And for Meyerson, entity is essential to explanation and should not be dissolved in relations or processes. More importantly, a historical fact is that the notion of ontology is almost always connected with that of substance. This connection constitutes an underpinning in the discourse of physical sciences and cannot be passed over in the examination of the foundations of physics.

Then what is substance? Substance is always characterized by a constellation of essential or primary qualities. These qualities exist in space and time and are conserved in the changes of their spatial and temporal locations, and to these all other qualities can be reduced. Since the nature of reality can only be discussed in terms of its symbolic representations, ontology in general, and substance in particular, as a model of reality, is a piece of science itself and cannot be separated from science. Thus the understanding of what are the primary qualities is different in different theories, and each theory determines its own kind of substance. However, a generally shared assumption, since the time of Leibniz, holds that substance must be fundamental (in contrast with epiphenomena), active or the source of activity, and self-subsistent, meaning that the existence of substance is not dependent upon the existence of anything else. One of the main conclusions of this book is that a conceptual revolution generally turns a previous substance into an epiphenomenon, and thus changes our conception of what is the basic ontology of the world.

In classical physics, Descartes took space or extension as substance. Newton's case was much more complicated. In addition to substance, his ontology also included force and space. And his substance referred not only to passive material particles, but also to active ether. For Leibniz, substance was a center of primitive activity. This activity was not the manifestation of a stuff or matter, but the activity itself was the substance, and matter was an appearance on the surface of this activity.

The dominant view after Leibniz was to take substance as inherently active objects, usually divided into different ontological categories: discrete individuals (such as visible massive particles and invisible atoms) and continuous plenum (such as the Cartesian extension and the classical field). An individual is a spatially bounded object and at least has some other properties. It is usually characterized as what can be identified, re-identified, and distinguished from other members of its domain.[4] Here identity is ensured by the conservation of essential qualities, and distinguishability has its origin in impenetrability, which presupposes a spatial bound of the object. The concept of individual is usually connected to that of particle because both have to be discrete, but it is narrower than the latter owing to its requirements of distinguishability and impenetrability. In quantum theory, quantal particles are identifiable but neither re-identifiable nor distinguishable from their like. They are thus not individuals but can still be accounted as particles mainly because of the conservation of rest-mass, charge, and spin.

This is one example of the theory dependence of our conception of substance. Another interesting example is that of the ontological status of energy. Traditionally, energy was thought to be one of the most important features of substance since it indicated that its carrier was active, and, as the measure of ability to act, it was conserved. However, as a measurable property rather than a self-subsistent object, energy itself was usually not to be regarded as substance. For example, when Carl Neumann claimed that potential energy was primary and able to propagate by itself, Maxwell maintained that energy could only exist in connection with material substance.[5] For the same reason, energeticism, according to which energy as pure activity is the basis of physical reality, was usually accused of being phenomenalism because of its rejection of substance. Yet it can be interpreted otherwise. What if energy is taken as substance with the new feature of being always active, always changing its form while keeping its quantity constant? Then energeticism would seem to be a precursor of James's functionalism and Whitehead's ontology of process.

These two examples suggest that an ontological assumption is fundamental not only to a specific theory, but also to a research programme. Let us have a closer look, from this perspective, at the genesis of the field theory programme. The electromagnetic field was taken to be responsible for continuously transmitting electromagnetic force through space. The substantiality of the field in 19th century physics is a subject for debate. Sometimes it is argued that Maxwell established the substantiality of the field because he proved the presence of energy in the field. But this claim is questionable. For Maxwell, the field was not an object but merely a state of the mechanical ether that obeyed Newton's laws of motion. This means that for Maxwell the field was not self-subsistent and hence could not be substantial. What the presence of energy in the field established was just the substantiality of the ether rather than that of the field.

Sometimes it is also argued that the removal of the mechanical ether entails the removal of the substantiality of the field, and this thereby supports the claim that spacetime points are the basic ontology of field theory.[6] In my opinion, however, it is precisely the removal of the mechanical ether that establishes the non-material substantiality of the field. The reason for this is that in this case the field becomes the only possible repository of the field energy, and the field energy presupposes a substance as its repository. As to spacetime points, the reason why they cannot be viewed as the basic ontology of field theories is that in the framework prior to the general theory of relativity, it is impossible to argue that they are active or the source of activity, while in the framework of the general theory of relativity, they are not self-subsistent since they are always occupied and, most importantly, individuated by the gravitational field.[7] In my opinion, the assumption that the continuous substantial field is the basic ontology of the world must be taken as the first of the basic tenets of field theories, though this was not always the case in the history of physics.

The field is distinct from the individual by its continuity, in contrast with the discreteness of the individual, and by the superimposability between its different portions, in contrast with the impenetrability between different individuals. To be sure, the field is also capable of displaying a form of discreteness through its periodicity introduced by boundary conditions (see section 6.5). Yet this kind of discrete existence is different from that of individuals, which exist permanently while the discreteness of the field exists only transiently.[8]

There is something deeper than these apparent distinctions when the electromagnetic field is taken to be a basic ontology of the world, rather than merely a mathematical device or a state of the mechanical ether.[9] The field is a new kind of substance, with Leibniz's primitive force as its precursor. It differs from both material individuals and the mechanical ether by its non-mechanical behavior. The introduction of this new non-mechanical ontology initiated a new programme, the field programme. The field programme differs from the mechanical programme in its new ontology and new mode of transmitting action by the field. The emergence of Lorentz's theory of the electron marked the end of the mechanical programme in two senses. First, it discarded the mechanical ether so that the transmission of the electromagnetic interaction could not be explained within the mechanical programme. Second, it introduced an independent substantial electromagnetic field that was not reducible to the mechanical ontology. And this prepared the way for the further advancement of the field programme (see chapter 2).

It is not so difficult to accept the importance of ontological assumptions as conceptual foundations for scientific research. But the Kuhnians would argue that the historical fact that any ontology posited by a theory is always replaced by a different and often contradictory ontology posited by a later theory seems to have convincingly indicated that our ontological assumptions have nothing to do with the real world. As I shall argue in the last chapter, however, the historical replacement of ontologies, at least in the context of twentieth century field theories, is not without a pattern and a direction. That is, an old ontology always turns out to be an epiphenomenon and can be derived from a new and more fundamental ontology. And this certainly has lent support to a realist interpretation of theoretical ontology.

Two caveats should be immediately added. First, this pattern is not always realized in a unilinear way but often through a dialectical synthesis. Second, ontological reduction is only one dimension of scientific development. Because of the existence of objective emergence, as I shall show in section 11.4 in the context of gauge theory, different levels of the world have their relatively autonomous ontologies, which are not reducible to an ultimate substratum. This entails a kind of ontological pluralism.

1.5 History and the history of science

History is not just a collection of past events but consists of movements with causes, patterns, and directions. If we onsider the huge amount of information buried in past events, historiography cannot be simply a copy of the real history but must be selective. Material is selected according to its historical significance, which is determined by a historian's interpretation. Since our interpretation of the past is gradually shaped in the light of the present, and of the future toward which we are moving, and also because historiography begins with the handing down of tradition, which means the carrying of the lesson of the past into the future, historiography is a dialogue between the past and the present whose purpose is to promote our understanding of the past in the light of the present and our understanding of the present and future in the light of the past; that is, to try to understand the transition from tradition to future.

Thus a good piece of historiography has to provide a plausible or convincing hypothesis about the cause, pattern, and direction of a historical movement, so that the past events become understandable, and a perspective or a constructive outlook can enlarge our understanding of the movement, thereby opening the way for further enquiries. Since some causes are merely accidental and others are generalizable, central to the hypothesis concerning our understanding of a past event is the question 'what should be taken as the major causes?' In the case of the history of science, this is a point at issue. The disagreement in this regard between the social constructivists and the intellectual historians of science has caused some tension between them.

It is true that science as a form of culture cannot be divorced from society. First, the language that science uses is a product of social communication. Second, problems that science tries to solve, motivations, and material and technical resources for solving problems are all provided by society. Third, institutions for mobilizing the resources for scientific investigations are supported and constrained by the socio-economic structures at large. All these considerations point to the social character of scientific activities. Moreover, a scientist can only acquire his ideas, metaphors, and conceptual schemes for solving problems and interpreting his results from his cultural environment, in which tradition is adapted to the current situation. This circumstance accords scientific knowledge a certain social, cultural, and historical specificity. At this *trivial* level, there is almost a unanimous consensus on the social character of science.

In the social constructivist account of science, 'social' refers to social relations among scientists as well as between scientists and society at large, including groupings of scientists, scientific institutions, and socio-economic structures. The social also refers to non-scientific forms of culture, such as religion and magic. But the special feature of social constructivism is that it defines the social in contrast with the individual and the intellectual. In this non-trivial sense of social constructivism, however, its account of scientific activities is incomplete, biased, and seriously flawed. The reason for its being flawed is that the constructivist account omits or even deliberately rejects the fact that all scientific activities are severely constrained and persistently guided by the aim of knowledge of nature. In fact the aim of knowledge of nature is built into the very concept of science as a particular social institution. Having missed this crucial point, the social constructive account of scientific activities cannot be accepted as it is presented now. For this reason, I take conceptual history, which summarizes humanity's persistent intellectual pursuit, rather than social history as the main body of the history of science, although I also admit that on the basis of a conceptual history, a social history may be interesting, if it successfully provides a more complete picture of how science as a pursuit of knowledge is actually developed.

Sometimes the history of physics is written mainly in terms of its empirical conquests. Yet it can be written in a different way. The development of fundamental physics necessarily involves great changes in ontological assumptions. The history of physics shows that most great physicists have intended their research to be the search for a true world picture, which serves to prepare the way for a conceptual revolution. This has not been an accidental side feature of the growth of physics but rather the central feature of its most important advance. Thus the history of physics is in a sense the history of the expressions of world-view, whose core consists of ontological assumptions encapsulated in the descriptive schemes. More specifically, what I plan to do in this volume is, with a historical and critical exposition of 20th century field theories, to detect a pattern in, and the direction of, the evolution of the changing worldview suggested by these theories; that is, to unfold the ontology shift and ontological synthesis that have occurred in these theories.

1.6 An outline of the main story

In my account, except for an intellectual precursor of Faraday's speculation of field as the conditions of space, the field programme, in the sense in which the field is taken to be a basic ontology of the world, started with Lorentz's theory of the electron, and became widely accepted after the advent of Einstein's special theory of relativity. However, both theories presupposed an independent ontology of space or spacetime points as the support of the fields. So in a reductionist sense the field programme at this stage was not a complete programme.

I begin my main story with the next step in the development of the field programme, that is, with Einstein's general theory of relativity (GTR). In GTR the gravitational fields transmitting interactions are inseparably bound up with the geometrical structures of spacetime. For this reason, I call GTR and subsequent developments in its direction the geometrical programme.

The interpretation of GTR is a controversial issue. It depends on whether matter, field, or spacetime point is taken to be its basic ontology; it also depends on the understanding of the relations between matter, fields, and spacetime. Historically, there were three interpretations. First, Einstein himself, following Mach, took ponderable bodies as the only physical reality that fully determined the gravitational field and the geometrical structures of spacetime. The discovery of the vacuum solutions of the gravitational field equations made this interpretation untenable (see section 4.3). Then, Hermann Weyl and Arthur Eddington developed a view according to which the geometrical structures of spacetime were taken as the physical reality to which the gravitational fields were to be reduced. Gravitation in this view was interpreted as a manifestation of the curvature of the spacetime manifold; that is, gravitation was geometrized. I call this the strong geometrical programme (see section 5.2). Einstein himself never endorsed this programme. Finally, in his unified field theory, Einstein took the gravitational field as a part of a total field that represented the ultimate physical reality, with spacetime being its structural quality. In a sense this was an expansion of Einstein's gravitization of the geometry of spacetime, originating from his point-coincidence argument of 1915 in justifying the principle of general covariance. I call this the weak geometrical programme (see sections 4.2, 5.1–3). Although the mathematical formalism remains the same, the ontological priority in different interpretations is different.

GTR constitutes the starting point of the geometrical programme. In the further developments of the programme, four generalizations were made: (i) the unification of gravity and electromagnetism; (ii) the unification of the mass and spin effects by introducing torsion of the spacetime manifold; (iii) the unification of gravity and quantum effects; (iv) the unification of matter and fields. Most of these generalizations were made in the strong geometrical programme. I shall show, however, that the attempts, within the strong geometrical programme, to incorporate quantum effects inevitably led to the very collapse of the programme (see section 5.3). This justifies my claim that the strong geometrical programme is inappropriate. But the weak geometrical programme was also unsuccessful in its attempts to extend its scope to the cases of electromagnetic, weak, and strong interactions.

Moreover, there is a serious flaw in the geometrical programme which has its seed in GTR itself. That is, the equations of GTR cannot avoid singular solutions or singularities, either in the big bang of an expanding universe, or in collapsing stars or black holes. The unavoidable occurrence of singularities implies that GTR must break down in the case of sufficiently strong gravitational fields. However, there is another kind of generalization of the geometrical programme, which is related to the recent developments of gauge theory within the framework of quantum field theory, and by which the singularity difficulty can be overcome. This promising attempt will be discussed in part III.

Another important variant of the field programme in twentieth century physics started with the quantum electrodynamics of Jordan and Dirac, the quantum field programme. Paradoxically, the field programme seems to be undermined by quantum theory. First, quantum theory puts a limit to the continuous distribution of energy throughout space, and this conflicts with the field ontology. Second, it also violates the principle of separability, according to which distant systems with zero interaction energy should be physically independent of one another. Finally, in quantum theory there is no room for particles to have continuous spatio-temporal paths during their quantum jumps or between their creation and annihilation, and this conflicts with the mode of transmitting interactions in the field programme. So how can I claim that quantum field theory (QFT) should be viewed as a variant of the field programme? The answer to this depends on my interpretation of the basic ontology and of the mode of transmitting action in QFT.

Then what is the ontology of quantum theory? The question is difficult to answer in general since the situation in non-relativistic quantum mechanics is different from that in QFT. In non-relativistic quantum mechanics, this problem is closely related with the interpretation of the wave function. De Broglie and Schrödinger held a realistic interpretation of the wave function. They assumed a field ontology and rejected the particle ontology because, they argued, quantal particles obeying quantum statistics showed no identity, and thus were not classically observable individuals. But this realistic view ran into severe difficulties: it implied a non-physical multi-dimensionality in many-body problems, a non-realistic superposition of states (such as the famous Schrödinger cat), and the collapse of wave functions in measurement, which has defied any explanation up to the present time. These difficulties indicated that the classical field was not an appropriate candidate for the ontology of non-relativistic quantum mechanics.

In his probabilistic interpretation, Max Born rejected the reality of the wave function, deprived it of energy and momentum, and assumed a particle ontology. However, Born did not claim a classical particle ontology. There were a number of difficulties, such as quantum statistics and the double slit experiment, that prevented him from doing so. As a response to these difficulties, Werner Heisenberg interpreted the wave function as a potential (see section 7.1). It should be noticed that this significant concession to a realistic field ontology had its roots in the dilemma faced by the probability interpretation with a particle ontology, and prepared the way for a great ontological shift in QFT.

In sum, the situation in non-relativistic quantum mechanics is this. In addition to a number of mathematical devices that cannot be interpreted realistically, the conceptually incoherent fusion of a mathematical structure (wave equations) based on a field ontology and a physical interpretation (the probability interpretation) based on a particle ontology makes it extremely difficult to find a coherent ontology, in terms of classical particles or fields, for this theory. Two ways out are possible. An instrumentalist solution to the difficulty is to take all the theoretical entities only as an analogical construction for economically describing observed phenomena, but not as a faithful description of what the quantum world really is. But this is only an evasion rather than a solution to the interpretation problem in quantum theory. Or one can try to remould our conception of substance, not restricting ontology to the dichotomy of classical particle and field. The remoulding process in the early history of QFT is examined in section 7.3, and can be summarized as follows.

A significant ontology shift, crucial to reclassifying substance, emerged when fermion field quantization was introduced. Before that, there were two quantization procedures for the many-body problem: second quantization and field quantization. These procedures presuppose, respectively, a particle ontology and a field ontology. The second quantization procedure actually is only a transformation of representation in a quantum system of particles, having nothing to do with the field ontology. In this formulation, the particles are permanent, their creation or destruction is just an appearance of jumps between different states, and the probability field is only a mathematical device for calculation. By contrast, the field quantization procedure starts with a real field represented by a collection of field oscillators, and exhibits the particle properties of the field by using the operators of creation and destruction, which can be interpreted as the excitation and de-excitation of field quanta.

However, when the quantization condition in non-relativistic quantum mechanics is applied to the field oscillators, what is quantized, however, is merely the motion (energy, momentum, etc.) of the fields, but not the fields themselves. There is no question of actually turning the field into a collection of particles, which would entail assuming a particle ontology, unless it is assumed that the quantization of motion entails the quantization of the carriers of motion, that is, of the fields themselves. But this is a significant metaphysical assumption, requiring a radical change of our conception of substance.

The two approaches of quantization were not contradictory but parallel, starting from different ontological assumptions. It is interesting to notice that Dirac, one of the original proposers of both approaches, in his early works (1927b, c) conflated one with the other, although he showed an inclination for the particle ontology, and for reducing a field into a collection of particles. This inclination suggests that the radical metaphysical assumption mentioned above was made by Dirac, although only unconsciously.

The introduction of fermion field quantization by Jordan and Wigner initiated further radical changes in the interpretation of quantum field theory (QFT). First, a realistic interpretation replaced the probability interpretation of the wave function: the wave function in their formulation had to be interpreted as a kind of substantial field, otherwise the particles as the quanta of the field could not get their substantiality from the field, leaving unsolved some of the original difficulties faced by Schrödinger's realistic interpretation. Second, the field ontology replaced the particle ontology: the material particle (fermion) was no longer regarded as having an eternally independent existence, but as being a transient excitation of the field, a quantum of the field, thus justifying my claim that QFT started a major variant of the field programme, namely, the quantum field programme.

But the reversion to the view of a real field evidently leaves a gap in the logic of models of reality and destroys the traditional conception of substance. A new conception of ontology is introduced. This new ontology cannot be reduced to that of the classical particle ontology because the field quanta lack a permanent existence and individuality. It also cannot be reduced to that of the classical field ontology because the quantized field has lost its continuous existence. It seems that 20th century field theories suggest that the quantized field, together with some non-linear fields (such as solitary waves), constitute a new kind of ontology, which Michael Redhead (1983) calls ephemeral. Among the founders of QFT, Jordan consciously made the advance to the new ontology and remoulded the conception of substance, while Dirac and Heisenberg unconsciously did so.

The new ontology of QFT was embodied in the Dirac vacuum. As an ontological background underlying the conceptual scheme of quantum excitation and renormalization, the Dirac vacuum was crucial for calculations, such as Weisskopf's calculation of the electron self-energy and Dancoff's discussion on the relativistic corrections to scatterings. The fluctuations existing in the Dirac vacuum strongly indicate that the vacuum must be something substantial rather than empty. On the other hand, the vacuum, according to the special theory of relativity, must be a Lorentz-invariant state of zero energy and zero momentum. Considering that energy has been loosely thought to be essential to substance in modern physics, it seems that the vacuum could not be taken as a kind of substance. Here we run into a profound ontological dilemma, which hints at the necessity of changing our conception of substance and of energy being a substantial property.

The mode of conveying interactions in QFT is different from that in the classical field programme in two aspects. First, interactions are realized by local couplings among field quanta, and the exact meaning of coupling here is the creation and annihilation of the quanta. Second, the actions are transmitted, not by a continuous field, but by discrete virtual particles that are locally coupled to real particles and propagate between them.[10] Thus the description of interactions in QFT is deeply rooted in the concept of localized excitation of operator fields through the concept of local coupling.

Yet the local excitation entails, owing to the uncertainty relation, that arbitrary amounts of momentum are available. Then the result of a localized excitation would not only be a single momentum quantum, but must be a superposition of all appropriate combinations of momentum quanta. And this has significant consequences. First, the interaction is transmitted not by a single virtual momentum quantum, represented by an internal line in a Feynman diagram, but by a superposition of an infinite number of appropriate combinations of virtual quanta. This is an entailment of the basic assumption of a field ontology in QFT. Second, the infinite number of virtual quanta with arbitrarily high momentum lead to infinite contributions from their interactions with real quanta. This is the famous divergence difficulty. Thus QFT cannot be considered a consistent theory without this serious difficulty being resolved. Historically, the difficulty was first circumvented by a renormalization procedure.

The essence of the original renormalization procedure is the absorption of infinite quantities into the theoretical parameters of mass and charge. This is equivalent to blurring the exact point model underlying the concept of localized excitations. While quantum electrodynamics meets the requirement of renormalizability, Fermi's theory of weak interactions and the meson theory of the strong nuclear force fail to do so. This difficulty, however, can be removed by using the idea of gauge invariance. Gauge invariance is a general principle for fixing the forms of fundamental interactions, on the basis of which a new programme, the gauge field programme, for fundamental interactions develops within the quantum field programme.

Gauge invariance requires the introduction of gauge potentials, whose quanta are responsible for transmitting interactions and for compensating the additional changes of internal degrees of freedom at different spacetime points. The role gauge potentials play in gauge theory is parallel to the role gravitational potentials play in GTR. While the gravitational potentials in GTR are correlated with a geometrical structure (the linear connection in the tangent bundle), the gauge potentials are correlated with a similar type of geometrical structure, that is, the connection on the principal fibre bundle. Deep similarity in theoretical structures between GTR and gauge theory suggests the possibility that the gauge theory may also be geometrical in nature.

Recent developments in fundamental physics (supergravity and modern Kaluza–Klein theory) have opened the door for associating gauge potentials with the geometrical structures in extra dimensions of spacetime (see sections 11.2 and 11.3). Thus it seems reasonable to regard the gauge field programme as a synthesis of the geometrical programme and the quantum field programme if we express the gauge field programme in such a way that interactions are realized through quantized gauge fields (whose quanta are coupled with material fields and are responsible for the transmission of interactions) that are inseparably correlated with a kind of geometrical structure existing either in internal space or in the extra dimensions of spacetime.

It is interesting to notice that the intimate link between the ontology of the quantum field programme and that of the geometrical programme became discernible only after the gauge field programme took the stage as a synthesis of the two. This fact suggests that the concept of synthesis is helpful for recognizing the continuity of theoretical ontologies in their structural properties across conceptual revolutions. The synthesis of scientific ideas requires a transformation of previous ideas. When we extend the concept of synthesis to the discussion of basic ontologies of theoretical systems, we find that ontological synthesis also requires a transformation of previous conceptions of ontology and, as a general feature, makes substance into an epiphenomenon, and thus accompanies a change of basic ontology (see section 12.4).

This feature suggests that the concept of ontological synthesis has captured some characteristics of conceptual revolutions, whose result is the birth of new research programmes based on new ontologies. On this view, a direct incorporation of the old ontology of a prerevolutionary programme into the new ontology of the postrevolutionary programme is very unlikely. Nevertheless, some of the discovered structural relations of the world, such as external and internal symmetries, geometrization, quantization, etc., embodied in the old ontology will certainly persist across the revolutions. I suggest that here is a coherent direction for ontological development, towards the true structure of local, though enlargeable, domains of investigation. Thus the occurrence of conceptual revolutions by no means implies that the development of theoretical science, as Kuhn suggests, is radically discontinuous or incommensurable. Rather, it is continuous and commensurable, compatible with a certain kind of scientific realism.

On the other hand, by drawing a new concept of ontological synthesis from the historical analysis of 20th century field theories, we also find that the growth of science does not necessarily take a unilinear form of continuity and accumulation, thus avoiding the untenable 'convergent realism'. The idea of ontological synthesis as a dialectical form of continuity and accumulation of world structures is more powerful than the idea of convergence in explaining the mechanism of conceptual revolutions and the patterns of scientific progress. Since the very idea of scientific rationality lies in the intention of obtaining more and more knowledge of the real structures of the world, then on the synthetic view of scientific growth, the conceptual revolution is one way of realizing this rationality.

The future direction of fundamental research suggested by the synthetic view is different from that suggested by Kuhn's view of incommensurability. For some Kuhnians, the direction of scientific research is dictated mainly by social factors, having little to do with intellectual factors. For others, a set of intellectual factors are important within a particular paradigm, yet play no role, or are even incomprehensible, in a new paradigm. On the synthetic view, however, as many internal structures of earlier theories as possible have to be preserved.

The suggestion for future research by the synthetic view is also different from that made by the unilinear view. According to the unilinear view, the existing successful theory must be taken as a model for future development. Yet on the synthetic view, scientists should be advised to keep their minds open to all kinds of possibilities, since a new synthesis beyond the existing conceptual framework is always possible. Looking at the future development of field theory from such a perspective, it is quite possible that its future does not exclusively rely on the researches within the gauge field programme, which are trying to incorporate the Higgs mechanism and Yukawa coupling into the programme, but also on the use of the results produced by S-matrix theory, whose underlying ideas, such as those of ontology and the nature of forces, are radically different from those of the field programme.

Notes

1. See Zilsel (1942).
2. See, for example, Gross (1992) and Weinberg (1992).
3. See Cantor and Hodge (1981).
4. See Strawson (1950).
5. The above interpretation of the ontological status of energy will play an important role in the discussion of chapter 7.
6. See, for example, Redhead (1983).
7. For the ontological primacy of the gravitational fields over spacetime points, some arguments will be given in section 4.2.
8. In some non-linear field theories, such as the theory of solitary waves, this distinction is blurred. Then impenetrability becomes the most important criterion for individuality.
9. It happened first in Lorentz's theory of the electron in which the ether had no mechanical properties and was synonymous with the absolute rest space, and then in Einstein's special theory of relativity in which the concept of an ether was simply removed from its theoretical structure.
10. The propagation of virtual particles is only a metaphor, because no definite spatial trajectory of an identifiable virtual particle can be traced out. The ontological status of virtual particles is a difficult question. Some preliminary discussion can be found in Brown and Harré (1988). The difficulty in conceptualizing the propagation of virtual particles is covered up by Feynman with his path-integral formulation (see Feynman and Hibbs, 1965).

2

The rise of classical field theory

Although the developments that I plan to explore began with Einstein's general theory of relativity (GTR), without a proper historical perspective, it would be very difficult to grasp the internal dynamics of GTR and subsequent developments as further stages of a field programme. Such a perspective can be suitably furnished with an adequate account of the rise of the field programme itself. The purpose of this chapter is to provide such an account, in which major motivations and underlying assumptions of the developments that led to the rise of the field programme are briefly outlined.[1]

2.1 Physical actions in a mechanical framework

As we mentioned in chapter 1, two intellectual trends, the mechanization and mathematization of the world that occurred in the early modern period, effectively changed people's conceptions of reality and causality. According to mechanical philosophers, such as Descartes and Boyle, the physical world was nothing but matter in motion. According to the Neoplatonists, such as Kepler and Henry More, the physical world was mathematical in its structure. As a synthesis of the two, the inner reality of the physical world appeared as merely material bodies with their motions governed by mathematical laws. Here, matter can take either the form of plenum, as in the case of Descartes, or the form of corpuscles, as in the case of Gassendi, Boyle, and Newton. The difference between the two mechanical systems led to different understandings of physical action, as we shall see in a moment.

The mechanization of the world also implied that the true nature of phenomena, the essence and cause of all changes and effects, can be found in the motion of material bodies in space. For example, Boyle argued that the essence of colour was nothing but the displacement of molecules. But Kepler, inspired by Neoplatonism, maintained that the cause of phenomena, the reason 'why they are as they are' lay in the underlying mathematical structures; whatever was true mathematically must be true in reality. The difference in the conception of causality, namely, the efficient cause versus the formal cause, also led to different understandings of physical action, which, as we shall see, were intertwined with different understandings stemming from different mechanical systems.

An issue central to our interests is how to explain physical actions, such as gravity, electricity, and magnetism, that are transmitted apparently at a distance. In the Cartesian system, since there was no empty space, the apparent action at a distance had to be mediated by some subtle forms of matter, some ethereal substances, such as fire, air, and various effluvia through impact or pressure. A typical example of this kind of mechanical explanation was Descartes's theory of magnetism. According to Descartes, the effluvia of subtle matter circulated through the body of a magnet and the surrounding space in a closed loop. Since the flux of the effluvia rarefied the air between magnet and iron, they were forced together by the pressure of the external air. In this way, Descartes reduced the magnetic action, which was apparently transmitted across a distance, to a contact action caused by the motion of material particles.

In an atomistic system where atoms were separated by void space, it was still possible, as argued by Boyle, to explain apparent action at a distance by contact action, in terms of local motion of some mediating substances with impact or pressure receiving from and exerting upon bodies that acted upon one another. But the admittance of the void space entailed, logically, the possibility of accepting action at a distance as a reality. In history, the notion of action at a distance was widely accepted mainly because Newton's theory of gravity enjoyed empirical successes in astronomy while Descartes's vortex theory of gravity, based on the concept of contact action, failed, as was soon realized by Huygens. The difficulties in explaining such physical phenomena as cohesion, elasticity, and magnetism in terms of ultimate action by contact also helped its acceptance.

But another important factor that also contributed to its wide acceptance was the formal conception of causality fostered by the Neoplatonists. According to this conception, if certain phenomena could be subsumed under certain mathematical laws of force, then they were explained, and thus were intelligible and real. Hence the equation of Newton's second law of motion, $F = ma$, seemed to have established a causal relation, namely, force is the cause of a body's acceleration. Since the mathematical law was regarded as the true cause of the phenomena, there was no need to search for the agent of force. Some of

Newton's disciples and successors, particularly the British empiricist philosophers Locke, Berkeley, and Hume took this conception very seriously. Newton himself, deep inside, felt that the ability of bodies to act where they were not was unintelligible, although sometimes he was also inclined to take this formal conception of causality as a refuge.

The notion of action at a distance was criticized by Leibniz. By appealing to a metaphysical principle of continuity, according to which causes and effects were continuously connected to one another, Leibniz rejected the idea of corpuscles moving in a void space, because it meant discontinuous changes of densities of bodies at their boundaries. For him, a body never moved naturally except when touched and pushed by another body, and the occult quality of attraction could only happen in an explicable manner by the impulsion of subtle bodies.

Leibniz's criticism of action at a distance was rejected by Roger Boscovich, who based his rejection, ironically, also on the principle of continuity. If the ultimate particles of matter were finite, Boscovich argued, there would be a discontinuous change of densities at their boundaries; and if they came into contact, their velocities would change discontinuously and an infinite force would be required. Without invoking an infinite regression of elastic parts, Boscovich concluded that the primary elements of matter must be simple points of no extent, with the capacity to exert forces on one another with magnitudes depending on their mutual distances. That is, the impact of bodies must ultimately involve forces at a distance, which can be represented by a continuous function of the distance between particles.

How to interpret Boscovich's force function is an interesting question. Is it only an abstract relation? Or should we take it as representing a continuous reality? If it is only a relation, then how can it be transformed into an entity so that it can propagate independently through space? If it represents a continuous reality, then what is its nature and constitution, and how can we depict the transmission of action through this continuous medium? These questions are interesting, not only because of the intrinsic merits of Boscovich's ideas (replacing extended particles by mathematical points having mechanical properties, and taking force as a quasi-entity existing somehow independently of material particles), but also because of their impact on later philosophers and physicists, such as Kant and Faraday. As to Boscovich himself, he was well in the positivist tradition of mathematical physics. He defined force solely in terms of a mathematical function describing the change of motion, without any intention of examining the ultimate nature of force. For Boscovich, as for Newton, force did not denote any particular mode of action, nor any mysterious quality, but only the propensity of masses to approach and recede.

2.2 The continuous medium

Although the formal conception of causality, which had its roots in Neoplatonism, was reinforced by the new philosophical trends of empiricism and positivism, the search for an efficient cause of the apparent actions at a distance persisted. This search had its roots in the mechanical world view. Ultimately, mechanical explanation required identifying permanent substances that underlay mechanisms, with which the cause was effectively transmitted, step by step, to the effect. A good example of this irresistible urge can be found in Newton's speculation on the ether.

The idea of an ether had a long history, and Newton was not the initiator. It was so central to the mechanical world view that from the early rise of mechanical philosophy, it had become quite popular in the literature. The reason for this was that an ethereal substance could provide a medium for transmitting motion across a distance and, more importantly, a variety of ethereal fluids possessing non-mechanical qualities could be deployed for explaining various non-mechanical phenomena, such as cohesion, heat, light, electricity, magnetism, and gravitation. For example, Descartes and his followers had proposed several effluvia for explaining optical, electric, and magnetic phenomena.

As subtle media for transmitting actions or mediating agents for non-mechanical effects, various ethereal substances (such as optical and gravitational ether, electric, magnetic, and caloric fluids, fire and phlogiston, the changes in whose distribution or states would lead to observable changes in ordinary bodies) were supposed to be of a different nature from mechanism: they were rare, invisible, and intangible, capable of penetrating gross bodies and permeating all space. By appealing to such a notion of ether, Newton tried to give a causal explanation of gravity. Without invoking the Cartesian concept of pressure and impact exerted by contiguous particles, Newton interpreted gravitation as the effect of forces of repulsion exerted by the particles of a rarefied medium, dispersed unevenly in the gravitating bodies and throughout the void space, and operating on ponderable bodies by repulsive forces.

Newton took the ether as the cause of gravity and the agency of attraction, and derived his idea of an ether from the analogy of nature: just like the light rays, Newton thought that the ethereal fluids were also formed of hard and impenetrable particles. But whereas bodies could move and float in the ether without resistance, the ether particles, as the constituents of a subtle form of matter, were supposed to be much rarer, subtler, and more elastic, having no *vis inertiae* and acting by laws other than those of contact action. However, these specific features do not qualify Newton's ether as non-mechanical because it was still constituted of material particles. Moreover, the action at a distance of gravity was only replaced by repulsive forces that also acted at a distance separating the ether particles.

Nevertheless, there was an important feature in Newton's ether that signaled a departure from mechanical philosophy. That is, although consisting of material particles, which were inert in their nature, Newton's ether possessed repulsive forces that were active and could not be reduced to inertia. Thus, as an active agent of gravity, or more generally, as an active substance, Newton's ether, as realized by Kant, was the metaphysical principle of the constitution of forces.

The idea of an ether as an active substance, which was buried in Newton's speculative writings, was articulated more clearly by Leibniz. For Leibniz, the whole world, ordinary bodies as well as rays of light, contained thin fluids, the subtle matter. Even space was not devoid of some subtle matter, effluvia, or immaterial substance. All these ultimately were constituted by a force plenum. The force plenum was an active substance because force as the cause of motion embodied activity. Thus, for Leibniz, nature's dynamism was innate in the force plenum, and all particulars were aspects of one dynamic unity. As suggested by Peter Heimann and J. E. McGuire,[2] in the developments from Descartes through Boyle, Newton, Leibniz, and later scientists, we find a decline of inert extension and a rise of active forces in the ontology of substance. That is why I take activity as part of the definition of substance in modern science.

However, a profound difficulty, which we have already encountered in Newton's ether, is that any continuous medium, whether active or not, if it consists of material particles, cannot provide an ontological basis for an alternative mode of transmitting actions to the mode of action at a distance. The reason, as pointed out by Kant, is that every particle with a sharp boundary always acts upon something outside itself, and thus acts where it is not. Thus a real alternative to action at a distance requires a new conception of continuous medium.

Except for some speculations about force plenum, fire, or effluvia, a scientifically important proposal for such a conception did not occur until Thomas Young's wave theory of light in 1802, in which an elastic fluid was proposed as the medium for a new mode of transmission of action across a distance: a step by step transmission by the undulatory motions of the fluid. This luminiferous ether soon took the form of elastic solid in Fresnel's formulation of 1816.

At first, the luminiferous ether was still conceived as consisting of point-centers of force or particles and forces acting between them (Navier, 1821; Cauchy, 1828), and the continuous transmission of action was thus reduced to the action among the contiguous centers or particles. Thus action at a distance was not replaced by a real alternative, since Kant's argument applies here, but was only to be transferred to a lower level.

The concept of a really continuous and deformable medium, not consisting of dimensionless points or extended but rigid bodies with sharp boundaries, was developed in the 1840s by George Stokes and William Thomson in their mathematical studies of the luminiferous ether. In their new conception of a continuous medium, the action between the infinitesimal portions of the medium, into which we might conceive the medium to be divided, was transmitted by a redistribution of strain, not of parts, through the whole medium, and the notion of force was replaced by that of pressure and stress acting over infinitesimal portions of the medium. What characterizes their mathematical studies of continuous media is that these were purely formal procedures to calculate the changes of infinitesimal portions of the continuous medium, without any reference to the hidden connection of the medium system. A concept crucial to their studies was that of the potential, previously developed by Laplace, Poisson, and Green, which was intimately connected with transmitting action in a continuous medium.

Originally, a potential in Laplace's theory of gravitation, or in Poisson's electrostatics for that matter, was not defined as an actual physical property of the space, but only a potential property: a force would be exerted at a point in a space if a test mass (or charge) were introduced into the space at that point. That is, the gravitational (or electrostatic) potential had no property except the potentiality for exerting an attractive force on masses (charges) introduced into the space where the potential was defined. There were two ways of treating the potential. If treated with an integral formulation, in which all actions between particles at a distance were integrated, a potential exhibited no continuous transmission of action through a physically continuous medium, but was used only as an *ad hoc* device, with which the force at every point of a space could be defined as the resultant of sources acting at a distance. Then the concept of a potential seemed to be irrelevant to the study of continuous media.

But a potential was also a quantity satisfying a certain partial differential equation. Since the differential laws, first invented by Newton, related immediately adjacent states of a system, they entered physics well suited to expressing the action exerted between contiguous portions of a continuous medium and, in the case of partial differential equations, to representing the evolution of a continuous system. More specifically, in the differential equation $\nabla^2 V = 4\pi\rho$, the second derivative of a potential V in the neighborhood of any point was only related to the density at that point in a certain manner, and no relation was expressed between the value of the potential at that point and the value of ρ at any point at a finite distance from it. Thus formulated, the notion of the potential could be transformed into the notion of the continuous medium with physical reality. This continuous medium would provide a physical basis for an alternative mode of transmitting action, both for action at a distance and for action by contact. In this mode, each point of the medium was characterized by certain mathematical functions representing physical quantities, such as energy, presented there, and the apparent action at a distance was explained by the changes in the states of stress in the medium.

Turning the continuous potential into such a physically continuous medium was in harmony with the aim of searching for efficient causes of the apparent actions at a distance. But how could it be possible to take a mathematical device that was supposed to be continuous as a representation of a physical medium if the medium actually was discontinuous? One possibility is to take the discontinuity as being smoothed out in a physical process. Although physically imaginable, metaphysically it is unsatisfying. It also misses the point that there exists a mathematical model that is able to deal with action in a strictly continuous medium, and in which the principle of continuity is satisfied.

Some mathematical physicists in the 1840s treated the problem in a different manner. Stokes, for example, took the continuity, or fluidity or elasticity for that matter, not as a macroscopic effect of constituent particles, but as a condition that attained perfection only in a perfectly continuous substance. He also took the motion of ordinary continuous medium as determinable from the equations of perfectly continuous substance. Here we find a juncture at which the Neoplatonist conception of reality, 'whatever is true mathematically is true in reality', came into play and helped mathematical physicists to shape their novel conception of a continuous substance. Conceptually, this move had profound consequences for the development of field theory.

2.3 The mechanical ether field theory

The rise of field theory in the 19th century was the result of several parallel developments. Most important among them were, first, the success of the wave theory of light, mainly due to the work of Young and Fresnel, which assumed an elastic luminiferous ether as an ontological basis for explaining optical phenomena; and second, the investigations of electricity and magnetism, started by Faraday and elaborated by Thomson, Maxwell, and many other physicists, which led to a recognition that the luminiferous ether might also be the seat of electromagnetic phenomena and thus play an explanatory role in the theory of electromagnetism, similar to its role in the wave theory of light.

The electric and magnetic phenomena can be quite successfully described within the framework of action at a distance. Earlier examples, before the work by Faraday and Maxwell, can be found in the formulations of the French physicists Coulomb, Poisson, and Ampère; later cases can be found in the formulations of the German physicists Weber and Helmholtz. Thus the motivation for developing a different framework was mainly that of explanation rather than description. That is, the purpose was to find the underlying cause of the apparently distant actions, and to explain them in terms of transmission of action in a continuous medium.

But such an attempt had existed in the history of physics for a long time, certainly much earlier than Faraday and Maxwell coming onto the stage. If we take the concept of the field as signifying a region of space considered, with respect to the potential behaviour of test bodies moved about in it, as an agent for transmitting actions, or as a medium whose local pressure and stress was the cause of the behaviour of the test bodies, then, as pointed out by John Heilbron (1981), the electricians of the 1780s already had this concept, though not the word.

However, the concept of the field developed in the eighteenth century and the early nineteenth century was essentially an *ad hoc* device rather than a representation of physical reality. The reason was obvious: the field was conceived only as the field of a body's action without other properties, and thus had no independent existence. Taking this into consideration, we find that the novelty of the developments initiated by Faraday lay in transforming the *ad hoc* device into a representation of physical reality, by arguing and demonstrating that the medium had other properties in addition to that related with a body's potential behavior. Most important among the additional properties were continuity and energy.

The concept of the electromagnetic field as a new type of continuous medium was first derived from the analogies with fluid flow and, in particular, with elastic media. In 1844, Faraday modified Boscovich's hypothesis of material particles being point-centers of converging lines of force, and suggested that the electric and magnetic actions between neighboring particles were transmitted, not at a distance, but through lines of force. He further suggested that the electric and magnetic phenomena could be characterized by these lines of force, whose directions represented the directions of the forces. In contrast with, for example, Poisson's idea of the potential, Faraday's modification of Boscovich's idea appeared to be the first conception of a field theory of electromagnetism: while Poisson's potential was completely determined by the arrangement of the particles and charges, and thus could be eliminated from the description of electromagnetism, Faraday proposed an independent existence for continuous lines of forces, through which the electro-magnetic forces could be continuously transmitted.

At first, Faraday argued against identifying these abstract lines of force with some space-filling elastic media, such as the luminiferous ether consisting of subtle particles. However, influenced by Thomson, who took the luminiferous ether seriously, Faraday adopted a more sympathetic attitude towards the idea of an ether. He was willing to regard magnetic force as a function of the ether, although he admitted that he had no clear idea about the actual physical process by which the lines existed and transmitted the actions. For some time, Faraday was not quite sure about the ontological status of the lines of force: they may be a mental representation of the intensity and direction of the force at a point, or a kind of physical reality, either as a 'state of mere space', or as a state of a substantial, though not ponderable, medium which 'we may call ether' (Faraday, 1844).

For the continuous lines of force to be a physical reality, Faraday proposed four criteria: they had to (i) be modified by the presence of matter in the intervening space; (ii) be independent of the bodies on which they terminate; (iii) be propagated in time; and (iv) exhibit a limited capability of action, which amounted to assuming energy being seated in the intervening space. The last criterion was a necessary condition for the continuity and reality of the lines of force, or of the field as it was called later. It is not difficult to see that if the continuity of motion was assumed, the field had to possess some energy: if a particle were placed in the intervening space at a point previously empty of matter, it might acquire certain kinetic energy, and this energy had to come from the surrounding space; otherwise, the conservation of energy would be violated (if action were propagated between bodies in a finite time, and if its energy were not present in the intervening space in the interval, it would no be conserved during that interval). All these criteria would guide later transformation of the concept of the field from an *ad hoc* device to a representation of physical reality by Thomson, and particularly by Maxwell.

Maxwell would not have been able to proceed without Thomson's pioneering work. By mathematical analogy, Thomson suggested a formal equivalence, first, in 1845, between electromagnetic phenomena and Fourier's work on heat conduction, and then, in 1847, between electrostatics and electromagnetism on the one hand and the equilibrium conditions of an elastic solid on the other. He used Stokes' solutions to the general equation of equilibrium of an elastic solid for the case of pure strain to represent electrostatic action, and solutions for the case of pure rotation to represent electromagnetic action. In 1849 and 1850, Thomson applied this line of reasoning more systematically and argued that a theory of magnetism could be developed by investigating continuous distributions of magnetic material which he called the field.

At first, Thomson did not specify the details of the field, only suggesting, on the basis of Faraday's discovery of magneto-optical rotation, that the lines of magnetic force could be identified with the luminiferous ether, a rotational and elastic continuous medium. From this idea of a mechanical ether, Thomson proposed several theories in which the ether was a dynamic substance, underlying the unity of nature and differentiating itself into various particulars of nature. From the 1850s, Thomson began to be interested in speculating about the hidden machinery represented by his mathematical results. In 1856 he gave a mechanical representation of his formal results of 1847. Knowing about Helmholtz's work of 1858 on vortex motion, which showed that a certain vortex motion would have a kind of permanence, Thomson argued that, on the assumption that the ether was a perfect fluid throughout space, these permanent vertex rings of the fluid could be identified as ordinary atoms. This was the first model connecting the discrete atom with a continuous plenum, explaining matter by the ether, and avoiding particles acting at a distance or by contact.

Thomson's ideas about the ether were extremely influential in the second half of the 19th century. However, all the models he proposed were mechanical in character. Although he greatly helped Maxwell with his work on the continuous medium and the potential, Thomson himself did not properly understand Maxwell's theory of the electromagnetic field. In 1884 he claimed:

I never satisfy myself until I can make a mechanical model of a thing. If I can make a mechanical model I can understand it. As long as I cannot make a mechanical model all the way through I cannot understand: and that is why I cannot get the electromagnetic theory.

(Thomson, 1884)

Starting with Thomson's mathematical analogy, Maxwell went beyond it and into a physical analogy. From the similarity between electromagnetic and continuous phenomena, he argued that electric and magnetic forces must involve actions through a continuous medium. To develop his theory of the electromagnetic field, Maxwell translated the solutions of formally equivalent problems in the mechanics of a continuous medium and potential theory into the language of electricity and magnetism. From the resultant wave equations, Maxwell inferred that 'light consists in the transverse undulations of the same medium which is the cause of electric and magnetic phenomena' (1861/2). However, this identification of the magneto-electric and luminiferous media was not yet an electromagnetic field theory of light, for the 'transverse undulations' that constituted the light waves were not given any definite interpretation in terms of electromagnetic variables. What Maxwell had accomplished was not a reduction of light to electricity and magnetism, but rather a reduction of both to the mechanics of a single ether.

Maxwell's idea of an ether, like Thomson's, was also completely mechanical in character. In his 'On physical lines of force' of 1861/2, Maxwell successfully embodied Faraday's idea of continuous transmission of force across a distance in terms of a limiting case of a chain of material particles within a mechanical model. Soon afterwards, beginning in 1864, Maxwell retreated from this kind of specific account of the mechanical ether. His 1864 paper 'A dynamical theory of the electromagnetic field' was a turning point for his transition from designing specific models to adopting an abstract and generalized dynamic (Lagrangian) approach. To avoid any unwarrantable assumption concerning the specific mechanism of the ether, and still maintain his commitment to the mechanical world view, Maxwell derived his famous wave equations by using the Lagrangian variational principle (1873). He felt that this dynamical theory was mechanical because Newtonian mechanics can be reformulated by using the Lagrangian formalism.

In a sense Maxwell was right. His interpretation of the laws he discovered was mechanical because his ether was a Newtonian dynamical system. There can be non-mechanical interpretations of his laws if the underlying dynamical system is taken to be non-Newtonian. But Maxwell was not ready for such a radical revision of the concept of the field. He gave a mathematical formulation of the field-theoretical conception of electric and magnetic action. But he did not fully articulate his concept of the field, because his field was only conceived as a state of an unknown dynamical system, the ether. When he was specific, his ether appeared as a limit of a dielectric of polarized particles. As a state of such a mechanical substance, the field did not have an ontologically independent status, neither in the sense of being irreducible, nor in the sense of being coequal with matter.

It is often claimed that one of Maxwell's most important contributions to field theory was that he established the reality of the field energy with his wave equations. It is true that he had indeed said that 'in speaking of the energy of the field, however, I wish to be understood literally'. It is also true that the solutions of his equations showed the time delay in transmitting the electromagnetic actions, which implied that some physical processes must be taking place in the intervening space. Thus what Faraday's lines of force represented were something really existing in this intervening space. But does this mean that Maxwell took the field as an ontologically independent physical reality that carries energy transmitted from one body to another? Far from it. With respect to what was propagating, Maxwell said:

We are unable to conceive of propagation in time, except either as the flight of a material substance through space, or as the propagation of a condition of motion or stress in a medium already existing in space.

(1873)

With respect to where the energy resided he said:

How are we to conceive this energy as existing in a point of space, coinciding neither with the one particle nor with the other? In fact, whenever energy is transmitted from one body to another in time, there must be a medium or substance in which the energy exists after it leaves one body and before it reaches the other.

(1873)

The purpose of this argument was to show that a medium in which the propagation took place and the energy resided was indispensable. Thus what Maxwell's famous 'energy arguments' established was not the reality of the energy of the field, but rather the reality of the energy of the ether.

Maxwell left no room for doubt that his conception of a mechanical medium as the seat of the field was incompatible with another understanding of the field, which 'is supposed to be projected from one particle to another, in a manner quite independent of a mechanical medium' (Maxwell, 1873). What Maxwell's arguments were directed against was Carl Neumann's view of the field, which was proposed in his theory of potential energy in 1868. Neumann's conception of the field went beyond the framework of a mechanical ether field and was well in the tradition of the concept of the force–energy field.

Neumann was not the first to contribute to the formation of the concept of the force–energy field in German physics. Although German physics in the nineteenth century generally favored the concept of action at a distance, there was a profound intellectual tradition in Germany which in its spirit was in favor of a field theory. What I mean here is the Leibnizian–Kantian tradition of rationalist metaphysics, which was influential among 19th century German intelligentsia through the natural philosophy of Schelling and Hegel.

From the principle of unity of substance and activity, Leibniz derived his concept of substance as an intensive continuum of force. This substance was a dynamic plenum, rather than an extensive plenum of matter, which pervaded the whole universe, underlay the unity of nature, and differentiated itself into the varied particulars of nature. Similarly, Kant wrote:

The elementary system of the moving forces of matter depends upon the existence of a substance (the primordially originating force) which is the basis of all moving forces of matter . . . .. There exists a universally distributed all penetrating matter within the space it occupies or fills through repulsion, which agitates itself uniformly in all its parts and endlessly persists in this motion.[3]

The above ideas emerged repeatedly in German physics. For example, Julius Mayer maintained the notion of a metaphysical force substance which was independent of matter but still had the same reality status as matter. To describe both light and heat transmitted through space, Mayer required not only a material ether to carry light waves, but also an independently transmitted and apparently immaterial force of heat. Mayer's view of a space-filling force field was resurrected in the new trend of energeticism, in which energy was taken to be the basis of all reality, in the last decade of the 19th century.

Another stimulus to the development of field theory in Germany came from Gauss's speculation (1845) that there must be some forces which would cause the electric actions to be propagated between the charges with a finite velocity. His pupil Bernhard Riemann, earlier than Neumann, made some efforts to develop mathematical expressions for the propagation of the potentials in time. Initially, Riemann (1853) attempted to analyze the processes of gravitation and light in terms of the resistance of a homogeneous ether to change of volume (gravitation) and shape (light). He reasoned that gravitation might consist of a continuous flow of an imponderable space-filling ether into ponderable atoms, which would depend on the pressure of the ether immediately surrounding the ponderable atoms, and the pressure in turn would depend on the velocity of the ether. Thus no action at a distance would exist. Riemann conjectured that the same ether would also serve to propagate the oscillations that were perceived as light and heat. A serious problem with Riemann's speculations concerned the possible ways in which new ether could be continuously created.

A few years before Maxwell established his electrodynamics, Riemann's more sophisticated unified ether field theory took shape. Riemann assumed that the cause of both motion and change of motion of a body at any point should be sought in 'the form of motion of a substance spread continuously through the entire infinite space' (1858). He called this space-filling substance the ether, and proposed certain motions of a homogeneous ether which could reproduce the partial differential equations of gravitation and light propagation. The first of the equations was a continuity equation about the flux of the ether. The second equation was a wave equation for transverse oscillations in velocity W (not oscillation in displacement) propagating at the speed of light. The combination of the two motions for gravity and light produced a well-behaved velocity function, which confirmed the possibility of unifying the two processes.

In a paper presented to the Göttingen Society of Science in 1858, Riemann proposed an electromagnetic theory of light:

I have found that the electrodynamic effects of galvanic current may be explained if one assumes the action of one electrical mass on the rest does not occur instantaneously but propagates to them with a constant velocity (equal to the velocity of light c within the limits of observational error). The differential equation for the propagation of electrical force, according to this assumption, will be the same as that for the propagation of light and radiant heat.

(Published in 1867 posthumously)

With the help of the concept of 'variations of density in the ether', Riemann proposed replacing Poisson's equation for the electrostatic potential $\nabla^2 V + 4\pi\rho = 0$ by the equation $\nabla^2 V - 1/c^2 \partial^2 V/\partial t^2 + 4\pi\rho = 0$, according to which the changes of potential would be propagated outward from the charges with a velocity c (1861a). Thus he found it possible to incorporate electricity and electromagnetism into his programme of unifying the physical sciences.

In characterizing Riemann's unification programme, we find that Riemann sought to replace discrete forces acting at a distance with continuous retarded actions between neighboring elements in the ether, with states and processes in the ether. In this way he provided forces with an ontological basis, and made the states of the ether and their variations the reality of force. This was a shift of considerable importance for the concept of energy, because the forces in this scheme can be understood as potential energy when forces were spread throughout space with the variations of the ether's states. Riemann's programme was also dynamic in character. He maintained that 'the laws of ether motion must be assumed for the explanation of phenomena', and the foundation for the dynamics of the ether was the principle of continuity of motion and the maximum–minimum condition on the potential (1858). These ideas were similar to those of the dynamical theories of Thomson, Maxwell, and, some time later, of Fitzgerald, Lodge, and Larmor.

Riemann's idea of an ether as a means of spreading forces in space transformed force into energy states of the ether, and thus reinforced a more general recognition of energy as an independent and conserved entity, as in the work of Neumann. To describe the apparent force acting at a distance between particles of Weber's electric ether, Neumann developed Riemann's earlier theory of propagated forces and retarded potentials in a new way. In a paper presented to the Göttingen Society of Science (1868), Neumann considered 'potential [energy] as primary, as the characteristic motive power, while conceiving force as secondary, as the form in which that power manifests itself'. He then demonstrated that a potential propagating with the velocity of light would entail all of the known laws of electromagnetic action. In this way, he made Weber's ether largely superfluous for the description of the propagation of light. Although he might have had in mind a material basis (such as an ether field) for propagating his potential, he made no reference to it, and thereby undermined the necessity of such a material foundation for a force–energy field.

By ignoring the unobservable ether, taking potential energy as fundamental and propagating through space, Neumann gave to it the independent status of an energy field, somewhat similar to Mayer's space-filling force field. So with good reasons the later energeticists looked back to Neumann as well as to Mayer as pioneers of their viewpoint.

Now we can see what the point at issue was in the disagreement between Maxwell and Neumann: could there be an ontologically irreducible physical substance that was different from material substance, and was responsible for continuously transmitting the electromagnetic action? Adhering to the mechanical conception of substance, Maxwell, in his criticism of Neumann, denied this possibility, and thus failed in this respect to contribute to the new understanding of the nature of the electromagnetic field.

The dominance of Maxwell's, and particularly Thomson's, mechanical view of the ether as a universal substratum in the thinking of British physicists soon led to the rise of a mechanical world view. Previously, there were electric particles of two kinds, bearing specific forces; magnetic particles of two kinds with their specific forces; and heat particles and light particles. As a result of the works of Thomson and Maxwell, however, this variety of particles and forces soon disappeared from the literature, giving way to a variety of motions in the ether: the phenomena of electricity and magnetism were explained as the motions and strains of the ether; heat was also viewed as nothing but patterns of motion in the ether. The underlying assumption was that all natural phenomena should be explainable by the dynamics of the universal ethereal substratum.

Ironically, however, the later developments of British physics following Maxwell's dynamical theory of the electromagnetic field of 1864 can be characterized as a process of demechanization and incorporation of the German, especially the Leibnizian, ideas of a non-mechanical but primordial substance. For example, W. K. Clifford was influenced by Riemann. Heinrich Hertz was influential in Britain after his discovery of electromagnetic waves, which made him and the general public believe in the reality of the field. In addition, Leibniz's idea of a primordial substance can be found in almost every British physicist's work, including those of Thomson and Larmor.

Echoing Thomson's dynamical programme of a universal ether, that the whole physical world was emergent from the vortex motion of the ether, Lodge said:

One continuous substance filling all space: which can vibrate as light; which can be sheared into positive and negative electricity; which in whirls constitutes matter; and which transmits by continuity, and not by impact, every action and reaction of which matter is capable. This is the modern view of the ether and its functions.

(1883)

However, Lodge maintained that 'the properties of ether must be somewhat different from that of ordinary matter'. Thus he hinted at the non-mechanical nature of the ether.

Fitzgerald, who was concerned with the ether theory through the 1880s and 1890s, also held the view that 'all of nature was to be reduced to motions in the universal plenum' (1885). According to Fitzgerald, there were two kinds of vortex motion. First, there were localized vortex atoms that constituted matter. Second, there were vortex filaments strung out throughout the universal fluid, and conferring on the fluid as a whole the characteristic properties of ether – its being the bearer of light and various forces. That is, both matter and ether were emergent from the universal plenum as a result of its satisfying exactly MacCullagh's system of dynamical assumptions, which, as Stokes pointed out, appeared to be unrealizable by any material dynamical system.[4] Here we find another hint of a non-mechanical ether.

Larmor was the last of the great British ether theorists. He synthesized the works of his predecessors and carried the existing trends to their logical conclusion. This synthesis was achieved in the context of a dynamical programme that represented the extreme development of the traditional mechanical ether theory, and thus paved the way for going beyond it. Here the dynamical programme means the Lagrangian formulation. Traditionally, in the Lagrangian approach to the ether theory, the abstract treatment would be followed, sooner or later, by an attempt to give a specific account of the hidden machinery that gave rise to the effects abstractly treated by the Lagrangian formalism – either in a realistic way, or at least by means of an illustrative model. But Larmor suggested that

the concrete model is rather of the nature of illustration and explanation, by comparison of the intangible primordial medium with other dynamical systems of which we can directly observe the phenomena.

(1894)

Thus, a crucial change in outlook was brought about by having the tail wag the dog: whereas the abstract Lagrangian approach had previously been regarded as a step towards the ultimate theory, which would display the ether as a specific mechanical system, now the abstract theory was in Larmor's view to be regarded as the final and ultimate theory, while specific models were significant only for heuristic or pedagogical reasons.

Larmor soon went even further in relinquishing the fundamental tenets of the British ether theory. In 1894 he gave up the vortex atom as the basis of mechanical matter. In 1900 he claimed that 'matter may be and likely is a structure in the ether, but certainly ether is not a structure of the matter', thus parting company with the material ether tradition of Thomson. 'Chiefly under the influence of Larmor', commented E. T. Whittaker (1951), 'by the close of the century, it came to be generally recognized that the ether is an immaterial medium, not composed of identifiable elements having definite locations in absolute space.'

This recognition helped physicists to take the idea of an energy field more seriously. The convertibility and conservation of various forms of energy indicated that energy embodied the unity of all natural power. Thus an energy field seemed to be a suitable candidate for the universal substratum. Although the foundation of the conservation of energy was still conceived to be mechanical energy, the non-mechanical implications of an energy field made the penetration of this conception into physicists' consciousness a necessary step for the transition from the conception of a mechanical ether field to a conception of an independent non-mechanical electromagnetic field.

2.4 The electromagnetic field theory

In the mechanical ether field theory, the electromagnetic field was more than an intellectual device for explaining apparent actions at a distance in terms of a continuous medium transmitting local actions. Rather, because the electromagnetic field possessed some properties other than those associated only with the transmission of actions (most important among them were energy and the property that the transmission of action through the field had the velocity of light and took time), the electromagnetic field could be conceived to be a representation of physical reality. However, since the electromagnetic field was conceived by the British physicists only as a state of the mechanical ether, rather than an independent object, I have called their theory the mechanical ether field theory instead of electromagnetic field theory.

Aside from continuity, the distinction between the ether and ordinary matter was quite vague. This was particularly so in the unified ether theory of matter of Thomson and Larmor, in which the material particles were conceived either as smoke rings or vortical atoms in the plenum, as suggested by magneto-optical rotation, or as centers of rotational strains in the elastic ether. Since both ether and ordinary matter were characterized by mechanical terms, there was no clear-cut distinction between the two. Of course, the electromagnetic field obeyed Maxwell's equations, while the material particles obeyed Newton's laws of motion. Yet, since the electromagnetic field was only a state of the mechanical ether that was supposed to obey Newton's laws, it was always assumed by the ether theorists that the electromagnetic laws could be reduced to the mechanical laws, although the failure to design a convincing mechanical model had postponed the task of reduction for the future. Moreover, since the relation between charge and ordinary matter was out of the scope of the ether field theory, the interactions between ordinary matter and the ether could only be dealt with in mechanical rather than electromagnetic terms. This further disqualifies the theory as an electromagnetic field theory.

The dynamical approach to electromagnetism, which was developed by the British physicists from Maxwell to Larmor and by Hertz and Lorentz on the Continent, and in which no mechanical details about the ether were postulated, was crucial for the rise of the electromagnetic field theory. Hertz, for example, was still committed to the idea of a mechanical ether, which could be represented in terms of the motion of hidden masses, and whose parts were connected by a mechanical structure. However, since Hertz detached the equations from any specific mechanical interpretation, and thus completely separated the mathematical formalism from its representation by mechanical models, his axiomatic approach fostered the development of the idea that the electromagnetic field had no mechanical properties.

A crucial step in developing an electromagnetic field theory was taken by Lorentz. In contrast with the British physicists who were interested in the mechanical constitution of the ether underlying the electromagnetic field, Lorentz gave his attention to the electric constitutions of material bodies and of the ether that was thought to be perfectly transparent to the uncharged matter. He first separated the ether completely from matter, and then dealt with the interactions between the ether and matter using his theory of the electron. In his ontological dualism, the material bodies were supposed to be systems of electrified particles (ions or electrons) that were embedded in the ether, and whose properties, such as charge or even mass, were electromagnetic rather than mechanical in nature. His ether was divested of all mechanical properties, and thus was separated from matter completely. The electromagnetic fields in this scheme were taken to be the states of the ether. Since the ether had no mechanical properties and its properties were just the same as in a void space, which underlay both the electromagnetic field and matter, the electromagnetic field enjoyed the same ontological status as matter. That is, it represented a physical reality independent of matter rather than a state of matter, and possessed, just as matter did, energy, and thus qualified as a non-mechanical substance.

Lorentz's conception of the ether was subject to several constraints and open to different interpretations and assessments.[5] Since the conception was crucial to his discussion of the relationship between the ether and matter, and also crucial to his general view of the physical world, it merits a few more words. Lorentz took the stationary character from Fresnel's conception of the ether, but modified it. For him, the ether was everywhere locally at rest. This means 'that each part of this medium does not move with respect to the other and that all perceptible motions of celestial bodies are relative motions with respect to the ether' (Lorentz, 1895). Thus the ether served as a universal reference frame, which gave correct measurements of length and time and played the same role played by absolute space in Newton's mechanics.

Moreover, Lorentz's ether was, in a special sense, substantial. The substantiality of the ether was required by his contraction hypothesis, which in turn was required by his commitment to the stationary ether. Lorentz tried to explain away the negative result of the Michelson–Morley experiment of 1887 by assuming that, in addition to the electric and magnetic forces, the molecular forces responsible for the dimension of a body would also be affected when the body moved through the ether, and thus the resultant contraction of the body's dimension would eliminate the effects to be expected within Fresnel's framework of the stationary ether. As a cause of a physical phenomenon, the alteration of the molecular force, which was supposed to be caused by some unknown physical interaction between the ether and the molecules of a body, the ether must be substantial so that it could interact with the molecules. However, since Lorentz's ether was perfectly transparent to uncharged matter, and no assumption was made about its structure, its substantiality inferred from the hypothesis of physical contraction was not enough to differentiate it from the vacuum.

In his discussion of the relationship between matter and the ether, Lorentz generalized Maxwell's dynamical approach and assumed that the dynamical system consisted of two parts: matter constituted by charged particles (electrons), and the electromagnetic fields that represented the states of the ether. Thus the relationship was reduced to a purely electromagnetic one, the interactions between electrons and the fields, no mechanical connection between the ether and matter being assumed to exist. More precisely, the presence and motion of a charged particle changed the local state of the ether and thus altered the field with which the charged particle acted on other charged particles at a later time. That is, the field resulted from the presence of charges and the change of their distribution, and acted on charges by exerting a certain 'pondero-motive' force, the so-called Lorentz force.

Lorentz's discussion of the relationship between matter and the ether amounted to the electrodynamics of moving electrons in the electromagnetic field. In his formulation, the kinetic and potential energies of the field were rederived and turned out to be the same as given by Maxwell's results. The only addition was his equation for the Lorentz force. Thus his electrodynamics appeared as a synthesis of Maxwell's concept of the electromagnetic field, in which the actions were propagated at the speed of light, and Continental electrodynamics, in which the electric actions were explained in terms of forces acting between the charged particles.

A special feature of Lorentz's electrodynamics was that while the charged particles experienced the ponderable force from the ether (through the field), the ether and the field experienced no reaction from the charged particles. This was necessary for maintaining his stationary ether hypothesis. Poincaré criticized Lorentz for his violation of Newton's third law of motion, the equality of action and reaction. But for Lorentz the ether was not a system of masses capable of motion and interacting by forces.[6] Thus, for Lorentz, not only were the ether and fields non-mechanical, but the charged particles that constituted material bodies were also non-mechanical in the sense that they violated Newton's third law of motion.

It is true that Lorentz believed in the absolute character of space and time, and retained the ether as an absolute reference frame. In spite of this, however, Lorentz began a major revision of the foundations of physics, because he rejected the universal validity of Newton's laws of motion, rejected the mechanical foundations of the electromagnetic field, and envisaged the foundations of physics on a purely electromagnetic ontology and concepts. Not only was the validity of the electrodynamical equations assumed to be fundamental without underlying mechanical explanation, but the laws of nature were reduced to properties defined by the electromagnetic equations. In particular, the laws of mechanics were viewed as special cases of the universal electromagnetic laws. Even the inertia and mass of a material body, according to Lorentz, had to be defined in electromagnetic terms, and could not be assumed constant. Thus a fundamental principle of Newton's mechanics was rejected.

Lorentz's electrodynamics was well received in the 1890s. In an antimechanistic cultural climate, which was testified to by Mach's philosophical criticism of mechanics and the rise of energeticism, Lorentz's work was influential and played a decisive role in the cultural debate between the electromagnetic versus mechanistic world views. Einstein, who had taken it as his mission to carry the field theory programme started by Faraday, Maxwell, and Lorentz to its completion, in his final years once commented on the importance of the step taken by Lorentz in developing the field theory:

It was a surprising and audacious step, without which the later development [of the field theory programme] would not have been possible.

(Einstein, 1949)

Notes

1. For details, see Cantor and Hodge (1981), Doran (1975), Harman (1982a, b), Hesse (1961) and Moyer (1978).
2. See McGuire (1974).
3. Cf. Wise (1981), from which I have quoted.
4. See Stein (1981).
5. For a different interpretation and assessment, see, for example, Nersessian (1984).
6. As pointed out by Howard Stein, the Lorentz force was not derived from Maxwell's stress defining the flow of momentum, otherwise there would have been no violation of the third law that amounted to anything but the conservation of momentum. In that case, since part of the flow transferred some momentum to the moving bodies through the Lorentz force, the amount of the momentum of the field must be changed. Since the mid-19th century, the conversion and conservation of energy was widely accepted, which implied the acceptance of non-kinematic forms of energy. And this made plausible non-kinematic forms of momentum, which can be ascribed to the electromagnetic fields without definite masses moving with definite velocities as their bearers. Thus the apparent conflict between mechanics and electrodynamics is not irreconcilable. For details, see Stein (1981).

Part I

The geometrical programme for fundamental interactions

The rise of classical field theory had its deep roots in the search for an efficient cause of apparent actions at a distance. In the case of electromagnetism, Thomson and Maxwell in their ether field theory succeeded in explaining the distant actions by introducing a new entity, the electromagnetic field, and a new ontology, the continuous ether. The field possessed energy and thus represented physical reality. But as a state of the mechanical ether, it had no independent existence. In Lorentz's electrodynamics, the field was still a state of the ether. However, since Lorentz's ether was deprived of all material properties and became synonymous with a void space, the field enjoyed an independent ontological status on a par with matter. Thus in physical investigations there emerged a new research programme, the field theory programme based on a field ontology, in contrast with the mechanical programme based on a particle ontology (together with space and force).

The new programme acquired a fresh appearance in Einstein's special theory of relativity (STR), in which the superfluous ontology of the Lorentz ether was removed from the theoretical structure. But in some sense the field theory programme was not yet completed. In Lorentz's electrodynamics as well as in STR, the fields had to be supported by a space (or spacetime). Thus the ultimate ontology of these field theories seemed not to be the fields, but the space (or spacetime), or more exactly the points of spacetime. As we shall see in chapter 4, this hidden assumption concerning the ultimate ontology of a field theory was not without consequences. With the development of the general theory of relativity (GTR), however, particularly with his point-coincidence argument (see section 4.2), Einstein showed that spacetime had no independent existence, but only existed as structural properties of the gravitational fields. Perhaps for this reason, Einstein regarded GTR as the completion of the classical field theory programme. I take GTR as the starting point of my discussion of 20th century field theories, because it was the first theory in which the field was the only ontology responsible for transmitting forces.

As the agent for transmitting forces, the gravitational fields in GTR functioned as geometrical structures of spacetime. In fact, GTR started a sub-programme within the field theory programme, the geometrical programme. Further developments along this line attempted to associate other force fields (such as the electromagnetic fields) or even matter and its properties (such as spin and quantum effects) with the geometrical structures of spacetime. The sub-programme was successful in some areas, such as relativistic astronomy and cosmology, and failed in other areas. Conceptually, however, the investigations within this sub-programme have provided a basis for a geometrical understanding of gauge field theory, and for a synthesis of quantum field theory (QFT) and GTR (see section 11.3).

3

Einstein's route to the gravitational field

Einstein in his formative years (1895–1902) sensed a deep crisis in the foundations of physics. On the one hand, the mechanical view failed to explain electromagnetism, and this failure invited criticisms from the empiricist philosophers, such as Ernst Mach, and from the phenomenalist physicists, such as Wilhelm Ostwald and Georg Helm. These criticisms had a great influence on Einstein's assessment of the foundations of physics. His conclusion was that the mechanical view was hopeless. On the other hand, following Max Planck and Ludwig Boltzmann, who were cautious about the alternative electromagnetic view and also opposed to energeticism, Einstein, unlike Mach and Ostwald, believed in the existence of discrete and unobservable atoms and molecules, and took them as the ontological basis for statistical physics. In particular, Planck's investigations into black body radiation made Einstein recognize a second foundational crisis, a crisis in thermodynamics and electrodynamics, in addition to the one in the mechanical view. Thus it was 'as if the ground had been pulled out from under one, with no firm foundation to be seen anywhere, upon which one could have built' (Einstein, 1949).

Einstein's reflections on the foundations of physics were guided by two philosophical trends of the time: critical scepticism of David Hume and Mach, and certain Kantian strains that existed, in various forms, in the works of Helmholtz, Hertz, Planck, and Henri Poincaré. Mach's historico-conceptual criticism of Newton's idea of absolute space shook Einstein's faith in the received principles, and paved for him a way to GTR. But a revised Kantian view, according to which reason played an active role in the construction of scientific theory, exerted a perhaps even more profound and persistent influence upon Einstein's thinking, and virtually shaped his style of theorizing. For example, Poincaré's emphasis on 'the physics of the principle', (1890, 1902, 1904) together with the victory of thermodynamics and the failure of 'constructive efforts', convinced Einstein that 'only the discovery of a universal

47

formal principle could lead to assured results' (Einstein, 1949). A universal principle discovered by Poincaré and Einstein was the principle of relativity. This principle applied to both mechanics and electrodynamics, and thus served as a firm foundation for physics as a whole. An extension of the principle of relativity, from the case of inertial reference frames, which was dealt with in STR, to the general case of non-inertial reference frames, incorporated gravitational fields into the scheme through another principle, the equivalence principle. And this led Einstein to GTR, a theory of gravitational fields.

3.1 Guiding ideas

It is well known that Einstein's critical spirit was nurtured by his reading the writings of Hume and Mach when he was young. While Hume's skepticism and his relational analysis of space and time might penetrate into young Einstein's unconsciousness, Mach's influence was more tangible. Although Mach himself liked to insist that he was neglected, his empirico-criticism actually exerted enormous influence from the 1880s on. Einstein (1916d) once commented that 'even those who think of themselves as Mach's opponents hardly know how much of Mach's view they have, as it were, imbibed with their mother's milk'. What circumstances made Mach so influential?

First, the revival of empiricism and the spread of a phenomenalist attitude among practicing scientists, as a reaction to the speculative natural philosophy, created a receptive climate for Mach's empirico-criticism. For example, Boltzmann (1888) wrote of Gustav Kirchhoff that he 'will ban all metaphysical concepts, such as forces, the cause of a motion'. Second, the mechanical view, according to which physical reality was characterized by the concepts of material particles, space, time, and forces, failed to explain electromagnetism. In the domain of electromagnetic phenomena, after Hertz's experiment in which the electromagnetic wave was detected, the wave theory prevailed, in which physical reality was characterized by the continuous wave field. A mechanical explanation of the field, based on the concept of the ether, was undermined by the negative result of the Michelson–Morley experiment. As a response to the failure, there emerged a more radical view of energeticism, in addition to the electromagnetic one. The energeticists rejected the existence of atoms, and believed in the ultimate continuity of nature; they took the laws of thermodynamics as their foundations, and energy as the only ultimate reality, and thus presented an anti-mechanical view.

In justifying their radical criticism of the mechanical view, some energeticists (Ostwald, Helm, and J. T. Merz) appealed to positivism: hypothetic entities, such as atoms and the ether, were to be omitted because their properties were not accessible to direct observation. Taking such a position, Merz (1904), for example, issued a call 'to consider anew the ultimate principles of all physical reasoning, notably the scope of force and action of absolute and relative motion'. It is not difficult, therefore, to see why Mach's empirico-criticism, which provided an epistemology for the phenomenologically oriented interpretation of new branches of physics, was so influential.

But Mach's major strength, as far as Einstein was concerned, came from his historico-conceptual analysis of mechanics. As a follower of Hume's empirical skepticism, Mach wanted to make clear the empirical origin of suspicious concepts by tracing their roots in the history of the concepts. According to Mach's general theory of knowledge, scientific constructs, such as concepts, laws, and theories, were just tools for economically describing experience and factual information. His study of the history of science revealed, however, that some constructs, though originally provisional, in time received metaphysical sanctions (because of successful applications) and became virtually unassailable. Thus insulated from the ongoing scene of action of science, where criticism, revision, and rejection were the rule, these constructs acquired the status of logical necessity, or received acceptance as self-evident and intuitively known truths. Mach accepted no such invulnerable truths.

Mach's major book, *The Science of Mechanics*, first published in 1883, was well known for his devastating critique of what he called the 'conceptual monstrosity of absolute space'. It was a conceptual monstrosity because

no one is competent to predicate things about absolute space and absolute motion; they are pure things of thought, pure mental constructs, that cannot be produced in experience. All our principles of mechanics are . . . experimental knowledge concerning the relative positions and motions of bodies.

(Mach, 1883)

Mach's criticism of absolute space consisted of four points: (i) space had no independent existence; (ii) it was only meaningful to talk of space in terms of the material objects in the universe and their properties and relations; (iii) motion was always relative to another object; (iv) the inertial forces arose in objects rotating or accelerating with respect to a material frame of reference. This criticism displayed Mach's iconoclasm and courage against the intrusion of empirically ungrounded a priori metaphysics in science; it also showed that Mach's empiricism was a powerful weapon for a critical reevaluation of classical physics.

Einstein read Mach's *Science of Mechanics* around the year of 1897. He recalled that 'the book extended a deep and persisting impression upon me, . . .

owing to its physical orientation towards fundamental concepts and fundamental laws' (1952c), and that the book shook his dogmatic faith that mechanics was 'the final basis of all physical thinking' (1949). According to Einstein, GTR came 'from skeptical empiricism of somewhat the kind of Mach's' (with Infeld and Hoffmann, 1938), and 'the whole direction of thought of this theory [GTR] conforms with Mach's, so that Mach quite rightly is considered as a forerunner of general relativity theory' (1930b).

While Humean–Machian skepticism liberated Einstein from the dogmatic faith in the universality of the principles of classical physics (mechanics, and thermodynamics as well as electrodynamics), in his search for a new conception of the firm foundations of physics, Einstein was mainly guided by a revised version of Kantianism. It is true that with the rise of classical field theory and the discovery of non-Euclidean geometries, many a priori principles (such as the axioms of Euclidean geometry and those about absolute space and action at a distance) that Kant took as indispensable for the construction of experience were discarded, and a widespread conviction was that Kant's apriorism in general, and his view of space and geometry in particular, were no longer tenable.

Yet in terms of the structure and cognitive content of scientific theories, Kantian rationalism, in the sense that (i) universal principles prescribed by reason were constitutive of our experience of the world, and were able to provide a consistent and coherent world picture that can serve as the foundation of physics, and (ii) all empirical knowledge of the actual world must conform with these principles, survived the developments in physics and mathematics. Although the universal formal principles were no longer taken to be fixed a priori, but as historically evolving or mere conventions, the legitimacy of theoretical constructions was powerfully argued for by many physicists and mathematicians as prominent as Planck and Poincaré, whose works Einstein studied carefully.

It is well known that Einstein's reading of Poincaré's book *Science and Hypothesis* (1902), before his formulation of STR, was an experience of considerable influence on him.[1] Some influences were tangible, others much subtler. In the former category we find Poincaré's denial of absolute space and time, and of the intuition of simultaneous events at different places; his objection to absolute motion; his reference to the principle of relative motion and to the principle of relativity. In order to locate the subtler influence of Poincaré upon Einstein, we have to take Poincaré's conventionalist epistemology seriously.[2]

The major epistemological problem for Poincaré was how objective knowledge and its continuous progress were possible in spite of apparently disruptive changes in mathematics and physics (Poincaré, 1902). In addressing this quasi-Kantian problem, Poincaré, taking into consideration the discovery of non-Euclidean geometries, the crisis of the mechanical view, and his own idea of the physics of the principles (see below), rejected the orthodox Kantian positions on (absolute) space, (Euclidean) geometry, and (Newtonian) physics. For Poincaré, science as a whole is empirical rather than a priori in nature.

Yet Poincaré was also fully aware that without being conceived in terms of an intellectual system, the mute occurrences would never become empirical facts; and that without understanding, no sense-appearances would become scientific experience, just as Kant (1783) asserted. Thus, according to Poincaré, a scientific theory must involve, in addition to empirical hypotheses, fundamental or constitutive hypotheses or postulates. Unlike the former, the latter, as the basic language of a theory, were the result of a conventional choice, not contingent on experimental discoveries. Empirical facts may guide the choice or change of language, but they play no decisive role in this respect. They could decisively affect the fate of empirical hypotheses, but only within a chosen language.

Then what was the objective content of science in terms of which the continuous progress of science could be defined? Poincaré's answer to this question can be summarized with three notions: the physics of the principles, the relativity of ontology, and structural realism.

According to Poincaré, unlike the physics of the central force, which desired to discover the ultimate ingredients of the universe and the hidden mechanisms behind the phenomena, the physics of the principles, such as the theories of analytic dynamics of Lagrange and Hamilton and Maxwell's theory of electromagnetism, aimed at formulating mathematical principles that systematized experimental results achieved on the basis of more than two rival theories, expressed the common empirical content and mathematical structure of these rival theories, and thus were neutral to different theoretical interpretations but susceptible to any of them.

The indifference of the physics of the principles to ontological assumptions was approved by Poincaré, because it fitted rightly into his conventionalist view of ontology. Based on the history of geometry, he accepted neither an a priori ontology that was fixed absolutely and had its roots in our mind, nor an intrinsic ontology that can be discovered by empirical investigations. For Poincaré, ontological assumptions were just metaphors, they were relative to our language, and thus were changeable.

But in the transition from an old theory with its ontology to a new one, some structural relations expressed by mathematical principles and formulations, in addition to empirical laws, might remain true if they reflected physical reality.

Poincaré is well known for his search for the invariant forms of physical laws. But the philosophical motivation behind this drive may be more interesting. Influenced by Sophus Lie, Poincaré took the invariants of groups of transformations, which described the structural features of collections of objects, as the foundation of geometry as well as physics (as a quasi-geometry). Here we have touched on the core of Poincaré's epistemology, namely, structural realism. Two features of this position are worth noticing. First, we can have objective knowledge of the physical world. Second, this knowledge is relational in nature. We can grasp the structures of the world, but we can never reach the things-in-themselves. This further justifies the claim that Poincaré's position was only a revised version of Kantianism.

Looking at the crisis of physics, which became apparent after the Michelson–Morley experiment of 1887, from such a perspective, Poincaré realized that what was needed was a theoretical reorientation, or a transformation of the foundations of physics. However, for Poincaré, what was crucial for the progress of physics was not the change of metaphor, such as the existence or non-existence of the ether, but the formulation of powerful mathematical principles that reflected the structures of the physical world. Among these principles that of relativity stood prominently in his mind, and he contributed substantially to its formulation.

In contrast with his frequent references to Mach, Einstein rarely mentioned Poincaré's name. Yet the salient affinity between Einstein's rationalist style of theorizing and Poincaré's revised Kantianism escapes nobody who is familiar with Einstein's work. An illuminating text, in this regard, is Einstein's Herbert Spencer Lecture at Oxford, entitled 'On the method of theoretical physics' (1933), where he summarized his position as follows: 'The empirical contents and their mutual relations must find their representation in the conclusions of the theory.' But the structure of the theoretical system 'is the work of reason', 'especially the concepts and fundamental principles which underlie it . . . are free inventions of the human intellect'. 'Experience may suggest the appropriate mathematical concepts, but they most certainly cannot be deduced from it . . . the creative principle resides in mathematics, . . . I hold it true that pure thought can grasp reality' with more unity in the foundations. As a genetic empiricist, as Poincaré was, Einstein acknowledged that the intuitive leap in the creation of concepts and principles must be guided by experience of the totality of sensible facts, and that 'experience remains, of course, the sole criterion of physical utility of a mathematical construction'. However, since 'nature is the realization of the simplest conceivable mathematical ideas, . . . pure mathematical constructions . . . furnish the key to the understanding of natural phenomena'.

Emphasizing the affinity, however, should not blind us to the difference in theorizing between Einstein and Poincaré. The main difference lies in the fact that on the whole, Einstein was more empirical than Poincaré. This is particularly true in terms of their assessments of the structure of specific theories. In the case of STR, Einstein was not as indifferent as Poincaré was to the existential status of the ether. Not only did Einstein take the hypothetic physical entity of the ether more seriously than Poincaré, but his willingness to remove the superfluous yet not logically impossible concept of the ether also reflected the empiricist influence of Mach's principle of the economy of thought. In the case of GTR, concerning the geometrical structure of physical space, Einstein significantly deviated from Poincaré's geometrical convention-alism and adopted an empirical conception of the geometry of physical space. That is, he took the geometry of physical space as being constituted and determined by gravitational fields (see section 5.1).

3.2 The special theory of relativity (STR)

In addition to Einstein, the special theory of relativity (STR) is inseparably associated with the names of Lorentz (transformations), Poincaré (group), and Minkowski (spacetime). Minkowski's work (1908) was a mathematical elabora-tion of Einstein's theory. Einstein acquainted himself with the problems that were central and ideas that were crucial to the formulation of STR by carefully studying the works by Lorentz (e.g. 1895) and Poincaré (e.g. 1902). Poincaré's STR (1905) was a completion of Lorentz's work of 1904, and the latter was a result of Poincaré's criticism (1900) of Lorentz's earlier work. Yet one of Lorentz's motivations in his (1892, 1895, 1899) was to solve the puzzle posed by the negative result of the Michelson–Morley experiment. Although it can be argued otherwise,[3] historically at least, the Michelson–Morley experiment was the starting point of a sequence of developments which finally led to the formulation of STR.

What was central to the developments was the concept of the ether. Lorentz took Fresnel's stationary ether as the seat of electromagnetic fields, which together with the electron were taken to be the ultimate constituents of the physical world (the so-called electromagnetic world view). Although the ether was considered by Lorentz to be endowed with a certain degree of sub-stantiality, it had nothing to do with energy, momentum, and all other mechan-ical properties except that it was at absolute rest. Thus the real function of the immobile ether in Lorentz's theory was only as an embodiment of absolute space, or as a privileged frame of reference, with respect to which the states of physical objects, whether they were at rest or in motion, could be defined, and in which the time was taken to be the true time and the velocity of light was constant.

Thus unlike the mechanical ether, the Lorentz ether allowed electromagnetic fields to enjoy an ontological status on a par with matter. Yet the retreat from Newton's conception of relative space (all inertial frames of reference, with a linear uniform motion with respect to each other at a constant velocity, were equal in the description of mechanical phenomena, no one having privilege over the others) posed a big puzzle for Lorentz's electromagnetic world view. The concept of an immobile ether, by which stellar aberration could be explained in terms of the partial entrainment of light waves by bodies in its motion relative to the ether (see Whittaker, 1951), was simply incompatible with the principle of relativity valid in mechanics and implied that there were effects of bodies' absolute motion with respect to the ether. The Michelson–Morley experiment showed that no such effects were detectable by optical (or equivalently by electromagnetic) procedures. This suggested that the principle of relativity seemed also valid in electrodynamics. If this were the case, the ether would be deprived of its last function as the privileged frame of reference and become superfluous, although not in contradiction with the rest of electro-dynamics.

Committed to the privileged status of the ether, Lorentz tried to explain the apparent relativity in electrodynamics, namely the persistent failure with optical means to detect the effects of bodies' absolute motion with respect to the ether, by assuming that the real effect of absolute motion was compensated by other physical effects, the contraction of the moving bodies in the direction of bodies' motion through the ether, and the apparent slowing down of clocks in the motion through the ether. Lorentz took these effects as caused 'by the action of the forces acting between the particles of the system and the ether', and represented it with his newly proposed rules for second order spatial transformation (Lorentz, 1899). Here the empiricist tendency and *ad hoc* nature of Lorentz's position are obvious: the observed relativity in the electromagnetic processes was not taken to be a universally valid principle but as a result of some specifically assumed physical processes, whose effects canceled each other, and thus in principle could never be observed.

Poincaré (1902) criticized Lorentz's *ad hoc* hypothesis and encouraged him (i) to search for a general theory whose principle would explain the null result of the Michelson–Morley experiment; (ii) to show that the electromagnetic phenomena occurring in a system could be described in terms of equations of the same form whether the system was at rest or in uniform translational motion; and (iii) to find transformation rules between frames of reference which were in relative uniform motion, so that with respect to these transformations the forms of the equations were invariant.

Lorentz's (1904) was designed to agree with Poincaré's principle of relativity according to which 'optical phenomena depend only on the relative motions of the material bodies' (Poincaré, 1899). Mathematically, except for some minor errors, Lorentz succeeded in presenting a set of transformation rules, the so-called Lorentz transformations, under which the equations of electrodynamics were invariant. But the spirit of Lorentz's interpretation of the Lorentz transformations failed to agree with the principle of relativity.

The main reason for the failure was that it treated different reference systems in different ways. In system S, whose coordinate axis was fixed in the ether, true time was defined; in system S', which was in relative motion to S, although the transformed spatial quantities were taken as representing a real physical change, the transformed temporal coordinate (local time) was taken only as an auxiliary quantity for mathematical convenience, without the same significance as true time. Thus Lorentz's interpretation was asymmetrical, not only in the sense that the spatial and temporal transformations had different physical status, but also in the sense that the transformations between two systems were not taken to be reciprocal in nature, and hence did not form a group and did not ensure strict invariance of the laws of electrodynamics. Moreover, Lorentz restricted his transformations to the laws of electrodynamics, and retained the Galilean transformations for mechanical systems.

For these reasons, Lorentz (1915) admitted that 'I have not demonstrated the principle of relativity as rigorously and universally true', and it was 'Poincaré . . . [who] obtained a perfect invariance of the equations of electrodynamics and formulated the "postulate of relativity", terms which he was the first to employ'.

Here, Lorentz referred to two facts. First, during the period from 1895 to 1904 Poincaré formulated the principle of relativity in terms of (i) the impossibility of detecting absolute motion by any physical method, and (ii) the invariance of physical laws in all inertial frames of reference.[4] If we also take notice of Poincaré's postulate of the constancy of the speed of light (1898),[5] then as convincingly argued by Jerzy Giedymin (1982; see also Zahar, 1982), the non-mathematical formulation of the principle of relativity and the postulate of the limiting and constant speed of light should be credited to Poincaré.

Second, in his (1905), Poincaré assumed from the outset that 'the principle of relativity, according to which the absolute motion of the earth cannot be demonstrated by experiment, holds good without restriction'. From such an understanding of relativity as a universal principle, Poincaré presented precise Lorentz transformations, interpreted them symmetrically by imposing group requirements, and expressed them in the form of a transformation group, under which the laws of electrodynamics were invariant. Thus Poincaré formulated STR, simultaneously with Einstein, in its mathematical form.

Poincaré's achievement was not an accident but the result of a long intellectual journey, begun with his abandonment of absolute space and time, and continued with his criticism of absolute simultaneity (1897, 1898, 1902),[6] which was based on his postulate of the constancy of the speed of light. The journey was driven by his increasing skepticism of the electromagnetic world view in general and of the existence of the ether in particular, and the skepticism was guided by his conventionalist epistemology. In *Science and Hypothesis* Poincaré declared:

We do not care whether the ether really exists; that is a question for metaphysicians to deal with. For us the essential thing is that . . . [it] is merely a convenient hypothesis, though one which will never cease to be so, while a day will doubtless come when the ether will be discarded as useless.

(1902)

Poincaré was indifferent to the hypothesis of the ether because, in line with his view of the physics of the principles (see section 3.2), his major interest was in the content common to two or more rival theories, which was statable in the form of a general principle. Thus what was important to him was not the hypothesis of the ether itself, but its function within the theory for which it might be responsible. Since it was increasingly clear that the hypothesis of the ether was in conflict with the principe of relativity, in his St Louis Lecture (1904), Poincaré dismissed the suggestion that the ether should be viewed as part of the physical systems to which the principle of relativity applied.

There is no doubt that Einstein had drawn inspiration and support for his first work on relativity from the writings of Poincaré. Yet Einstein's simultaneous formulation of STR had its own merit: it was simpler in physical reasoning and saw more clearly the physical implications of the principle of relativity, although Poincaré's formulation was superior in its mathematical elaboration (see next section).

Einstein is famous for solving the ether puzzle by cutting the Gordian knot. At the beginning of his (1905c) he said:

The unsuccessful attempts to discover any motion of the earth relative to the 'light medium', suggest that the phenomena of electrodynamics as well as of mechanics possess no properties corresponding to the idea of absolute rest. They suggest rather that . . . the same laws of electrodynamics and optics will be valid for all frames of reference for which the equations of mechanics hold good. We will raise this conjecture (hereafter called the Principle of Relativity) to the status of a postulate . . ..

In the same sentence Einstein added his second postulate: 'light is always propagated in empty space with a definite velocity c which is independent of the state of motion of the emitting body', without giving an explicit reason. While the first postulate was firmly entrenched in the tradition of mechanics, the second was accepted only as a consequence of the wave theory of light, in which the luminiferous ether was the medium for light to travel in. Einstein did not refer in his second postulate to the ether, but to any inertial frame. Since the Galilean transformations could not explain the constancy of the speed of light, the Lorentz transformations with their apparently local time had to play a role in relating phenomena between inertial systems.

Under the Lorentz transformations, the equations of electrodynamics remained unchanged, but the laws of mechanics changed because, in Lorentz's theory of the electron, they were transformed according to the Galilean transformations. Thus Einstein found that central to the ether puzzle was the notion of time employed by Lorentz, whose theory was burdened with *ad hoc* hypotheses, unobservable effects, and asymmetries. The transformation rules for the laws of electrodynamics and mechanics depended on two different notions of time: a mathematical notion of local time for the former, and a physical notion of true time for the latter. But Einstein's extension of the principle of relativity to electrodynamics implied the equivalence of inertial systems related by the Lorentz transformations. That is, the local time was the true time, and the true times in different inertial systems differed because each of them depended on its relative velocity. This meant that there was no absolute time and simultaneity, and the notion of time had to be relativized.

The relativization of time, together with the rejection of interpreting Lorentz's spatial transformations as physical contraction caused by the *ad hoc* molecular forces, led Einstein to the conclusion that no state of motion, not even a state of rest, can be assigned to the ether. Thus the ether, as a privileged frame of reference for the formulation of the laws of electrodynamics, proved to be superfluous and amounted to nothing but the empty space with which no physical processes were associated.

The removal of the ether undermined both the mechanical and the electromagnetic world views, and provided a foundation for the unity of physics. In the case of electrodynamics, Einstein's two postulates were in harmony with the new kinematics, which was based on the relativistic notion of space and time and associated with the Lorentz transformations. For mechanics and the rest of physics, the old kinematics, which was based on the notion of absolute time and absolute simultaneity (although the notion of space was relative) and associated with the Galilean transformations, was in conflict with Einstein's postulates. The unity of physics based on Einstein's two postulates required an adaptation of mechanics and the rest of physics to the new kinematics. The result of the adaptation is a relativistic physics, in which simultaneity depends on the relative state of moving systems, the transformation and addition of velocities obey new rules, and even mass is no longer an invariant property of a physical entity, but is dependent on the state of motion of the system with respect to which it is described, and thus is changeable.

3.3 The geometrical interpretation of STR

According to Felix Klein (see Appendix A1), the geometry of an n-dimensional manifold is characterized as the theory of invariants of a transformation group of an n-dimensional manifold. Klein's elegant and simple approach was promptly accepted by most mathematicians of the time. A surprising application of this idea was the discovery by Poincaré and Hermann Minkowski that STR was none other than a geometry of a four-dimensional spacetime manifold,[7] which can be characterized as the theory of invariants of the Lorentz group.[8]

In his 1905 paper, Poincaré introduced several ideas of great importance for a correct grasp of the geometrical import of STR. He chose the units of length and time so as to make $c = 1$, a practice that threw much light on the symmetries of relativistic spacetime. He proved that Lorentz transformations form a Lie group. He characterized the group as the linear group of transformations of four-dimensional spacetime, which mixed spatial and temporal coordinates but left the quadratic form $s^2 = t^2 - x^2 - y^2 - z^2$ invariant. Poincaré further noted that if one substituted the complex valued function (it) for t, so that (it, x, y, z) were the coordinates in four-dimensional space, then the Lorentz transformation was simply a rotation of this space about a fixed origin. Besides, he discovered that some physically significant scalars and vectors, e.g. electric charge and current density, can be combined into Lorentz invariant four-component entities (which were subsequently called 'four-vectors'), thus paving the way for the now familiar four-dimensional formulation of relativistic physics.

In spite of all these contributions by Poincaré, however, the geometrical interpretation of STR was mainly developed by Minkowski. In his (1907), Minkowski grasped the most profound significance of the principle of relativity as bringing about 'a complete change in our ideas of space and time': now 'the world in space and time is in a certain sense a four-dimensional non-Euclidean manifold'. In (1908), he noted that the relativity postulate (about the Lorentz invariance of physical laws) 'comes to mean that only the four-dimensional world in space and time is given by phenomena'.

According to Minkowski, physical events invariably include place and time in combination. He called 'a point of space at a point of time, that is, a system of values x, y, z, t' 'a world point'. Since everywhere and everywhen there was substance, he fixed his 'attention on the substantial point which is at the world point'. Then he obtained, as an image of 'the everlasting career of the substantial point, a curve in the world, a world-line'. He said: 'In my opinion physical laws might find their most perfect expression as reciprocal relations between these world-lines.' For this reason, he preferred to call the relativity postulate 'the world postulate', which 'permits identical treatment of the four coordinates x, y, z, t'. He declared: 'Henceforth space by itself, and time by itself, are doomed to fade away into mere shadows, and only a kind of union of the two will preserve an independent reality.'

The change in the ideas of space and time brought about by STR included the relativization of time, which made inertial systems equivalent (only as different partitions of physical events into classes of simultaneous events) and the ether superfluous. As a result, the absolute space embodied in the ether was replaced by the relative spaces of inertial systems, although non-inertial systems were still defined with respect to an absolute space. In Minkowski's reformulation of STR, the relative space and the relative time of each inertial system can be obtained by projections of a single four-dimensional manifold, 'the world', as Minkowski called it, or as it is now usually called, Minkowskian spacetime.

Minkowski took STR as having endowed the four-dimensional manifold with a fixed kinematic structure that was independent of dynamical processes. This captured the essence of STR as a new kinematics. The kinematic structure appeared as a chrono-geometry characterized by the Minkowskian metric $g_{ij} = \text{diag}(1, -1, -1, -1)$, which uniquely determined the inertial structure of Minkowskian spacetime (defined by a free-falling system and characterized by affine connection).[9] Minkowski represented physical entities as geometrical objects at a world-point (he called them spacetime vectors). Since the representation was independent of frames of reference, it was uniquely determined. Thus the four-dimensional manifold was conceived as a stage of physical events, and physical laws were expressed as relations between geometrical objects.

In this way Minkowski reduced the Lorentz invariance of physical laws to a chrono-geometrical problem in the four-dimensional manifold. Minkowskian chrono-geometry is characterized by the Lorentz invariance of the Minkowskian interval defined as $s^2 = g_{ij}(x^i - y^i)(x^j - y^j)$ (here i, $j = 0, 1, 2, 3$), and this invariance fixed g_{ij} as $\text{diag}(1, -1, -1, -1)$ globally. The fact that Euclidean distance is always non-negative while the corresponding Minkowskian interval can be positive, negative or even zero for non-zero $(x^i - y^i)$ implies that the chronogeometry characterized by the Lorentz (or Poincaré) group is in effect non-Euclidean.

In Minkowski's interpretation, it becomes clearer that STR is a theory of space and time, characterized by a kinematic group, the Poincaré group, defined in the four-dimensional Minkowskian spacetime manifold. The Poincaré group and its invariants determine the chrono-geometrical structure, which in turn determines the inertial structure of Minkowskian spacetime. The last point is true, however, only when there is no gravitation. But gravitation was going to come to the focus of Einstein's attention soon after his work on STR in (1905c).

3.4 The introduction of gravitational fields: the principle of equivalence

STR eliminated the absolute space associated with the ether as a privileged frame of reference, only to reestablish the relative spaces associated with the inertial frames of reference as a class of privileged frames of reference, or the static Minkowskian spacetime as a privileged frame of reference. If the spirit of the principle of relativity was to argue for the invariant forms of physical laws under the changes of the frames of reference, namely, to deny the existence of any privileged frame of reference, then, Einstein asked himself, 'Is it conceivable that the principle of relativity is also valid for systems that are accelerated relative to one another?' (1907b).

The pursuit of this research agenda finally led to the discovery of the general theory of relativity, which in Einstein's mind always referred to the extension of the principle of relativity, or to the elimination of privileged frames of reference. Historically, and also logically, the starting point of this pursuit was the exploration of the proportionality between inertial and gravitational masses and the formulation of what was later to be called the equivalence principle.

In examining the general meaning of mass–energy equivalence ($M = m_0 + E_0/c^2$, where M is the mass of a moving system and E_0 is its energy content relative to the rest system), which was one of the consequences of the relativity of simultaneity, Einstein found that 'a physical system's inertial mass and energy content appear to behave like the same thing', both of them being variable. More importantly, he also found that in the measurement of mass variation, it was always implicitly assumed that it could be carried out with the scale, and this amounted to assuming that mass–energy equivalence

holds not only for the inertial mass, but also for the gravitational mass or, in other words, inertia and gravity are under all circumstances exactly proportional . . . The proportionality between inertial and gravitational masses is valid universally for all bodies with an accuracy as precisely as can currently be determined; ... we must accept it as universally valid.

(Einstein, 1907b)

On the basis of the proportionality or equivalence between inertial and gravitational masses, Einstein suggested a further equivalence of profound significance, with the help of a gedanken experiment.

Consider two systems Σ_1 and Σ_2. Σ_1 moves in the direction of its x axis with a constant acceleration γ relative to Σ_2 that remains at rest. In Σ_2 there is a homogeneous gravitational field in which all bodies fall with an acceleration $-\gamma$. Einstein wrote:

As far as we know, the physical laws referred to Σ_1 do not differ from the laws referred to Σ_2; this is due to the fact that all bodies undergo the same acceleration in the gravitational field. Hence, in the present state of our experience, we have no inducement to assume that the systems Σ_1 and Σ_2 differ in any respect. We agree therefore to assume from now on the complete physical equivalence of a gravitational field and the corresponding acceleration of the reference frame.

(1907b)

Einstein referred to this statement as the equivalence principle (EP for short) in (1912a), and called this idea 'the happiest thought of my life' in (1919).

For the extension of the principle of relativity, Einstein introduced, via EP, the gravitational field as a constitutive factor in the inertio-geometrical structure of spacetime. EP deprived the uniform acceleration of absoluteness, seemed to reduce the gravitational field to the status of relative quantity (see comments on this below), and thus made inertia and gravity inseparable. Yet the only empirical evidence Einstein cited for EP was that 'all bodies undergo the same acceleration in the gravitational field'. What this fact supports is just the so-called weak EP, which entails that mechanical experiments will follow the same course in a uniformly accelerated system and in a system at rest in a matching homogeneous gravitational field. But what Einstein claimed was the so-called strong EP, which assumed that two such systems were completely equivalent, so that it was impossible to distinguish between them by means of physical experiments of any kind. In the following discussion, EP always refers to the strong EP.

Conceptually, EP extended 'the principle of relativity to the case of the uniformly accelerated translation of the reference system' (Einstein, 1907b). The heuristic value of EP 'lies in the fact that it enables us to substitute a uniformly accelerated reference frame for any homogeneous gravitational field' (1907b). Thus 'by theoretical consideration of processes which take place relative to a system of reference with uniform acceleration, we obtain information as to how the processes take place in a homogeneous gravitational field' (1911). As we shall see in section 4.1, this interpretation of EP played a guiding role in Einstein's understanding of the (chrono-)geometrical structures of space(-time) in terms of gravitational fields.

Originally, the classical gravitational field (in the sense that it was similar to the Newtonian scalar gravitational field), introduced, through EP, as a coordinate effect, was of a special kind. It was defined on the relative space associated with a uniformly accelerated frame of reference, rather than on the four-dimensional spacetime. Thus the occurrence of the field depended on the choice of frame of reference, and could always be transformed away. Moreover, since EP was valid only for the uniformly accelerated frame in Minkowskian spacetime, the inertial structure of Minkowskian spacetime (the affine connection determined by the Minkowskian metric g_{ij}) and the classical gravitational field obtained through a coordinate transformation could only be taken as a special case of the general gravitational field that is a four-dimensional extension of the classical gravitational field.

It is often argued that since EP cannot be applied to the non-uniformly accelerated and rotational frames of reference for introducing arbitrary gravitational fields, it fails to relativize arbitrary acceleration in general, fails to eliminate the privileged frames of reference, and thus fails to be a foundation of GTR.[10] It is true that EP has only very limited applicability. It is also true that even the combination of EP and STR is not enough for the formulation of GTR. The reason is simple. What EP establishes is only the local equivalence of free-falling non-rotating frames of reference with the inertial frames of reference of STR. But the Minkowskian metric that characterizes the inertial systems of STR, as a reflection of the flatness of Minkowskian spacetime, is fixed. So it cannot serve as a foundation of GTR, in which the gravitational fields are of a dynamic nature, reflecting the curved nature of spacetime.[11]

However, the methodological importance of EP lies in the fact that if it is possible to prove (through a transformation of the coordinate system which was connected, e.g., with a change from an inertial frame of reference to a uniformly accelerated one) that the inertial structure of the uniformly accelerated frame of reference is indistinguishable from the classical gravitational field, then it is imaginable that the inertial structures of rotational and arbitrarily accelerating frames of reference (obtainable by non-linear transformations of the coordinates) might also be equivalent to general gravitational fields: those fields that correspond to the inertial structures of rotational frames would be velocity dependent, and those that correspond to the inertial structures of arbitrarily accelerating frames would be time dependent. If this were the case (we shall call this assumption the generalized version of EP), then all kinds of frames of reference would have no intrinsic states of motion, and all of them could be transformed to inertial ones. In this case, they are equivalent, and the difference between them comes solely from different gravitational fields.

Thus what EP suggests is not the dependence of the gravitational field on the choice of frame of reference. Rather, it suggests a new way of conceiving the division between relative spaces and the gravitational fields. If the origin of structures responsible for the inertial effects (affine connections) lies in the gravitational fields rather than relative spaces, then the latter possess no intrinsic quality and all their structures have to be conceived as being constituted by gravitational fields. Here we find that the internal logic of extending the principle of relativity inevitably led to a theory of gravitation.

The significance of EP for GTR can be established in two steps. First, GTR for Einstein meant that all Gaussian coordinate systems (and the frames of reference to which they were adapted) were equivalent. The generalized EP had shifted all physically meaningful structures from relative spaces associated with frames of reference to the gravitational fields, so the frames of reference had no intrinsic structures and could be represented by arbitrary coordinate systems. None of them had any privilege over the others. In the sense that the generalized EP took away the physical significance of any coordinate systems, and thereby made them equivalent, it was a necessary vehicle for Einstein to use to achieve the goal of GTR. Second, in an arbitrary coordinate system, the Minkowskian metric g_{ij} became changeable. In view of EP, it could be taken as a gravitational field. Although it is only a special kind of gravitational field, the Minkowskian metric g_{ij} provided a starting point for the extension to a four-dimensional formulation of an arbitrary gravitational field. In this sense, EP also provided Einstein with an entrance to the theory of gravitation, the physical core of GTR.

It is interesting to notice that, as in his postulation of the principle of relativity and the principle of the constancy of the speed of light, Einstein in his formulation of EP and his exploration of a more speculative generalized EP provided us with no logical bridge from empirical evidence to universal principles. People are often puzzled by the fact that Einstein's principles were not 'deduced from experience by "abstraction"'. Yet as Einstein later confessed: 'All concepts, even those which are closest to experience, are from the point of view of logic freely chosen conventions' (1949). From the discussion of Einstein's exploitation of EP, we may get some flavor of how deep and how powerful the influence of Poincaré's conventionalism was upon Einstein's style of theorizing. The great success of Einstein's research may encourage philosophers of science to explore the source of the fruitfulness of the conventionalist strategy.

Notes

1. See, for example, Seelig (1954), p. 69.
2. It is interesting to notice that Poincaré's epistemological position has become the subject of controversy in recent years. Interested readers can find a disagreement between Gerald Holton (1973) and Arthur Miller (1975) on the one hand, who took Poincaré the theoretical physicist as mainly an inductivist, and Jerzy Giedymin (1982) and Elie Zahar (1982) on the other, who took Poincaré's conventionalist position as a decisive factor in his independent discovery of the special theory of relativity.
3. For example, Holton (1973) presents powerful arguments for the irrelevance of the Michelson–Morley experiment to Einstein's formulation of STR.
4. For example, in 1895 Poincaré asserted the impossibility of measuring 'the relative motion of ponderable matter with respect to the ether'. In 1900, he told the Paris International Congress of Physics that 'in spite of Lorentz', he did not believe 'that more precise observations will ever reveal anything but the relative displacements of material bodies' (1902). In his 1904 St Louis lecture, he formulated the principle of relativity as one 'according to which the laws of physical phenomena must be the same for a stationary observer as for an observer carried along in a uniform motion of translation, so that we do not and cannot have any means of determining whether we actually undergo a motion of this kind' (1904).
5. 'Light has a constant speed . . .. This postulate cannot be verified by experience, . . . it furnishes a new rule for the determination of simultaneity' (Poincaré, 1898).
6. 'There is no absolute time . . .. Not only have we no direct intuition of the equality of true time intervals, but we do not even have a direct intuition of the simultaneity of two events occurring at different places' (Poincaré, 1902).
7. Time was explicitly referred to as a fourth dimension of the world in some 18th century texts. In his article 'Dimension' in the *Encyclopedie*, D'Alembert suggested thinking of time as a fourth dimension (see Kline, 1972, p. 1029). Lagrange, too, used time as a fourth dimension in his *Mechanique Analytique* (1788) and in his *Theorie des Functions Analytiques* (1797). Lagrange said in the latter work: 'Thus we may regard mechanics as a geometry of four dimensions and analytic mechanics as an extension of analytic geometry.' Lagrange's work put the three spatial coordinates and the fourth one representing time on the same footing. However, these early involvements in four dimensions were not intended as a study of chronogeometry proper. They were only natural generalizations of analytic work that was no longer tied to geometry. During the 19th century this idea received little attention, while *n*-dimensional geometry became increasingly fashionable among mathematicians.
8. The extended (non-homogeneous) Lorentz group is also called the Poincaré group, which is generated by the homogeneous Lorentz transformations and translations.
9. Cf. Torretti (1983).
10. See Friedman (1983).
11. The same idea can be used against the infinitesimal notion of EP. For detailed discussion on this point, see Torretti (1983), Friedman (1983), and Norton (1989).

4

The general theory of relativity (GTR)

In comparison with STR, which is a static theory of the kinematic structures of Minkowskian spacetime, GTR as a dynamical theory of the geometrical structures of spacetime is essentially a theory of gravitational fields. The first step in the transition from STR to GTR, as we discussed in section 3.4, was the formulation of EP, through which the inertial structures of the relative spaces of the uniformly accelerated frames of reference can be represented by static homogeneous gravitational fields. The next step was to apply the idea of EP to uniformly rotating rigid systems. Then Einstein (1912a) found that the presence of the resulting stationary gravitational fields invalidated the Euclidean geometry. In a manner characteristic of his style of theorizing, Einstein (with Grossmann, 1913) immediately generalized this result and concluded that the presence of a gravitational field generally required a non-Euclidean geometry, and that the gravitational field could be mathematically described by a four-dimensional Riemannian metric tensor $g_{\mu\nu}$ (section 4.1). With the discovery of the generally covariant field equations satisfied by $g_{\mu\nu}$, Einstein (1915a–d) completed his formulation of GTR.

It is tempting to interpret GTR as a geometrization of gravity. But Einstein's interpretation was different. For him, 'the general theory of relativity formed the last step in the development of the programme of the field theory . . . . Inertia, gravitation, and the metrical behaviour of bodies and clocks were reduced to a single field quality' (1927). That is, with the advent of GTR, not only was the theory of spacetime reduced to the theory of gravitational-metric fields, but for any theory of other fields, if it is to be taken as consistent, the interactions between these fields and the metric fields should be taken seriously. This claim is justified by the fact that all fields are defined on spacetime, which is essentially a display of the properties of metric fields and is dynamical and, quite likely, capable of interacting with other dynamical fields.

When Einstein claimed the dependence of spacetime upon the metric fields, what he meant was more than the describability and determination of the spacetime structures by the metric fields. Rather, in his reasoning towards general covariance, Einstein finally took a metaphysical position whereby the existence of spacetime itself was ontologically constituted by the metric field (section 4.2). Thus the dynamical theory of the metric field is also a dynamical theory of spacetime as a whole, which, initially, was conceived as a causal relationship between the metric fields and the distribution of matter and energy (section 4.3), and various solutions to the field equations can serve as foundations for various cosmological models (section 4.4). Here, we find that GTR provides us, literally, with a world picture. According to this picture, the gravitational fields individuate the spacetime points and constitute various spacetime structures for supporting all other fields and their dynamical laws.

4.1 The field and geometry

Conceptually, EP implies, or at least heuristically suggests, the possibility of an extension of the principle of relativity from inertial systems to arbitrary systems. The extension can be achieved by introducing gravitational fields as causal agents responsible for different physical contents of relative spaces associated with different physical frames of reference. To obtain a physical theory of general relativity, however, Einstein had first to free himself from the idea that coordinates must have an immediate metrical significance. Second, he had to obtain a mathematical description of the gravitational field that was solely responsible for the metrical properties of spacetime. Only then could Einstein proceed to find the correct field equations. A crucial link in this chain of reasoning, as John Stachel (1980a) pointed out, was Einstein's consideration of the rotating rigid disk.

An outstanding and mysterious feature of GTR is its dual use of the tensor $g_{\mu\nu}$ as the metric tensor of the four-dimensional Riemannian spacetime manifold, and also as a representation of the gravitational field. However, Einstein did not mention, until 1913, the metric tensor as a mathematical representation of the gravitational field, nor mention any need for a non-flat (non-Euclidean) spacetime, when he was still concerned only with the static gravitational field, based on the consideration of EP in terms of uniformly accelerated frames.

It is true that even in the context of STR, spacetime as a whole, as proclaimed by Minkowski, had to be conceived as a non-Euclidean manifold. Yet a three-dimensional relative space, associated with an inertial system and represented by a congruence of parallel time-like worldlines, was still Euclidean in nature. Also, the relative spaces (associated with the uniformly accelerated frames of reference in the application of EP) provided no clear indication of their geometrical nature: whether they were Euclidean in nature or otherwise. Only when Einstein tried to extend his investigations of the gravitational field from the static case to the non-static but time-independent case, the simplest one of which occurred in the rotating rigid disk problem, was he convinced that a non-Euclidean spacetime structure was needed for describing the gravitational field.

According to STR, rigid bodies cannot really exist. But it was possible to define, as an idealization: (i) rigid inertial motion, (ii) a rigid measuring rod, and (iii) a dynamical system capable of exerting rigid inertial motion.[1] Assuming that this was the case, Einstein found that

in a uniformly rotating system, in which, on account of the Lorentz contraction [the measuring rod applied to the periphery undergoes a Lorentzian contraction, while the one applied along the radius does not], the ratio of the circumference of a circle to its diameter would have to differ from π.

(1912a)

Thus a basic assumption of Euclidean geometry was not valid here.

In the final version of GTR, Einstein deployed the rotating disk argument to show that

Euclidean geometry does not apply to (a rotating system of coordinates) K'. The notion of coordinates ... which presupposes the validity of Euclidean geometry, therefore breaks down in relation to the system K'. So, too, we are unable to introduce a time corresponding to physical requirements in K', indicated by clocks at rest relatively to K'.

(1916b)

But according to the extended version of EP,

K' may also be conceived as a system at rest, with respect to which there is a gravitational field. We therefore arrive at the result: the gravitational field influences and even determines the metrical laws of the spacetime continuum. If the laws of configuration of ideal rigid bodies are to be expressed geometrically, then in the presence of a gravitational field the geometry is not Euclidean.

(1921b)

A mathematical treatment of non-Euclidean geometry with which Einstein was familiar was Gauss's theory of two-dimensional surfaces. Unlike a Euclidean plane on which the Cartesian coordinates signify directly lengths measured by a unit measuring rod, on a surface it is impossible to introduce coordinates in the same manner so that they have a simple metrical significance. Gauss overcame this difficulty by first introducing curvilinear coordinates arbitrarily, and then relating these coordinates to the metrical properties of the surface through a quantity g_{ij}, which was later called the metric tensor. 'In an analogous way we shall introduce in the general theory of relativity arbitrary coordinates, x_1, x_2, x_3, x_4' (Einstein, 1922). Thus, as Einstein repeatedly claimed (e.g. 1922, 1933, 1949), the immediate metric significance of the coordinates was lost if one (i) admitted non-linear transformations of coordinates (as first suggested by the example of the rotating rigid disk and demanded by EP), which invalidated Euclidean geometry; and (ii) treated a non-Euclidean system in a Guassian way.

Another heuristic role of Gauss's theory of surfaces was its suggestion concerning the mathematical description of the gravitational field. 'The most important point of contact between Gauss's theory of surfaces and the general theory of relativity lies in the metrical properties upon which the concepts of both theories, in the main, are based' (Einstein, 1922). Einstein first had the decisive idea of the analogy of mathematical problems connected with GTR and Gauss's theory of surfaces in 1912, without knowing Riemann's and Ricci's or Levi-Civita's work. Then he became familiar with these works (the Riemannian manifold and tensor analysis) from Marcel Grossmann, and obtained a mathematical description of the gravitational field (Einstein and Grossmann, 1913).

Three points were crucial for achieving this goal. First, Minkowski's four-dimensional formulation provided a concise description of STR, which was regarded by Einstein as a special case of the sought-for GTR. Second, the rotating disk problem suggested that the spacetime structures, when a gravitational field is present, are non-Euclidean in nature. Third, the metrical properties of a two-dimensional Gaussian surface were described by the metric g_{ij} rather than by arbitrary coordinates. Then, what was needed, as Einstein saw it when he approached Grossmann for help, was a four-dimensional generalization of Gauss's two-dimensional surface theory, or, and this amounts to the same thing, a generalization of the flat metric tensor of Minkowski's formulation of STR to a non-flat metric.

In the final version of GTR, Einstein connected the gravitational field with the metric $g_{\mu\nu}$. In a Mikowskian spacetime where STR is valid, $g_{\mu\nu} = \text{diag}(-1, -1, -1, 1)$. Einstein said:

A free material point then moves, relatively to this system, with uniform motion in a straight line. Then if we introduce new spacetime coordinates x_1, x_2, x_3, x_4, by means of any substitution we choose, the $g_{\mu\nu}$ in this new system will no longer be constant, but functions of space and time. At the same time the motion of the free material point will present itself in the new coordinates as a curvilinear non-uniform motion and the law of this motion will be independent of the nature of the moving particle. We shall therefore interpret this motion as a motion under the influence of a gravitational field.

We thus find the occurrence of a gravitational field connected with a space-time variability of the $g_{\mu\nu}$.

(1916b)

Precisely based on this consideration, Einstein suggested that 'the $g_{\mu\nu}$ describe the gravitational field' and 'at the same time define the metrical properties of the space measured' (*ibid.*).

4.2 The field and spacetime: general covariance

The achievement of mathematically describing gravitational (or more precisely the inertio-gravitational) structures by the metric tensor $g_{\mu\nu}$, which at the same time define the chronogeometrical structures characterizing the behavior of clocks and rods, opened a door for further investigations of the relationship between the gravitational field and spacetime structures. The clarification of this relationship, however, was not on Einstein's agenda when he was in collaboration with Grossmann in 1913. The immediate task for them, once a mathematical expression of the gravitational field had been found, was to establish equations for the fields as dynamical structures. The guiding idea in this pursuit remained the same as in Einstein's previous investigations: extending the relativity principle (RP) from inertial frames of reference to arbitrary ones so that the general laws of nature would hold good for all frames of reference.[2] It was this requirement of extending RP that led Einstein to the introduction of the gravitational field via the formulation of EP, and to taking $g_{\mu\nu}$ as the mathematical expression of the gravitational fields. Yet with $g_{\mu\nu}$ at hand, this requirement can be put in a mathematically more precise way: the equations satisfied by $g_{\mu\nu}$ should be covariant with respect to arbitrary transformations of coordinates. That is, the principle of general relativity required the field equations to be generally covariant.

The generally covariant equation for $g_{\mu\nu}$ was easily found by generalizing the Newtonian equation for the gravitational potential (Poisson's equation): the sources of the fields were obtained by relativistically generalizing the Newtonian mass density to the rank-two stress-energy tensor $T_{\mu\nu}$; then they were related to the metric tensor and its first two derivatives. The only generally covariant tensor that could be formed from the latter, as Grossmann realized, was the Ricci tensor $R_{\mu\nu}$. Einstein and Grossmann discovered and rejected the generally covariant equations that were based on the Ricci tensor, because these equations failed to reduce to the Newtonian limit for weak static gravitational fields (1913). Instead, Einstein derived a set of field equations that satisfied several apparently necessary requirements, but were invariant only under the restricted covariance group.[3]

At first, Einstein was disturbed by his failure to find generally covariant equations. But before long, he felt that it was unavoidable, and posed a simple philosophical argument against all generally covariant field equations. The argument referred to here is the well-known hole argument. Interestingly enough, by first posing and then rejecting the hole argument, the initial concern of which was causality or uniqueness or determinism, Einstein had greatly clarified his understanding, within the framework of GTR, of the relationship between the gravitational field and spacetime,[4] which at first was made mysterious by the double function of the metric tensor $g_{\mu\nu}$.

Here is Einstein's formulation of the hole argument:

We consider a finite portion Σ of the continuum, in which no material process occurs. What happens physically in Σ is then completely determined if the quantities $g_{\mu\nu}$ are given as functions of the coordinates x_ν with respect to a coordinate system K used for the description. The totality of these functions will be symbolically designated by $G(x)$.

Let a new coordinate system K' be introduced that, outside of Σ, coincides with K, within Σ, however, diverges from K in such a way that the $g'_{\mu\nu}$ referred to K', like the $g_{\mu\nu}$ (and their derivatives), are everywhere continuous. The totality of the $g'_{\mu\nu}$ will be symbolically designated by $G'(x')$. $G'(x')$ and $G(x)$ describe the same gravitational field. If we replace the coordinates x'_ν by the coordinates x_ν in the functions $g'_{\mu\nu}$, i.e., if we form $G'(x)$, then $G'(x)$ also describes a gravitational field with respect to K, which however does not correspond to the actual (i.e., originally given) gravitational field.

Now if we assume that the differential equations of the gravitational field are generally covariant, then they are satisfied by $G'(x')$ (relative to K') if they are satisfied by $G(x)$ relative to K. They are then also satisfied relative to K by $G'(x)$. Relative to K there then exist two distinct solutions, $G(x)$ and $G'(x)$, although at the boundary of the region both solutions coincide; i.e., what happens in the gravitational field cannot be uniquely determined by generally covariant differential equations.

(1914b)

Three points are crucial to the argument. First, the new tensor field $G'(x)$ (the 'dragged-along' field) can be obtained or generated from the original field $G(x)$ by a point (or active) transformation (or a diffeomorphism) on the manifold ('the continuum'), on which the field is defined. The transformation is so contrived that the metric field $G(x)$ is mapped onto a new metric field $G'(x)$, while the stress-energy tensor is mapped onto itself. Second, the general covariance (GC) of a set of field equations entails that if $G(x)$ is a solution of the equations, the actively transformed (or dragged-along) field $G'(x)$ is also a solution of the equations. Third, $G(x)$ and $G'(x)$ are the components of two physically different tensor fields at the same point in the same coordinate system (K). With these points in mind Einstein felt that GC made it impossible to uniquely determine the metric field out of the stress–energy tensor. That is, GC violated the law of causality, thus was unacceptable.

What underlay the hole argument against GC was the claim that $G(x)$ and $G'(x)$ represented different physical situations in the hole (Σ). Then what was the justification for such a claim? No justification was explicitly given. Tacitly, however, a consequential position concerning the nature of spacetime (or the physical identity of the points of the manifold as spatio-temporal events) was assumed. That is, the points of the manifold were assumed to be physically distinguishable events in spacetime even before the metric fields were dynamically specified. It was this assumption that entailed that mathematically different fields $G(x)$ and $G'(x)$ defined on the same point x should be taken as physically distinct fields. Otherwise, if the physical identity of the points of the manifold as spacetime events had to be established by the dynamical metric fields, the legitimacy of defining two different metric fields at the same spacetime point would be an open question, depending on one's understanding of physical reality (see below).

Then a serious question was whether this assumption was physically justifiable. In the context of Newtonian theory, since absolute (non-dynamical) space and time endowed the points of an empty region of a manifold with their spatial and temporal properties, and thus provided the points with preexisting physical identity, the assumption was justifiable. In the context of STR, an inertial frame of reference (consisting of, e.g., rods and clocks in rigid and unaccelerated motion, or various combinations of test particles, light rays, clocks, and other devices) played the similar role of individuating the points of an empty region of a manifold as absolute space and time did in Newtonian theory, and thus the assumption was also justifiable. In the context of GTR, however, there was no such absolute (non-dynamical) structure that could be used to individuate the points of the manifold, and thus the justification of this assumption was quite dubious. Yet to examine the justification of the underlying assumption of the hole argument thoroughly, more philosophical reflections were needed.

In his struggle with difficulties connected with the field equations of limited invariance, and in particular in his attempt to explain the precession of Mercury's perihelion in November 1915, Einstein developed a new idea of physical reality, which turned out to be crucial to his rejection of the hole argument. According to Einstein (1915e, 1916a, and 1949), with regard to the independent physical reality of space and time, there was an essential difference between STR and GTR. In STR, the coordinates (associated with frames of reference) had immediate physical (metrical) significance, because the points of an empty region of a manifold were individuated as spacetime points by means of the inertial reference frame. This meant that these points had preexistent spatio-temporal individuality independent of the dynamical metric field; or phrased differently, this meant that space and time had a physical reality independent of the dynamical metric field.

In GTR, however, all physical events were to be built up solely out of the movements of material points, the meeting of the points (i.e., the points of intersection of their world lines or the totality of spatio-temporal point-coincidences), which was dictated by the dynamical metric field, was the only reality. In contrast to what depended on the choice of the reference system, the point-coincidences were naturally preserved under all transformations (and no new ones were added) if certain uniqueness conditions were observed. These considerations had deprived the reference systems K and K' in the hole argument of any physical reality, and took 'away from space and time the last remnant of physical objectivity' (Einstein, 1916b).

Since the points of a manifold in GTR must inherit all of their distinguishing spatio-temporal properties and relations from the metric fields, a four-dimensional manifold could be defined as a spacetime only when it was structured by metric fields.[5] That is, in GTR there would be no spacetime before the metric field was specified or the gravitational equations were solved. In this case, according to Einstein, it would be senseless and devoid of any physical content to attribute two different solutions $G(x)$ and $G'(x)$ of a set of field equations to the same coordinate system or the same manifold. For this reason, Einstein stressed that

the (simultaneous) realization of two different g-systems (better said, of two different gravitational fields) in the same region of the continuum is impossible by the nature of the theory . . . . If two systems of the $g_{\mu\nu}$ (or generally, of any variables applied in the description of the world) are so constituted that the second can be obtained from the first by a pure space-time transformation, then they are fully equivalent.

(Einstein, 1915e)

Logically, the lack of structures in GTR other than the metric fields for individuating the points of a manifold as spacetime points justified the identification of a whole equivalence class of drag-along metric fields with one gravitational field. The identification had undermined the underlying assumption of the hole argument (the drag-along metric fields were physically different) and averted the logical force of the hole argument (GC would violate the law of causality), and had thus removed a philosophical obstacle in Einstein's way to the generally covariant equations. Moreover, since Einstein was convinced that the physical laws (equations) should describe and determine no more than the physically real (the totality of the spacetime coincidences), and that the statements of the physically real would not 'founder on any (single-valued) coordinate transformations', the demand for GC became most natural and imperative (*ibid.*). As a result, a definitive formulation of GTR, that is, the generally covariant form of the field equations (the Einstein field equations)

$$G_{\mu\nu} = -k(T_{\mu\nu} - \tfrac{1}{2}g_{\mu\nu}T) \tag{1}$$

or, equivalently,

$$G_{\mu\nu} - \tfrac{1}{2}g_{\mu\nu}G = -kT_{\mu\nu}, \tag{2}$$

was quickly obtained (Einstein, 1915d).[6]

For Einstein, GC was a mathematical expression of the extension of RP. This claim, however, was rejected by Eric Kretschmann (1917) and other later commentators. They argued that since any spacetime theory can be brought into a generally covariant form purely by mathematical manipulation, and these generally covariant theories included the versions of Newtonian spacetime theory that obviously violated RP, the connection between GC and RP expressed by Einstein was an illusion. In fact, they argued, GC, as a formal property of certain mathematical formulations of physical theories, was devoid of any physical content.

The Kretschmannian argument can be rejected in two steps. First, from the above discussion, we have seen clearly that GC, as it was understood and introduced in GTR by Einstein, was a consequence of a physically profound argument, the point-coincidence argument. The latter assumed that the physical reality of spacetime was constituted by the points of intersection of the world lines of material points, which, in turn, were specified by the dynamical metric fields. Thus, from its genesis, GC was less a mathematical requirement than a physical assumption about the ontological relationship between spacetime and the gravitational field: only physical processes dictated by the gravitational field can individuate the events that make up spacetime. In this sense, GC is by no means physically vacuous.

Second, it is true that it is always possible, by introducing some additional mathematical structures (such as the general metric tensor and curvature tensor), to bring a theory (such as Newtonian theory or STR) that initially was not generally covariant into a generally covariant form. To assert that the modified theory has the same physical content as the original one, however, one has to make physical assumptions that these additional structures are merely auxiliary without independent physical significance, that is, that they are physically empty and superfluous. In the case of STR, for example, a restriction (that the curvature tensor must be vanishing everywhere) has to be imposed so that the Minkowskian metric (together with the priviliged inertial frames of reference) can be recovered. This kind of restriction, necessary for recovering the physical content of theories prior to general relativity, makes the apparent GC in their formulations trivial and physically uninteresting. It is particularly instructive to notice that since the solutions of the trivial GC theories have to meet the restrictions imposed on the formulations, they cannot be generally covariant themselves.

By ruling out the cases of trivial GC, the connection between GC and the extension of RP seems deep and natural: both presuppose and endorse the relationalist view of spacetime, rejecting that spacetime has a physical reality independent of the gravitational field.[7]

4.3 Matter versus spacetime and the field: Mach's principle

Although Kretschmann failed to prove rigorously that RP, as expressed in GC equations, was only a requirement about the mathematical formulation without physical content, his criticism did prompt Einstein to reconsider the conceptual foundations of GTR. In March 1918, Einstein for the first time coined the expression 'Mach's principle' and listed it, in addition to (a) RP and (b) EP,[8] as one of three main points of view on which GTR rested:

(c) *Mach's principle:* The G-field is completely determined by the masses of the bodies. Since mass and energy are identical in accordance with the results of the special theory of relativity and the energy is described formally by means of the symmetric energy tensor ($T_{\mu\nu}$), the G-field is conditioned and determined by the energy tensor of matter.

(Einstein, 1918a)

In a footnote Einstein said:

Hitherto I have not distinguished between principles (a) and (c), and this was confusing. I have chosen the name 'Mach's principle' because this principle has the significance of a generalization of Mach's requirement that inertia should be derived from an interaction of bodies.

(Ibid.)

Einstein held that the satisfaction of Mach's principle (MP) was absolutely necessary because it expressed a metaphysical commitment concerning the ontological priority of matter over spacetime. He claimed that the field equations (1) as an embodiment of MP implied that no gravitational field would exist without matter. He also claimed that MP was a statement about the spacetime structure of the whole universe, since all the masses of the universe participated in the generation of the gravitational field (*ibid.*).

By stressing the importance of MP, Einstein tried to clarify the conceptual basis of GTR. The logical status of MP in GTR, however, is much more complicated than Einstein thought, as we shall see shortly. Einstein's effort, nevertheless, revealed his Machian motivation in developing and interpreting GTR. In fact, as early as 1913, Einstein had already proclaimed the Machian inspiration of his theory: he postulated that

the inertial resistance of a body can be increased by placing unaccelerated inertial masses around it; and this increase of the inertial resistance must fall away if those masses share the acceleration of the body.

(1913b)

and claimed that

this agrees with Mach's bold idea that inertia originates in an interaction of the massive particle under consideration with all the rest.

(Einstein and Grossmann, 1913)

In a letter to Mach on 25 June 1913, Einstein wrote enthusiastically:

Your helpful investigations on the foundations of mechanics will receive a brilliant confirmation. For it necessarily turns out that inertia has its origin in a kind of interaction, entirely in accord with your considerations on the Newton pail experiment.

(1913c)

In the obituary he wrote when Mach died in 1916, Einstein even claimed that Mach 'was not far from postulating a general theory of relativity half a century ago' (1916d).

As we indicated in section 3.1, Mach's ideas had exercised a profound influence upon Einstein's thinking. Most important among them was his criticism of Newton's notion of absolute space, which, according to Mach, was out of the reach of experience and had an anomalous causal function: it affected masses, but nothing can affect it. Newton justified his notion of absolute space by appealing to the notion of absolute motion with respect to absolute space, which generated observable inertial effects, such as centrifugal forces in the case of a rotating bucket. In arguing against Newton, Mach suggested instead that

Newton's experiment with the rotating vessel of water simply informs us, that the relative rotation of the water with respect to the sides of the vessel produces no noticeable centrifugal forces, but that such forces are produced by its relative rotation with respect to the mass of the earth and other celestial bodies.

(Mach, 1883)

Here the notion of absolute motion with respect to absolute space was replaced by motion relative to physical frames of reference.[9]

Both relative motion and physical frames of reference were observable. Yet it was still mysterious why relative motion would be able to produce inertial forces. Mach's suggestion, as was hinted immediately after the above quotation, was that inertial forces and inertial masses, as well as the inertial behavior of bodies in a relative space associated with the physical frame of reference, were determined by some kind of causal interaction between bodies and their environment:

No one is competent to say how the experiment would turn out if the sides of the vessel increased in thickness and mass until they were ultimately several leagues thick.

(Ibid.)

In other places, Mach asserted that the contribution to the production of inertial forces and to the determination of the inertial behavior of bodies by 'the nearest masses vanishes in comparison with that of the farthest' (Mach, 1872). This brings out another characteristic feature of Mach's idea: the local inertial system or local dynamics is determined by the distribution of matter in the whole universe.

When Einstein coined the expression 'Mach's principle' and asserted that the state of space was completely determined by massive bodies, he certainly had in his mind the whole complex of Mach's ideas: space as an abstraction of the totality of spatial relations of bodies (which were embodied in various frames of reference), the relativity of motion, and the determination of inertia by the causal interactions of bodies. Mach's idea of causal interactions, however, was based on the assumption of instantaneous action at a distance. The idea of distant actions explained why the distribution of distant masses could play an essential role in the local dynamics (such as inertial forces), but it was in direct contradiction with the field concept of action cherished dearly by Einstein. Thus, to take over Mach's ideas, Einstein had to reinterpret them in the spirit of field theory. There was no particular difficulty in this pursuit because Einstein's idea of gravitational field served perfectly well as the medium of causal actions among massive bodies. Thus, according to Einstein, 'Mach's notion finds its full development' in the notion of gravitational field (1929b).

There were other modifications of Mach's ideas in Einstein's discussion of MP, which were also related with Einstein's thinking within the field-theoretical framework. First, what the energy tensor in the Einstein field equations described was not just the energy of bodies, but also that of the electromagnetic fields. Second, in Einstein's generalization of Mach's ideas of inertia (that inertial masses, inertial forces, and the law of inertia had their origin in the causal interactions of bodies), the notion 'inertia' referred to inertial systems (as distinct frames of reference in which the laws of nature, including local laws of electrodynamics, took a particularly simple form) as well. Thus Mach's ideas of inertia provided inspiration not only for EP (if the causal interactions were taken to be gravitational interactions), but also for the extension of RP, by which the inertial frames of reference were deprived of their absolute character and became a special case of relative spaces characterized by metric tensors (which describe both the inertio-gravitational fields and the chrono-geometrical structures of spacetime).

As we noted above, Einstein's distinction of MP from RP was a response to Kretschmann's criticism. Originally, Einstein might have felt that RP as an extension of Mach's idea of the relativity of inertia was enough to characterize the physical content of GTR. Under the pressure of Kretschmann's criticism that RP in the form of GC was physically vacuous, however, Einstein felt that, to avoid confusion, it was necessary to clarify the conceptual foundations of GTR. And this could be done by formulating a separate principle, which was coined MP, to characterize some of the guiding ideas of GTR, concerning (i) the relationship between the inertial motions (dictated by the geometrical structures of spacetime) and the matter content of the universe; (ii) the origin and determination of spacetime structure; and (iii) the ontological status of spacetime itself, while at the same time maintaining the heuristic value of RP.

There is no doubt that Mach's ideas motivated and inspired Einstein's working on GTR. However, the logical status of MP in GTR is another matter. This is a complicated and controversial issue. One may follow Einstein and claim, loosely, that RP is a generalization of Mach's idea about the relativity of accelerated and rotational motions (and thus of physical frames), that EP arises from Mach's idea that inertia comes from (gravitational) interactions between massive bodies, and that the Einstein field equations implement Mach's idea that matter is the only real cause for the metric structures of space (time), which as an abstraction of material relations, in turn, dictate motions of bodies.[10] Einstein, however, went further and derived some testable non-Newtonian predictions from GTR. These predictions, if confirmed, would lend strong support to Mach's ideas about the relativity of inertia, and would entail that 'the whole inertia, that is, the whole g_{ij}-field, [is] to be determined by the matter of the universe' (Einstein 1921b). For this reason, they have been called Machian effects.

In fact, the following Machian effects were anticipated by Einstein as early as 1913, well before the formulation of MP, on the basis of his tentative theory (with Grossmann) of the static gravitational field. (A) The inertial mass of a body must increase when ponderable masses are piled in its neighborhood (Einstein, 1913b). (B) A body must experience an accelerating force when nearby bodies are accelerated (Einstein, 1913a). (C) A rotating hollow body must generate inside of itself a Coriolis field, which drags along the inertial system, and a centrifugal field (Einstein, 1913c). These effects (especially effect (C), namely the dragging along of inertial systems by rotating masses) were used by Einstein and later commentators as criteria for the Machian behavior of solutions of the field equations of GTR (Einstein, 1922). But the most characteristic aspect of Einstein's Machian conception was expressed in his assertion (D):

In a consistent theory of relativity there can be no inertia relative to 'space', but only an inertia of masses relative to one another. Hence if I have a mass at a sufficient distance from all other masses in the universe, its inertia must fall to zero.

(1917a)

Assertion (D), however, is contradicted by the existence of a variety of solutions of Einstein's field equations. The first among them is the Schwarzschild solution. In extending Einstein's (1915c), in which an approximate gravitational field of a static spherically symmetrical point-mass in a vacuum was obtained, Schwarzschild (1916a), with the assumption that the gravitational potentials are continuous everywhere except at $R = 0$ (R as well as θ and ϕ is a polar coordinate), obtained an exact solution for an isolated point-mass in otherwise empty spacetime (a star or the solar system):

$$ds^2 = (1 - \alpha/r)c^2\,dt^2 - dr^2/(1 - \alpha/r) - r^2(d\theta^2 + \sin^2\theta\,d\phi^2), \qquad (3)$$

where

$$r = (R^3 + \alpha^3)^{1/3}, \quad \text{with} \quad \alpha = 2Gm/c^2.$$

This solution implies that the metric at infinity is Minkowskian; thus a test body has its full inertia however far it may be from the only point-mass in the universe. It is obvious that this solution is in direct contradiction with assertion (D).

Einstein knew the Schwarzschild solution quite early and wrote, on 9 January 1916, to Schwarzschild: 'I liked very much your mathematical treatment of the subject' (1916b). Yet the clear anti-Machian implications of the treatment urged Einstein to reaffirm his Machian stand to Schwarzschild:

In the final analysis, according to my theory, inertia is precisely an interaction between masses, not an 'action' in which, in addition to the masses contemplated, 'space' as such participates. The essential feature of my theory is just that no independent properties are attributed to space as such.

(1916b)

To meet the challenge to the Machian conception, posed by the Schwarzschild solution and by a plausible boundary condition of the field equations[11] (which specifies that the metric at infinity has a Minkowskian character), Einstein, in his classic paper (1916c), appealed to the concept of 'distant matter which we have not included in the system under consideration'. These unknown masses could serve as the source of the Minkowski metric at infinity, thus providing a Machian explanation of Minkowskian boundary conditions, and as a major source of all inertia.

If a Machian conception is taken for granted, then there is a self-consistent justification for the existence of distant masses. It is observed that to a very high degree of accuracy inertia in our part of the universe is isotropic. But the matter in our immediate vicinity (planets, sun, stars) is not isotropic. Thus, if inertia is produced by matter, an overwhelming part of the effect must come from some isotropically distributed distant masses that lie beyond our effective observation.

Yet the price for saving the Machian position in this way was very high, as was pointed out by de Sitter.[12] First, a universal Minkowskian boundary condition is in contradiction with the spirit of GTR because it is not generally covariant; and the idea that various gravitational potentials are necessarily reduced to the Minkowskian value is an ingredient of a disguised version of absolute space. In general, any specification of boundary conditions, as Einstein himself soon realized in his letter to Besso at the end of October of 1916, is incompatible with GTR as local physics.[13] Second, distant masses are supposed to be different from absolute space, which is independent of reference systems and non-observable in principle. Yet just like absolute space, distant masses are also independent of reference systems and are non-observable in principle: to serve the function of determining gravitational potentials at infinity, these masses must be conceived as beyond any effective observation. Thus the philosophical appeal of the Machian programme, compared to Newton's idea of absolute space, has faded away.

By the time Einstein made his assertion (D), he was fully aware of these difficulties. The reason for his adherence to the Machian position, regardless of the Schwarzschild solution and various difficulties associated with the Machian conception of boundary conditions, is that he had developed a new strategy for implementing his Machian program (1917a). This strategy led to the genesis of relativistic cosmology, which has been rightly regarded as the most fascinating consequence of MP. The cosmological considerations also gave Einstein new confidence in Mach's idea, which was testified in his formulation of MP in (1918a) and his further discussion of the Machian effects in (1922). Before turning to cosmology, however, let us look at the Machian effects briefly.

Einstein thought that effect (A) was a consequence of GTR (1922). However, this was shown to be only an effect of the arbitrary choice of coordinates.[14] In fact, inertial mass in GTR, in which EP holds,[15] is an intrinsic and invariant property of bodies, independent of the environment. To express the generation of inertial mass by interactions with distant masses successfully, at least one more long-range force generated by a new (different from the $g_{\mu\nu}$) field is required. Yet this breaks EP and amounts to nothing but a modification of GTR, such as the failed Brans–Dicke theory.[16]

Under certain initial and boundary conditions, effects (B) and (C) can be derived from GTR.[17] Yet this cannot be simply construed as a demonstration of the validity of MP. The problem is that, in these derivations, a Minkowskian spacetime was tacitly assumed to hold at infinity, partially under the disguise of the method of approximation. But this assumption, as noted above, violates the spirit of MP on two counts. First, the boundary condition at infinity in this context assumes the role of Newton's absolute space and, in contradiction with MP, is immune to the influence of the massive bodies in question. Second, instead of being the only source of the total structure of spacetime, the presence of matter in this context merely modifies the latter's otherwise flat structure. Although the flat structure can be taken to be dictated by non-observable distant masses, the latter are philosophically no more appealing than absolute space.

The failure to confirm the Machian effects suggests that MP is neither a necessary presupposition nor a logical consequence of GTR. Nevertheless, two more options are still available for incorporating MP into GTR. First, we can modify GTR and bring it into a form compatible with MP. One attempt of this kind, the one appealing to non-observable distant masses, failed, as noted above. Another modification of GTR by the introduction of a cosmological constant will be discussed in the next section. The second option is to take MP as a selection rule for solutions of Einstein's field equations. That is, MP is taken as an external constraint on the solutions, rather than as an essential ingredient of GTR. But this makes sense only in the context of relativistic cosmology. Therefore it will also be treated in the next section.

There is another reason why MP was not endorsed, even by Einstein himself in his later years. MP presumes that material bodies are the only independent physical entities that determine the structures, or even the very existence, of spacetime. Einstein rejected this premise when he began to be fascinated by the idea of a unified field theory from the mid-1920s (see section 5.3). He argued that 'the [stress–energy tensors] T_{ik} which are to represent "matter" always presuppose the g_{ik}' (1954a). The reason Einstein gave for this is simple. In order to determine the stress-energy tensor T_{ik}, it is necessary to know the laws governing the behavior of matter. The existence of non-gravitational forces requires the laws of these forces to be added to those of GTR from outside. Written in a covariant form, these laws contain the components of the metric tensor. Consequently we will not know matter apart from spacetime. According to Einstein's later view, the stress–energy tensor is merely a part of the total field rather than its source. Only the total field constitutes the ultimate datum of physical reality, from which matter is to be constructed. Thus the ontological priority of matter over the field is explicitly and firmly rejected. From such a perspective, Einstein summed up his later position on MP:

In my opinion we ought not to speak about the Machian Principle any more. It proceeds from the time in which one thought that the 'ponderable bodies' were the only physical reality, and that all elements that could not be fully determined by them ought to be avoided in the theory.

(1954a)

Some clarifications on Einstein's position on MP are needed. In my opinion, there are in fact two MPs, MP_1 and MP_2. MP_1, held by Mach and the early Einstein and rejected by the later Einstein, says that ponderable bodies are the only physical reality that fully determines the existence and structure of spacetime. MP_2, held by Einstein in his unified field theory period, says that spacetime is ontologically subordinate to the physical reality represented by a total substantial field, among whose components we can find gravitational fields, and that the structures of spacetime are fully determined by the dynamics of the field. More discussion on Einstein's view of spacetime and its geometric structures will be given in section 5.1. Here it is sufficient to say that with respect to the relation between spacetime and the field, Einstein's later position was still Machian in spirit. The only difference between this position (MP_2) and Mach's (MP_1) is that Einstein took the field rather than ponderable matter as the ultimate ontology that gives the existence and determines the structures of spacetime.

In a letter to Max Born on 12 May 1952, Einstein wrote:

Even if no deviation of light, no perihelion advance and no shift of spectral lines were known at all, the gravitational field equations would be convincing, because they avoid the inertial system – that ghost which acts on everything but on which things do not react.

(1952d)

It is an undeniable fact that Einstein's successful exorcism of that ghost was inspired and guided by Mach's idea about the relativity of inertia, although in its original formulation, MP_1, this idea can be viewed neither as a necessary premise nor as a logical consequence of the field equations, nor has it been established as a useful selection rule, as we shall see in the next section.

4.4 The consistency of GTR: the genesis of relativistic cosmology

Since Einstein took the initial inspiration, core commitment, and major achievement of GTR to be the dismissal of absolute space (as well as the privileged status of the inertial system), to have a consistent theory he had to meet two challenges. One was posed by rotation, whose absolute character, as argued by Newton and his followers, required support from absolute spatial structures. The other was associated with the boundary conditions of the field equations, which functioned as a disguised version of absolute space. In order to deprive rotation of absoluteness, Einstein, following Mach, resorted to the ever-elusive distant masses. This strategy, however, was sharply criticized and resolutely rejected by de Sitter, as noted earlier. De Sitter also disagreed with Einstein on the boundary conditions, which were mingled with the idea of distant masses in the Machian context, and on what a consistent theory of general relativity should be. Historically, the Einstein–de Sitter controversy was the direct cause of the emergence of relativistic cosmology.[18]

After having established the field equations, Einstein tried to determine the boundary conditions at infinity, which was only a mathematical formulation of the Machian idea that the inertia of a mass must fall to zero when it is at a sufficient distance from all other masses in the universe. He clearly realized that the specification of boundary conditions was closely related with the question of 'how a finite universe can exist, that is, a universe whose finite extent has been fixed by nature and in which all inertia is truly relative'.[19]

On 29 September 1916, in a conversation with de Sitter, Einstein suggested a set of generally covariant boundary conditions for $g_{\mu\nu}$:

$$
\begin{matrix}
0 & 0 & 0 & \infty \\
0 & 0 & 0 & \infty \\
0 & 0 & 0 & \infty \\
\infty & \infty & \infty & \infty^2
\end{matrix}
\tag{4}
$$

De Sitter immediately realized two of its implications: its non-Machian nature and the finiteness of the physical world. Since the degenerate field of $g_{\mu\nu}$ at infinity represented a non-Machian and unknowable absolute spacetime, 'if we wish to have complete [consistent] four-dimensional relativity for the actual world, this world must of necessity be finite'.[20]

De Sitter's own position on boundary conditions was based on his general epistemological view that 'all extrapolations beyond the region of observation are insecure' (1917e). For this reason, one had to refrain entirely from asserting boundary conditions for spatial infinity with general validity; and the $g_{\mu\nu}$ at the spatial limit of the domain under consideration had to be given separately in each individual case, as was usual in giving the initial conditions for time separately.

In his 'cosmological considerations' Einstein acknowledged that de Sitter's position was philosophically incontestable. Yet, as he confessed immediately, such complete resignation on this fundamental question, namely the Machian question, was for him a difficult thing. Fortunately, when Einstein wrote this paper, he had already found a way out of this difficult situation, a consistent way of carrying through his Machian conception of GTR. Einstein's new strategy was to develop a cosmological model in which the absence of boundary conditions was in harmony with a sensible conception of the universe as a totality:

If it were possible to regard the universe as a continuum which is finite (closed) with respect to its spatial dimensions, we should have no need at all of any such boundary conditions.

(1917a)

Then, by assuming that the distribution of matter was homogeneous and isotropic, Einstein succeeded in showing that

both the general postulate of relativity and the fact of the small stellar velocities are compatible with the hypothesis of a spatially finite universe.

(Ibid.)

And this marked the emergence of relativistic cosmology, which was the result of philosophical reflections upon the relationship between matter and space-time.

However, in order to carry through this idea, Einstein had to modify his field equations, i.e., introduce a cosmological constant λ so that the field equations (1) were transformed into

$$G_{\mu\nu} - \lambda g_{\mu\nu} = -k(T_{\mu\nu} - \tfrac{1}{2}g_{\mu\nu}T). \tag{5}$$

The reason for introducing the λ term was not so much to have a closed solution – which was possible without the λ term, as realized by Einstein in the article, and as fruitfully explored later on by Alexander Friedman (1922) – as to have a quasi-static one, which was required, Einstein claimed, 'by the fact of the small velocities of the stars'. Mathematically, this first quantitative cosmological model is essentially the same as Schwarzschild's interior solution (1916b). The major difference is this. While the interior metric is reducible to the Minkowski metric at infinity, Einstein's universe is spatially closed.[21]

The physical meaning of the λ term is rather obscure. It can be interpreted as a locally unobservable energy density of the vacuum, or as a physical force of repulsion (negative pressure) that increases with distance, counteracting the effects of gravitation and making the distribution of matter stationary or even expanding. In any case, it is an obviously non-Machian source term for the gravitational field, representing an absolute element that acts on everything without being acted upon. With the help of the cosmological term, however, Einstein was able to predict a unique relation between the mass dentity ρ in the universe and its radius of curvature R,

$$k\rho c^2 = 2\lambda = 2/R^2, \tag{6}$$

by taking matter to be pressure-free. This fixing of a definite periphery of the universe had removed any necessity for a rigorous treatment of boundary conditions. And this was exactly what Einstein had hoped for.

If the Machian motivation was the only constraint in cosmological construction, then the introduction of the λ term was neither necessary nor sufficient. It was not necessary because a closed solution, though not static, could be derived from the original field equations (1) as noted above. Nor was it sufficient because the modified field equations still had non-Machian solutions. As was shown by de Sitter (1917a, b, c, d), even if the system were devoid of matter, a solution to the modified field equations could still be found when time was also relativized,

$$ds^2 = (1 - \lambda r^2/3)\, dt^2 - (1 - \lambda r^2/3)^{-1}\, dr^2 - r^2(d\theta^2 + \sin^2\theta\, d\phi^2), \tag{7}$$

with a radius of curvature $R = (3/\lambda)^{1/2}$, which was labeled by de Sitter system B, as in contrast with Einstein's cosmological model, which was labeled system A.

Einstein's first reaction to model B, as recorded in the postscript of de Sitter's (1917a), was understandably negative:

In my opinion it would be unsatisfactory to think of the possibility of a world without matter. The field of $g_{\mu\nu}$ ought to be conditioned by matter, otherwise it would not exist at all.

Facing this criticism, de Sitter reasoned that since in system B all ordinary matter (stars, nebulae, clusters, etc.) was supposed not to exist, Einstein and other 'followers of Mach are compelled to assume the existence of still more matter: the world-matter' (1917b). Even in system A, de Sitter wrote, since the cosmological constant was very small, as indicated by our knowledge of the perturbations in the solar system, the total mass of the hypothetical and non-observable world-matter must be 'so enormously great that, compared with it, all matter known to us is utterly negligible' (1917a).

Einstein did not accept de Sitter's reasoning. In his letter to de Sitter (1917d), Einstein insisted that there was no world-matter outside the stars, and that the energy density in system A was nothing more than a uniform distribution of the existing stars. But if the world-matter were just some kind of ideal arrangement of ordinary matter, de Sitter (1917b) argued, then system B would be completely empty by virtue of its equations (which would be a disaster from a Machian point of view), and system A would be inconsistent. The reason for the inconsistency was given as follows. Since ordinary matter in system A was primarily responsible for deviations from a stationary equilibrium, to compensate for the deviations, the material tensor had to be modified so as to allow for the existence of internal pressure and stress. This was possible only if one introduced the world-matter, which, unlike ordinary matter, was identified with a continuous fluid. If this were not the case, then the internal forces would not have the desired effect and the whole system would not remain at rest, as was supposed to be the case. De Sitter further argued that the hypothetic world-matter 'takes the place of the absolute space in Newton's theory, or of the "inertial system". It is nothing more than this inertial system materialized' (de Sitter, 1917c).

The exact nature of the world-matter aside, Einstein, in a letter to de Sitter on 8 August 1917 (1917e), presented an argument against de Sitter. Since dt^2 in model B was affected by a coefficient $\cos^2(r/R)$,[22] all physical phenomena at the distance $r = \pi R/2$ could not have duration. Just as in the near neighborhood of a gravitating massive point where clocks tend to stop (the g_{44} component of the gravitational potential tends to zero), Einstein argued, 'the de Sitter system cannot by any means describe a world without matter but a world in which the matter is completely concentrated on the surface $r = \pi R/2$ [the equator of the de Sitter space]' (1918b). Conceivably, this argument gave Einstein certain confidence in his Machian conception and helped him in formulating Mach's principle at the same time, various difficulties discussed above notwithstanding.

But de Sitter rejected Einstein's criticism on two grounds: (i) the nature of the matter on the equator, and (ii) the accessibility of the equator. In the margin of Einstein's letter (Einstein, 1917e) de Sitter commented that

if g_{44} must become zero for $r = \pi R/2$ by 'matter' present there, how large must the 'mass' of that matter then be? I suspect ∞! We then adopted a matter [which is] not ordinary matter . . ... That would be distant masses yet again . . ... It is a *materia ex machina* to save the dogma of Mach.

In (1917c, d) de Sitter also argued that the equator was not 'physically accessible' because a particle can only 'reach it after an infinite time, i.e., it can never reach it at all'.[23]

Different implications for the Machian conception of spacetime notwith-standing, systems A and B shared a common feature: both were static models of the universe. Conceptually, a gravitating world can only maintain its stability with the help of some physical force of repulsion that counterbalances the effects of gravitational attraction. The cosmological constant can be taken as such an agent for fulfilling this function.[24] Physically, this counterbalance is a very delicate business: at the slightest disturbance the universe will expand to infinity or contract to a point if nothing intervenes.[25] Mathematically, it was a Russian meteorologist, Alexander Friedmann (1922, 1924), who showed that systems A and B were merely limiting cases of an infinite family of solutions of the field equations with a positive but varying matter density: some of them would be expanding, and others contracting, depending on the details of the counterbalancing. Although Friedmann himself took his derivation of various models merely as a mathematical exercise, in fact he proposed a mathematical model of an evolving universe. Einstein first resisted the idea of an evolving universe, but Friedmann's rigorous work had forced Einstein to recognize that his field equations admitted dynamic as well as static solutions of the global spacetime structure, although he remained suspicious of the physical signifi-cance of an evolving universe (Einstein, 1923c).

The conceptual situation began to change in 1931. Unaware of Friedmann's work, a Belgian priest and cosmologist, Georges Lemaître (1925, 1927), working on the de Sitter model, rediscovered the idea of an evolving universe and proposed a model that was asymptotic to the Einstein universe in the past and to the de Sitter universe in the future. Lemaître's work (1927) was translated into English and republished in 1931. This, together with the discovery of Hubble's law (1929) – which strongly suggested an expanding universe – and Eddington's proof of the instability of the Einstein static universe (1930), made the concept of an evolving universe begin to receive wide acceptance. It was not surprising that in 1931 Einstein declared his preference for non-static solutions with zero cosmological constant (1931), and de Sitter voiced his confidence that 'there cannot be the slightest doubt that Lemaître's theory is essentially true, and must be accepted as a very real and important step towards a better understanding of Nature' (1931).

When Einstein, having travelled 'rather a rough and winding road', created relativistic cosmology in 1917, he wrote that his construction of a model for the whole universe 'offers an extension of the relativistic way of thinking' and 'a proof that general relativity can lead to a noncontradictory system'.[26] However, de Sitter challenged Einstein's claim and, with the help of his own works in 1917, showed that a cosmological model of GTR could be in direct contradiction with Mach's dogma. Thus, for GTR to be a consistent theory, according to de Sitter, Mach's dogma had to be abandoned. In his letter to Einstein on 4 November 1920, de Sitter declared: 'The [gravitational] field itself is real' rather than subordinate to ponderable matter (1920). This, in fact, indicated a direction in which Einstein himself had already been moving, starting with his Leiden Lecture (1920a), and ending with his later conception of a unified field theory.

One may assume that with the acceptance of the notion of an evolving universe, which has provided a framework within which more and more non-Machian solutions of the field equations[27] have been discovered, MP should have no place in a consistent theory of general relativity. But this is not the case. Some scholars argue that although MP is in contradiction with some solutions of the field equations, to have a consistent theory, however, we can, instead of giving up MP, take MP as a selection rule to eliminate non-Machian solutions. John Wheeler, for example, advocated (1964a) that MP should be conceived as 'a boundary condition to separate allowable solutions of Einstein's equations from physically inadmissible solutions'. For D. J. Raine, MP as a selection rule is not merely a metaphysical principle for regulation. Rather, he argues, since MP embraces the whole universe and is confronted by cosmological evidence, it can be tested by cosmological models and verified by empirical data. It can be argued that such a study of the empirical status of MP has a direct bearing on our understanding of what should be taken as a consistent GTR. Let us have a brief look at this issue.

MP as a selection rule can take many forms.[28] The most important one is concerned with rotation. Our previous discussion of boundary conditions of the field equations indicates that GTR is not incompatible with the idea that the material universe is only a small perturbation on absolute space. If absolute space exists, we expect that the universe may have some rotation relative to it. But if MP is right and there is no such thing as absolute space, then the universe must provide the standard of non-rotation, and could not itself rotate. Thus the rotating solutions, such as the Gödel universe (1949, 1952) or the Kerr metric (1963) are not allowed. In this particular case, empirical evidence seems to be in favor of MP because, as Collins and Hawking (1973a, b) interpreted, the observed isotropy of the microwave background (i) puts a limit on the possible rotation (local vorticity) of the universe (in the sense of a rotation of a local dynamical inertial frame at each point with respect to the global mass distribution), and (ii) shows that at no time in its history can the universe have been rotating on a time scale shorter than the expansion time scale. This, according to Raine, 'provides us with the strongest observational evidence in favor of the [Mach] Principle' (1975).

Raine further argues that MP is also well supported by other empirical data.

For example, (i) anisotropically expanding non-rotating model universes are non-Machian (because locally the shearing motions mimic a rotation of the matter relative to a dynamic inertial frame) and they are ruled out by the microwave observations of the extremely small ratio of the shear to the mean Hubble expansion; and (ii) the homogeneity of the Machian Robertson–Walker universe is supported by the observed zero galaxy correlation, which measures the tendency for pairs of galaxies to cluster together in associations that have a mean separation less than the average. Thus, according to Raine, these aspects of MP are also empirically verified to a high precision (1981).

Be that as it may, the usefulness of MP as a general selection rule is far from established in many cases without special symmetries because, as Wolfgang Rindler points out, 'Raine's methods are difficult to apply except in situations with special symmetries' (1977).

Notes

1. See Stachel (1980a).
2. Einstein usually talks about coordinate systems (mathematical representations) instead of frames of reference (physical systems). But judging from the context, what he really means when he talks about coordinate systems is in fact frames of reference.
3. For a detailed account of these physical requirements, see Stachel (1980b) or Norton (1984).
4. Cf. Stachel (1980b) and Earman and Norton (1987).
5. Or, more precisely, only when the points of a manifold were individuated by four invariants of the Riemann tensor constructed from the metric field. See Stachel (1994) and references therein.
6. Here $G_{\mu\nu}$ is the Ricci tensor with contraction G, and $T_{\mu\nu}$ is the stress–energy tensor with contraction T. The left hand side of equation (2) is now called the Einstein tensor.
7. Some commentators (Earman and Norton, 1987) tried to use the hole argument against the substantivalist view of spacetime, by incorporating formulations with trivial GC, even in the context of theories prior to general relativity. Their arguments collapsed because, as pointed out by Stachel (1994), they failed to recognize that the existence of non-dynamic individuating structures in theories prior to general relativity prevents them from avoiding the hole argument by identifying all the drag-along fields with the same gravitational field. Moreover, in theories prior to general relativity, spacetime is rigid in nature (any transformation that reduces to identity in a region must reduce to identity everywhere), and thus the hole argument (presupposing a transformation that reduces to identity outside the hole but non-identity inside the hole) is not applicable in these cases in the first place. For this reason, the nature of spacetime in the scientific context of theories prior to general relativity cannot be decided on the hole argument. For more details, see Stachel (1994).
8. 'Inertia and weight are identical in nature ... the symmetric "fundamental tensor" ($G_{\mu\nu}$) determines the metrical properties of space, the inertial behavior of bodies in it, as well as gravitational effects. We shall denote the state of space described by the fundamental tensor as the G-field' (Einstein, 1918a).
9. In Mach's discussion, physical frames of reference sometimes were represented by the fixed stars, on other occasions by the totality of the masses in the universe.
10. See, e.g., Raine (1975).
11. Since Einstein's field equations are second order differential equations, boundary conditions are necessary for obtaining solutions from the equations.

12. See de Sitter (1916a, b).
13. See Speziali (1972), p. 69.
14. See Brans (1962).
15. So it is always possible to define, in the framework of GTR, a local coordinate system along the world line of any particle with respect to which physical phenomena obey the laws of STR and show no effects of the distribution of the surrounding mass.
16. Brans and Dicke (1961).
17. See Lense and Thirring (1918); Thirring (1918, 1921).
18. For an excellent summary of the Einstein–de Sitter controversy, see Kerszberg (1989).
19. Einstein's letter to Michele Besso (14 May 1916), see Speziali (1972).
20. See de Sitter (1916a). The non-Machian nature of the very idea of boundary conditions was also clear to Arthur Eddington. On 13 October, while waiting for a reprint of de Sitter's (1916a) paper, he wrote to de Sitter: 'When you choose axes which are rotating relative to Galilean axes, you get a gravitational field which is not due to observable matter, but is of the nature of a complementary function due to boundary conditions – sources or sinks – at infinity . . .. That seems to me to contradict the fundamental postulate that observable phenomena are entirely conditioned by other observable phenomena' (Eddington, 1916).
21. Temporally, Einstein's static universe is not closed. It has a classic cosmic time, and thus has the name 'cylindrical universe'.
22. Some mathematical manipulations are needed to derive this statement from equation (7). See Rindler (1977).
23. Connected with these remarks are the questions of singularity and horizon. More discussions on these are given in section 5.4.
24. Cf. Georges Lemaître (1934).
25. Cf. Arthur Eddington (1930).
26. See Einstein's letter to de Sitter (1917c) and to Besso (Speziali, 1972).
27. Roughly speaking, most cosmologists take all vacuum solutions (such as the de Sitter space and the Taub–NUT space) and flat or asymptotically flat solutions (such as the Minkowski metric or the Schwarzschild solution) to be non-Machian. They also take all homogeneous yet anisotropically expanding solutions (such as various Bianchi models) or rotating universes (such as the Gödel model and the Kerr metric) to be non-Machian. Some people also take all models with a non-zero cosmological constant λ as non-Machian, because the λ term in these models must be treated as a non-Machian source term; but others show more tolerance towards the λ term. For a survey of the criteria and formulations for Machian or non-Machian solutions, see Reinhardt (1973), Raine (1975, 1981) and Ellis (1989).
28. For a survey of these forms, see Raine (1975, 1981).

5

The geometrical programme (GP)

Einstein's GTR initiated a new programme for describing fundamental inter-actions, in which the dynamics was described in geometrical terms. After Einstein's classic paper on GTR (1916c), the programme was carried out by a sequence of theories. This chapter is devoted to discussing the ontological commitments of the programme (section 5.2) and to reviewing its evolution (section 5.3), including some topics (singularities, horizons, and black holes) that began to stimulate a new understanding of GTR only after Einstein's death (section 5.4), with the exception of some recent attempts to incorporate the idea of quantization, which will be addressed briefly in section 11.3. Consider-ing the enormous influence of Einstein's work on the genesis and developments of the programme, it seems reasonable to start this chapter with an examination of Einstein's views of spacetime and geometry (section 5.1), which underlie his programme.

5.1 Einstein's views of spacetime and geometry

The relevance of spacetime geometry to dynamics

Generally speaking, a dynamical theory, regardless of its being a description of fundamental interactions or not, must presume some geometry of space for the formulation of its laws and interpretation. In fact a choice of a geometry predetermines or summarizes its dynamical foundations, namely, its causal and metric structures. For example, in Newtonian (or special relativistic) dynamics, Euclidean (or Minkowskian) (chrono-) geometry with its affine structure, which is determined by the kinematic symmetry group (Galileo or Lorentz group) as the mathematical description of the kinematic structure of space (time), determines or reflects the inertial law as its basic dynamical law. In these theories, the kinematic structures have nothing to do with dynamics. Thus dynamical laws are invariant under the transformations of the kinematic symmetry groups. This means that the kinematic symmetries impose some restrictions on the form of the dynamical laws. However, this is not the case for general relativistic theories. In these theories, there is no a priori kinematic structure of spacetime, and thus there is no kinematic symmetry and no restriction on the form of dynamical laws. That is, the dynamical laws are valid in every conceivable four-dimensional topological manifold, and thus have to be generally covariant. It should be noted, therefore, that the restriction of GC upon the form of dynamical laws in general relativistic theories is different in nature from the restrictions on the forms of dynamical laws imposed by kinematic symmetries in non-generally relativistic theories.

The nature of spacetime and the nature of motion

It has long been recognized that there is an intimate connection between the nature of spacetime and its geometrical structures on the one hand, and the nature of motion on the other, in terms of their absoluteness or its opposite. Here the notion of absolute motion is relatively simple: it means non-relative or intrinsic motion. But the notion of absolute space(time) is more complicated. It can mean rigid (non-dynamic), substantive (non-relational), or autonomous (non-relative). In traditional debates on the nature of space, all three senses of absolutism, as opposed to dynamism, relationalism, and relativism, were involved.[1] In contemporary discussions in the context of GTR, however, rigidity is not an issue any more. But the (stronger) substantivalist and (weaker) autonomist versions of absolutism are still frequently involved.

If there are only relative motions, then the notion of an absolute space(time) will have no justification at all. If, on the other hand, space(time) is absolute, then there must be some absolute motions. Yet the entailments here do not go the other way round. That is, the non-absolute nature of space(time) does not entail the absence of absolute motions, and absolute motions do not presuppose an absolute space(time). Such logical relations notwithstanding, the existence of absolute motions does require that spacetime must be endowed with structures, such as points and inertial frames of reference, which are rich enough to support an absolute conception of motion. Thus a difficult question has been raised, concerning the ontological status of these spacetime structures. Historically, this conceptual situation is exploited, often successfully, by abso-lutists. For this reason, there is a long tradition, among relationists, of rejecting the existence of absolute motions as a protective stratagem, even though it is not logically required.[2] Among relationists taking this stratagem we find Mach and Einstein.

Einstein's views of spacetime

Einstein's views can be characterized by his rejection of Newton's absolutist views of space and time, and by his being strongly influenced by developments in field theories and his finally taking physical fields as the ontologically constitutive physical basis of spacetime and its structures.

Einstein's understanding of Newton was shaped by Mach's interpretation. According to Mach, Newton held that space must be regarded as empty, and such an empty space, as an inertial system with its geometrical properties, must be thought of as fundamental and independent, so that one can describe secondarily what fills up space.[3] According to Mach, empty space and time in Newtonian mechanics play a crucial dual role. First, they are carriers of coordinate frames for things that happen in physics, in reference to which events are described by the space and time coordinates. Second, they form one of the inertial systems, which are considered to be distinguished among all conceivable systems of reference because they are those in which Newton's law of inertia is valid.

This leads to Mach's second point in his reconstruction of Newton's position: space and time must possess as much physical reality as do material points in order for the mechanical laws of motion to have any meaning. That is, 'physical reality' is said to be conceived by Newton as consisting of space and time on the one hand, and of permanently existing material points, moving with respect to the independently existing space and time, on the other. The idea of the independent existence of space and time can be expressed in this way. If there were no matter, space and time alone could still exist as a kind of potential stage for physical happenings.

Moreover, Mach also noticed that in Newton's equations of motion, the concept of acceleration plays a fundamental part and cannot be defined by temporally variable intervals between massive points alone. Newton's acceleration is conceivable or definable only in relation to space as a whole. Mach reasoned, and this reasoning was accepted by Einstein, that this made it necessary for Newton to attribute to space a quite definite state of motion, i.e., absolute rest, which is not determinable by the phenomena of mechanics. Thus, according to Mach, in addition to its physical reality, a new inertia-determining function was also tacitly ascribed to space by Newton. The inertia-determining effect of space in Newtonian mechanics must be autonomous, because space affects masses but nothing affects it. Einstein called this effect the 'causal absoluteness' of space[4] (1927).

Newtonian mechanics was replaced by STR, the emergence of which Einstein took as an inevitable consequence of the development of field theory by Faraday, Maxwell, and Lorentz. The field theory of electromagnetism is metaphysically different from mechanics because, as Einstein summarized in (1917b),

(a) 'in addition to the mass point', there arises 'a new kind of physical reality, namely, the "field"', and

(b) 'the electric and magnetic interaction between bodies [are] effected not by forces operating instantaneously at a distance, but by processes which are propagated through space at a finite speed'.

Taking (a) and (b) into consideration, Einstein in STR claimed that the simultaneity of events could not be characterized as absolute, and that in a system of reference accelerating relative to an inertial system, the laws of disposition of solid bodies did not correspond to the rules of Euclidean geometry because of the Lorentz contraction. From non-Euclidean geometry, and on the considerations of GC and EP, Einstein suggested in GTR that the laws of disposition of solid bodies were related closely with the gravitational field. These developments in field theories had radically modified the Newtonian concepts of space, time, and physical reality, and further developments in field theories required a new understanding of these concepts. It was in this interplay that Einstein developed his views of space, time, and physical reality.

According to Einstein, the notion that spacetime is physically characterless is finally disposed of because

the metrical properties of the spacetime continuum in the neighbourhood of separate spacetime points are different and conjointly conditioned by matter existing outside the region in question. This spatio-temporal variability [of the metric relations], or the knowledge that 'empty space' is, physically speaking, neither homogeneous not isotropic, ... compels us to describe its state by means of ten functions, [i.e.,] the gravitational potential $g_{\mu\nu}$ .... No space and no portion of space is without gravitational potential, for this gives it its metrical properties without which it is not thinkable at all. The existence of the gravitational field is directly bound up with the existence of space.

(1920a)

For Einstein, spacetime was also divested of its causal absoluteness. In GTR, the geometrical character of spacetime is constituted by metric fields, which, according to EP, are at the same time the gravitational fields. Thus spacetime, whose structures depend on physically dynamic elements ($g_{\mu\nu}$), is no longer absolute. That is to say, it not only conditions the behavior of inertial masses, but it is also conditioned, as regards its state, by them.

It is worth noticing that Einstein's views on spacetime, though consistently anti-Newtonian in nature and field-theoretically oriented, underwent a subtle change nevertheless. The change occurred in the mid-1920s, apparently concomitant with his pursuing a unified field theory (UFT) starting in 1923 (see section 5.3). In his pre-UFT period, the Newtonian concept of empty space and the causal absoluteness of space and time were rejected, but the physical reality of space and time was still somewhat retained:

Our modern view of the universe recognizes two realities which are conceptually quite independent of each other, even though they may be causally connected [via MP], namely the gravitational ether and the electromagnetic field, or – as one might call them – space and matter.

(1920a)

In contrast with this dualistic position on physical reality and a semi-absolutist view of space, Einstein's view in his UFT period was a thoroughly relationist one: 'space is then merely the four-dimensionality of the field' (1950b), 'a property of "things" (contiguity of physical objects)', and

the physical reality of space is represented by a field whose components are continuous functions of four independent variables – the coordinates of space and time, ... it is just this particular kind of dependence that expresses the spatial character of physical reality.

(1950a)

He went even further and asserted that

space as opposed to 'what fills space', which is dependent on the coordinates, has no separate existence, ... if we imagine the gravitational field, i.e., the function g_{ik}, to be removed, there does not remain a space of type (1) [$ds^2 = dx_1^2 + dx_2^2 + dx_3^2 - dx_4^2$], but absolutely nothing.

(1952a)

For Einstein, the functions g_{ik} describe not only the field, but at the same time also the metrical properties of spacetime. Thus

space of the type (1) ... is not a space without field, but a special case of the g_{ik} field, for which ... the functions g_{ik} have values that do not depend on the coordinates. There is no such thing as an empty space, i.e., a space without field. Space-time does not claim existence on its own, but only as a structural quality of the field.

(Ibid.)

As far as the ontological primacy of the gravitational field over spacetime is concerned, Einstein was a relationist as early as 1915, when he rejected the hole argument and regained GC (see section 4.2). But what was crucial for his change from a dualist to a monist view of physical reality was his new understanding of the nature and function of the $g_{\mu\nu}$ fields. If the function of the $g_{\mu\nu}$ fields is only to constitute spacetime ontologically and specify its metrical structures (and thus motions) mathematically, then whether we take spacetime as an absolute or a relational representation of the $g_{\mu\nu}$ fields will only be a semantic problem, without any essential difference. But if the $g_{\mu\nu}$ fields are to be taken as part of physically substantial fields, transformable into and from other physical fields, as claimed by UFT, then a dualist view of physical reality will not be sustainable, and a thoroughly relationalist view of spacetime will give no room to any autonomous view of spacetime, even though there is still room for defining absolute motions, which are not relative to absolute spacetime, but relative to the spacetime constituted by the totality of physical fields (rather than by some of them).

On 9 June 1952, Einstein summed up his view of spacetime as follows:

I wish to show that space-time is not necessarily something to which one can ascribe a separate existence, independently of the actual objects of physical reality. Physical objects are not in space, but these objects are spatially extended. In this way the concept 'empty space' loses its meaning.

(1952b)

In the last few years of his life, Einstein frequently maintained that the basis of the above view is the programme of field theory, in which physical reality, or the ultimate ontology, or the irreducible conceptual element in the logical construction of physical reality, is represented by the field. He insisted that only this programme can make a separate concept of space superfluous, because it would be absurd to reject the existence of empty space if one insisted that only ponderable bodies are physically real (1950b, 1952a, 1953).

Einstein's views of geometry

Closely related to Einstein's relationalist view of spacetime is his distinct view of geometry. This view was neither axiomatic nor conventionalist, but was practical in nature, trying to maintain a direct contact between geometry and physical reality. Einstein attached special importance to his view, 'because without it I should have been unable to formulate the theory of relativity', and 'the decisive step in the transition to generally covariant equations would certainly not have been taken' (1921a).

According to the axiomatic view, what is assumed is not the knowledge or intuition of the objects of which geometry treats, but only the validity of the axioms that are to be taken in a purely formal sense, i.e., as void of all content of intuition or experience. These axioms are free creations of the human mind. All other propositions of geometry are logical consequences of the axioms. 'It is clear that the system of concepts of axiomatic geometry alone cannot make any assertions as to the behaviour of real objects', because the objects of which geometry treats are defined by axioms only and have no necessary relations with real objects (Einstein, 1921a).

For Einstein, however, geometry 'owes its existence to the need which was felt of learning something about the behaviour of real objects' (1921a), 'geometrical ideas correspond to more or less exact objects in nature, and these last are undoubtedly the exclusive cause of the genesis of those ideas' (1920a). The very word 'geometry', which means earth-measuring, endorsed this view, Einstein argued, 'for earth-measuring has to do with the possibilities of the disposition of certain natural objects with respect to one another', namely, with parts of the earth, scales, etc. (1921a). Thus, to be able to make any assertion about the behavior of real objects or practically rigid bodies, 'geometry must be stripped of its merely logico-formal character by the coordination of real objects of experience with the empty conceptual framework of axiomatic geometry' (1921a), and becomes 'the science of laws governing the mutual spatial relations of practically rigid bodies' (1930a). Such a 'practical geometry' can be regarded 'as the most ancient branch of physics' (1921a).[5]

According to Einstein, the significance of the 'practical geometry' lies in the fact that it establishes the connections between the body of Euclidean geometry and the practically rigid body of reality, so that 'the question whether the geometry of the universe is Euclidean or not has a clear meaning, and its answer can only be furnished by experience' (1921a).

In Poincaré's view of geometry, there was no room for such a connection between the practically rigid body and the body of geometry. He pointed out that under closer inspection, the real solid bodies in nature were not rigid in their geometrical behavior, that is, their possibilities of relative disposition depended upon temperature, external forces, etc. Thus immediate connections between geometry and physical reality appeared to be non-existent. According to Poincaré, therefore, the application of a geometry to experience necessarily involved hypotheses about physical phenomena, such as the propagation of light rays, the properties of measuring rods, and the like, and had an abstract (conventional) component as well as an empirical component, just as in every physical theory. When a physical geometry was not in agreement with observations, agreement might be restored by substituting a different geometry or a different axiom system, or by modifying the associated physical hypotheses.[6] Thus it should be possible and reasonable, in Poincaré's conventionalist view of geometry, to retain Euclidean geometry whatever may be the nature of the behavior of objects in reality. For if contradictions between theory and experience manifest themselves, one could always try to change physical laws and retain Euclidean geometry, because intuitively, according to Poincaré, Euclidean geometry was the simplest one for organizing our experience.

Einstein could not reject Poincaré's general position in principle. He acknowledged that according to STR, there was no real rigid body in nature, and hence the properties predicated of rigid bodies do not apply to physical reality. But, still, he insisted that it was not a difficult task

to determine the physical state of a measuring-body so accurately that its behaviour relative to other measuring-bodies should be sufficiently free from ambiguity to allow it to be substituted for the 'rigid' body. It is to measuring-bodies of this kind that statements about rigid bodies must be referred.

(1921a)

Einstein also acknowledged that, strictly speaking, measuring rods and clocks would have to be represented as solutions of the basic equations (as objects consisting of moving atomic configurations), and not, as it were, as theoretically self-sufficient entities, playing an independent part in theoretical physics. Nevertheless, he argued that in view of the fact that 'we are still far from possessing such certain knowledge of theoretical principles of atomic structure as to be able to construct solid bodies and clocks theoretically from elementary concepts', 'it is my conviction that in the present stage of development of theoretical physics these concepts must still be employed as independent concepts' (1921a).

Thus Einstein provisionally accepted the existence of practically rigid rods and clocks, of which practical (chrono)geometry treated, in order to connect (chrono)geometry and physical reality, and to give (chrono-)geometrical significance to the $g_{\mu\nu}$ fields.

However, one would be greatly mistaken to think thereby that Einstein held a kind of external view of (chrono)geometry, external in the sense that (chrono) geometry was considered to be constituted by rods and clocks.[7] This is a mistaken view for three reasons. First, Einstein admitted non-linear transformations of coordinate in GTR, which entails that the immediate metrical significance of the coordinates is lost. Thus in the Gauss–Riemann sense that geometry of the spacetime manifold is independent of coordinates,[8] Einstein's view was intrinsic rather than external. Second, Einstein held that (chrono-) geometrical properties of spacetime were given by the gravitational fields alone (1920b). Thus in the sense that the (chrono-)geometrical structures of spacetime were presumed to exist independently of its being probed by rods and clocks, or even independently of the existence of rods and clocks, Einstein's view was internal rather than external. Third, Einstein emphasized that the introduction of two kinds of physical thing apart from four-dimensional spacetime, i.e., rods and clocks on the one hand, and all other things (e.g., material points and the electromagnetic fields) on the other, in a certain sense was inconsistent. Although he thought it was better to permit such an inconsistency at first, the inconsistency had to be eliminated at a later stage of physics, namely, in a unified field theory (1949).

However, if we accept that Einstein took (chrono)geometry to be onto-logically constituted and mathematically described by the $g_{\mu\nu}$ fields, then certain complications should not be ignored. First, there were those to do with the interpretation of the field equations. Since the field equations equate the contracted curvature tensor, expressed in terms of $g_{\mu\nu}$, which characterizes the (chono)geometry, with the stress–energy tensor, which specifies the matter distribution, in a symmetrical way, neither term can claim priority over the other. Thus what the field equations suggested was a mutual causal connection between the metrical fields and matter (including all non-metrical fields). Yet Einstein's early obsession with Mach's ideas made him claim that the metrical fields and the chronogeometry of spacetime itself had to be unilaterally and completely determined by matter (1918a). This seems to be in favor of the claim that Einstein held a kind of external view of (chrono)geometry. It is fair to say that to the extent that Einstein was sticking to MP, this claim is somewhat justifiable.

But with the impression he received from investigations of the vacuum field equation, which showed that the metric fields did not owe their existence to matter, and with his moving from MP to UFT, Einstein gradually realized that just as the electromagnetic field had won liberation and the position as an independent dynamic entity under Maxwell, the metric field should also be allowed at his hands to cast off its chains under Mach, and step onto the stage of physics as a participant in its own right, with dynamic degrees of freedom of its own. In fact, near the end of his life, Einstein clearly pointed out that in his UFT, 'the T_{ik} which are to represent "matter" always presuppose the g_{ik}', the metric tensor (1954a).

Second, there were considerations regarding the physical basis of chrono-geometry. Einstein held that the physical reality of spacetime was represented by the field (1950a); that spacetime and its chronogeometry claimed existence only as a structural quality of the physical field (1952a), of which gravitation and metric were only different manifestations; and that 'the distinction between geometrical and other kinds of fields is not logically founded' (1948a).[9] If we remember all these statements of Einstein's, then the position of taking Einstein as holding an internal or even an absolutist view of chronogeometry would not seem to be tenable.

It is well known that the core of Einstein's view of geometry is Riemann's celebrated thesis:

We must seek the ground of its metric relations outside it [i.e. actual things forming the groundwork of space], in binding forces which act upon it.

(Riemann, 1854)

Here what 'binding forces' means in Einstein's view is clearly the gravitational force, or, more precisely, the gravitational interactions mediated by the gravitational fields. Thus it seems possible to summarize Einstein's view of geometry as follows.

In so far as Einstein regarded the gravitational field as the only ontologically necessary and sufficient constitutive element for gravitational interactions and hence for chronogeometry, without any need of rods and clocks as necessary constitutive elements for chronogeometry, his view can be considered internal.

To the extent that Einstein maintained that chronogeometry is not ontologically irreducible but, rather, only a manifestation of the structural properties of the gravitational field, his view should be taken as a relationalist one.

5.2 The geometrical programme: strong and weak versions

In terms of its ontological commitment, the geometrical programme for fundamental interactions (GP), whose foundations were laid down by Einstein in GTR, is a variation of the field theory programme; in terms of its description of dynamics, geometrical terms play an exclusive part. Conceptually, two assumptions constitute its points of departure. First, EP, which assumes the inseparability of inertia and gravity, thus assumes the role of the gravitational field in constituting geometrical (i.e. inertial or affine) structures of spacetime (see section 3.4), and the role of geometrical structures in describing the dynamics of gravity. Second, the view of 'practical geometry', according to which geometry is directly connected with the physical behavior of practically rigid rods and clocks, and thus the geometrical structures of spacetime, as the basis of physical phenomena, is not given a priori but, rather, is determined by physical forces. For this reason, geometry should be regarded as a branch of physics, and the geometrical character of the physical world is not an a priori or analytic or conventional problem, but an empirical one. As we indicated at the end of section 5.1, this idea has its intellectual origin in Riemann's celebrated inaugural lecture of 1854.

Starting with these two assumptions, Einstein discovered that the trajectories of particles and rays in a gravitational field have the properties of geodesics in a non-Euclidean manifold, and that the occurrence of a gravitational field is connected with, and described by, the spatio-temporally variable metrical coefficients $g_{\mu\nu}$ of a non-Euclidean manifold. In contrast with classical field theory, in which the field is described in terms of potential functions that depend on coordinates in a pregiven Euclidean space or Minkowskian space-time, Einstein in GTR associated gravitational potentials directly with the geometrical character of spacetime, so that the action of gravity could be expressed in geometrical terms. In this sense, we may say that Einstein initiated the geometrization of the theory of gravity; hence the epithet 'geometrical programme'.

The basic idea of GP is this. Gravitational (or other) interactions are realized through certain geometrical structures, such as the curvature of spacetime. The geometrical structure influences the geodesics of the motion of matter, and is influenced by the energy tensor of matter. The latter influence is expressed by the field equations, and the former influence is expressed by the geodesic equations of motion, which were first postulated independently of the field equations (Einstein and Grossmann, 1913b), but were then proved as only a consequence of the field equations (Einstein, Infeld, and Hoffman, 1938).

GP has two versions: a strong version and a weak version. According to the strong version, (i) the geometrical structures of spacetime themselves are physically real, as real as matter and the electromagnetic fields, and have real effects upon matter; and (ii) the gravitational interaction is taken as a local measure of the effect of spacetime geometry on the motion of matter, as a manifestation of the spacetime curvature, and is expressed by the equation of geodesic deviation.

Einstein himself held this view in his earlier years. In fact, as late as 1920, he still identified the gravitational field with space (1920b) or, more precisely, with the geometrical structures defined on a manifold. With this view, Einstein transformed the geometrical structures of spacetime from rigidly given, unchangeable, and absolute entities into variable and dynamical fields interacting with matter. It is worth noting that the strong version of GP closely agrees with the Newtonian position in asserting the reality and activity of space, although it differs from the latter position in viewing space as a dynamic reality full of physical character, rather than taking space as empty and rigid as conceived by Newtonians.

The weak version of GP rejects the independent existence of spacetime structures, and takes them only as structural qualities of the fields. Einstein held this view in his later years when he pursued UFT. Logically speaking, however, the weak version does not have to presume UFT. What it does presume is that the geometrical structure of spacetime is ontologically constituted by physical (gravitational) fields.

Then a serious question facing the weak version of GP is how to justify this presumption. The justification is readily provided by Einstein's view of practical geometry. According to this view, the geometrical structure manifests itself only in the behavior of rods and clocks, and is determined wholly by the interactions between the gravitational field on the one hand, and rods and clocks on the other. Rods and clocks, in addition to their function as probes of the metric of spacetime, can also influence the gravitational field. But the field itself, according to Einstein, is sufficient to constitute the metric of spacetime. In empty spacetime, it is the purely gravitational field that is the only physical reality and constitutes metrics (Minkowskian or otherwise) for spacetime; in non-empty spacetime, it is the gravitational field interacting with matter (and/or the electromagnetic field) that plays the same constitutive role.

Einstein's theory of gravitation, as he himself repeatedly pointed out from the days of GTR to the end of his life, can only be regarded as a field theory, in which gravity as a causal process is mediated by physical fields.

I have characterized Einstein's research programme of gravitation as a geometrical, and also as a field-theoretical, one. The compatibility of these two characterizations is apparent. In both Einstein's earlier and later views, the fields and the geometry of spacetime, as the medium of gravitational inter-actions, are directly and inseparably connected to each other. In fact, the two views of GP that form two phases of the development of GP can also be considered as two views, or two phases of the development, of the field programme.

View 1: the geometrical structures of spacetime are taken as physically real and the gravitational fields are reducible to them. Apparently, this view is compatible with the dualism of matter and space/field.

View 2: the gravitational field is taken to be a physical substance, underlying the geometrical structures of spacetime and having spacetime as the manifestation of its structural quality. It is not so difficult to see that this view prepares the way for monism as was endorsed by Einstein's UFT.

Judging from Einstein's publications, we find that the transition from view 1 to view 2 took place in the first half of the 1920s, concomitant with his turn to the pursuit of UFT. At the beginning of the 1920s, Einstein still viewed space as an independent reality (1920a), with the possibility of its geometrical structures having some field properties (1922). But five years later, Einstein already regarded the metric relations as identical with the properties of the field, and declared that 'the general theory of relativity formed the last step in the development of the programme of the field theory' (1927).

In the transition, Einstein's view of 'practical geometry' played an important role. If the geometrical structure of spacetime manifests itself only in the behavior of rods and clocks, and is determined wholly by the gravitational interactions mediated by the gravitational fields, then it is quite natural, if one accepts activity, reality, and self-subsistence as the criteria of substance, to regard the gravitational field interacting with matter and with itself as a physical substance, and geometrical structure as its structural quality. It is worth noting that the two extremely important papers (1921a, 1925b) that had clarified Einstein's view of geometry happened to be written during the period of change.

Thus we have reached our most important conclusion. After he entered the second phase of the evolution of his ideas, what Einstein did was not to geometrize the theory of gravitation, but to gravitize the geometry of spacetime.[10] That is, he regarded geometry as the manifestation of gravitational interactions rather than the other way round. The geometrical structures of spacetime, after being gravitized, evolve with the evolution of gravitational interactions, and its laws of evolution are identifiable with the dynamical laws of the field, namely, Einstein's field equations.

What is involved in the above discussion is only Einstein's views about the relationship between the gravitational fields and spacetime with its geometrical structures. During the first two phases of Einstein's field theory, in addition to the reality of the fields, the reality of matter was presupposed. But with the development of the theory, and in particular, with the deeper investigations of the vacuum field equations, Einstein was forced to commit himself to consider the relationship between field and matter.

What had to be resolved was the following dilemma. On the one hand, Einstein's commitment to MP forced him to take matter as the only source of the field and the only cause of changes of the field. On the other hand, the existence of mathematically valid solutions to the vacuum field equations seemed to imply that the field itself was a kind of substance, not derived from matter. If he rejected MP and took the field as a primary substance, he had to explain why matter acted, as indicated by the field equations, as a source of the field, and a cause of its changes. Einstein's solution was quite simple. He took matter, which he incorrectly reduced to the electromagnetic field, and the gravitational field as two manifestations of the same substratum, the so-called non-symmetrical total field. Conceptually, this provided a substantial basis for the mutual transformability of the two. Thus he said, 'I am convinced, however, that the distinction between geometrical and other kinds of fields is not logically founded' (1948b).[11] With this solution Einstein carried GP on to a new stage, the stage of UFT.

In sum, GP, which was mainly based on GTR and Einstein's other works, in all its three phases is radically different from Newtonian physics in that it is a field theory. In the Newtonian programme, gravitation appears as an instantaneous action at a distance, and the gravitational force cannot be interpreted as an activity determined by, and emanating from, a single body, but has to be taken only as a bond between two bodies that interact across the void. In Einstein's programme, gravitation was considered as a local intermediary action, propagating not instantaneously but with the velocity of light, and the force was split up into the action of one body (excitation of the field determined by this body alone) and the reaction of another body (temporal change of its momentum caused by the field). Between the two bodies the field transmitted momentum and energy from one body to another (Einstein, 1929a, b).

In Einstein's opinion (1949), Mach's criticism of Newtonian theory was bound to fail because Mach presumed that masses and their interactions, but not fields, were the basic concepts. But

the victory over the concept of absolute space or over that of the inertial system became possible only because the concept of the material object was gradually replaced as the fundamental concept of physics by that of the field . . . up to the present time no one has found any method of avoiding the inertial system other than by way of the field theory.

(1953)

Einstein was right that his victory over Newtonian theory was a victory of the field theory. But, unfortunately, he failed to show in a convincing way, though he wanted to, that matter and fields (both electromagnetic and gravitational ones) can be described with a unified field.

Needless to say, Einstein's field theory is of a particular kind, in which the fields are inseparably bound up with the geometrical structures of spacetime. Then, what is its relation with a more fruitful field theory, quantum field theory? This interesting question, together with a number of historical and philosophical questions, will be addressed later in this volume.

5.3 Further developments

At about the same time as Einstein presented his final version of GTR, David Hilbert proposed a new system of the basic equations of physics (1915, 1917), synthesizing the ideas of Gustav Mie's electromagnetic field theory of matter (1912, 1913) and Einstein's general relativity. In his new theory, Hilbert used a variational principle and obtained, from Mie's world function $H(g_{\mu\nu}, g_{\mu\nu,\rho},$

$g_{\mu\nu,\rho\sigma}$, A_μ, $A_{\mu,\rho}$), fourteen equations for fourteen potentials: ten of these equations contained the variations with respect to the gravitational potentials $g_{\mu\nu}$, and were called the equations of gravitation; while four stemmed from the variations with respect to the electromagnetic potentials A_μ, and gave rise to the generalized Maxwell equations. With the help of a mathematical theorem, which says that in the process of applying the variational principle, four relations will be found among n fields, Hilbert claimed that

the four [electromagnetic equations] may be regarded as a consequence of the gravitational equations . . .. In this sense electromagnetic phenomena are gravitational effects.

(1915)[12]

Hilbert's unified theory of gravity and electromagnetism provided a strong stimulus to the early development of GP. In particular, his conceived connection between the metric $g_{\mu\nu}$ and electromagnetic potentials A_μ led him to a general view on the relationship between physics and geometry:

Physics is a four-dimensional pseudo-geometry whose metric $g_{\mu\nu}$ is related to the electromagnetic quantities, that is matter.

(Hilbert, 1917)

However, Hilbert said nothing about the geometrical foundations of physics, nor about the geometrical structures of spacetime. In addition, the geometrical correspondence of the electromagnetic potentials A_μ was also unclear in his theory.

What was crucial for the development of GP in terms of its mathematical underpinnings was Levi-Civita's introduction of the concept of the infinitesimal parallel displacement of a vector (Levi-Civita, 1917). From such a concept, the elements in Riemann's geometry, such as the Riemann tensor of curvature, can be obtained. In this way, Riemann's geometry was generalized. Levi-Civita also endowed the three-index Christoffel symbol ($\Gamma^\mu_{\rho\nu}$) with the meaning of expressing the infinitesimal transport of the line elements in an affinely connected manifold, with their lengths being preserved. Thus a parallel transport of vectors in a curved space was made equivalent to the notion of covariant differentiation: $a^\mu_{;\nu} = a^\mu_{,\nu} + (\Gamma^\mu_{\rho\nu})a^\rho$. Following Levi-Civita's idea, Hessenberg (1917) considered space as formed by a great number of small elements cemented together by parallel transport, i.e., as an affinely connected space.

The first extension of Einstein's theory in a fully geometrical sense was given by Hermann Weyl in (1918a, b). Strongly influenced by Einstein's earlier ideas, Weyl held that geometry was 'to be regarded as a physical reality since it reveals itself as the origin of real effects upon matter', and 'the phenomena

of gravitation must also be placed to the account of geometry' (1922). Furthermore, he wanted to set forth a theory in which both gravitation and electromagnetism sprang from the same source and could not be separated arbitrarily, and in which 'all physical quantities have a meaning in world geometry' (1922; cf. 1918a, b).

For this purpose, Weyl had to generalize the geometrical foundations of Einstein's theory. As a starting point, he carried the idea of field theory into geometry, and made the criticism that

Riemann's geometry goes only half-way towards attaining the ideal of a pure infinitesimal geometry. It still remains to eradicate the last element of geometry 'at a distance', a remnant of its Euclidean past. Riemann assumes that it is possible to compare the lengths of two line elements at different points of space, too; it is not permissible to use comparisons at a distance in an 'infinitely near' geometry. One principle alone is allowable; by this a division of length is transferable from one point to that infinitely adjacent to it.

(1918a)

Thus in Weyl's geometry a particular standard of length should be used only at the time and place where it is, and it is necessary to set up a separate unit of length at every point of space and time. Such a system of unit standards is called a gauge system. Likewise, the notion of vectors or tensors is a priori meaningful only at a point, and it is only possible to compare them at one and the same point. What then is their meaning in all spacetime? On this point, Weyl absorbed Levi-Civita's idea about infinitesimal parallel displacement and Hessenberg's idea about affinely connected space. His own original view was the assumption that the gauge system, just like the coordinate system, was arbitrary:

The $g_{\mu\nu}$ are determined by the metrical properties at the point p only to the extent of their proportionality. In the physical sense, too, it is only the ratio of the $g_{\mu\nu}$ that has an immediate tangible meaning.

(Weyl, 1922)

Weyl demanded that a correct theory must possess the property of double invariance: invariance with respect to any continuous transformation of coordinates, and invariance under any gauge transformation in which $\lambda g_{\mu\nu}$ is substituted for $g_{\mu\nu}$, where λ is an arbitrary continuous function of position. Weyl declared that 'the supervention of this second property of invariance is characteristic of our theory' (*ibid.*).

The arbitrariness of the gauge system entails that the length l of a vector at different points is changed infinitesimally and can be expressed as $dl^2 = l^2 \, d\phi$, where $d\phi$ is a linear differential form: $d\phi = \phi_\mu \, dx^\mu$ (Riemannian geometry is

the limiting case when ϕ_μ is zero). This entails that from the quadratic fundamental form (Riemannian line element) $ds^2 = g_{\mu\nu} dx^\mu dx^\nu$ we can obtain

$$g_{\mu\nu,\rho} - \Gamma_{\mu,\nu\rho} - \Gamma_{\nu,\mu\rho} = g_{\mu\nu}\phi_\rho. \tag{1}$$

Substituting $\lambda g_{\mu\nu}$ for $g_{\mu\nu}$ in $ds^2 = g_{\mu\nu} dx^\mu dx^\nu$, one can easily find that
$d\phi'_\mu = d\phi_\mu + d(\log \lambda)$. Then gauge invariance implies that $g_{\mu\nu} dx^\mu dx^\nu$ and
$\phi_\mu dx^\mu$ are on an equal footing with $\lambda g_{\mu\nu} dx^\mu dx^\nu$ and $\phi_\mu dx^\mu + d(\log \lambda)$. Hence
there is an invariant significance in the anti-symmetrical tensor $F_{\mu\nu} =$
$\phi_{\mu,\nu} - \phi_{\nu,\mu}$. This fact led Weyl to suggest interpreting ϕ_μ in the world
geometry as the four-potential, and the tensor $F_{\mu\nu}$ as the electromagnetic field.

From the above consideration, Weyl held that not only could the electromagnetic field be derived from the world geometry (1918a, b),[13] but the affine
connection Γ of spacetime also depended on the electromagnetic potentials ϕ_μ,
in addition to the gravitational potentials $g_{\mu\nu}$, as can be easily seen from
equation (1) (1922). The presence of ϕ_μ in the affine connection Γ of spacetime
suggested that the geometry of spacetime must deviate from a Riemannian one.
Thus Weyl's unified theory differed from Hilbert's in its richer geometrical
structures, and also in its clearer and closer connection between electromagnetism and the geometry of spacetime. It also differed from Einstein's theory in its
taking the world geometry as an ontologically fundamental physical reality,
and gravity and electromagnetism as derivative phenomena; that is, it belonged
to the strong version of GP.

Weyl's concept of the non-integrability of the transference of length, derived
from the local definition of the length unit, invited many criticisms. Most
famous among them was that of Einstein, who pointed out its contradiction
with the observed definite frequencies of spectral lines (Einstein, 1918b). In
response, Weyl suggested the concept of 'determination by adjustment' instead
of 'persistence' to explain this difficulty away (1921). But this was too
speculative and he himself did not work it out in detail.

In spite of all the difficulties this concept encountered, another concept
associated with (or derived from) it, that of (dilation-)gauge invariance, has
important heuristic significance. In his unified theory, Weyl established a
connection between the conservation of electricity and the dilation-gauge
invariance. He regarded this connection as 'one of the strongest general
arguments in favour of my theory' (1922). Although the original idea of
dilation-gauge invariance was abandoned soon after its occurrence because its
implication was in contradiction with observation, in the new context of
quantum mechanics, Weyl revived the idea in 1929 (see section 9.1) This time
the local invariance was that of the quantum phase in electromagnetism. Our
present view of electric charge and the electromagnetic field relies heavily on

this idea. In fact, the relation between the gauge invariance of physical laws and the geometrical structures of spacetime has been one of the most fascinating subjects in contemporary physics (see part III).

According to Arthur Eddington, however, Weyl's geometry still suffered from an unnecessary restriction. Eddington hoped to show that

in freeing Weyl's geometry from this limitation, the whole scheme becomes simplified, and new light would be thrown on the origin of the fundamental laws of physics.

(1921)

In passing beyond Euclidean geometry with the help of GTR, Einstein obtained gravity; in passing beyond Riemannian geometry with the help of the principle of gauge invariance, Weyl obtained electromagnetism. So one might ask what remained to be gained by further generalization. Eddington's answer was that hopefully we could obtain non-Maxwellian binding forces that would counter-balance the Coulomb repulsion and hold an electron together.

Eddington, like Weyl, started from the concept of parallel displacement and gauge invariance. The main difference between them lies in the fact that while Weyl took $dl^2 = l^2 \, d\phi$ and equation (1), Eddington took the right side of (1) as $2K_{\mu\nu,\rho}$ instead of the special form $g_{\mu\nu}\phi_\rho$. This led to $dl^2 = 2K_{\mu\nu,\rho}\xi^\mu\eta^\nu \, dx^\rho$. Thus with the expressions

$$S_{\mu\nu,\sigma} = K_{\mu\nu,\sigma} - K_{\mu\sigma,\nu} - K_{\nu\sigma,\mu} \quad \text{and} \quad 2k_\mu = S^\sigma_{\sigma\mu},$$

the generalized gauge invariant curvature tensor $G^*_{\mu\nu}$ is split into a symmetric and an anti-symmetric part:

$$G^*_{\mu\nu} = R_{\mu\nu} + F_{\mu\nu}, \quad \text{with} \quad F_{\mu\nu} = k_{\mu,\nu} - k_{\nu,\mu},$$

While $F_{\mu\nu}$ can be viewed as the electromagnetic field, the symmetric part

$$R_{\mu\nu} = G_{\mu\nu} + H_{\mu\nu}, \quad \text{with} \quad H_{\mu\nu} = k_{\mu,\nu} + k_{\nu,\mu} - (S^\sigma_{\mu\nu})_\sigma + (S^\beta_{\alpha\nu}S^\alpha_{\beta\mu} - 2k_\alpha S^\sigma_{\mu\nu})$$

includes the curvature $G_{\mu\nu}$ responsible for gravitation as well as $H_{\mu\nu}$, an expression for the difference between the whole energy tensor and the electromagnetic energy tensor. Eddington suggested that this difference must represent a non-Maxwellian electronic part of the whole energy tensor.

Although Eddington obtained a more general geometry, he insisted that the natural geometry of the real world was Riemann's geometry, not Weyl's generalized geometry, nor his own. He said:

What we have sought is not the geometry of actual space and time, but the geometry of the world structure, which is the common basis of space and time and things.

(Ibid.)

There is no doubt that the introduction of the notion of the geometry of the world structure was important both for the development of GP and for the evolution of the conceptions of geometry (see section 11.3).

Closely related in spirit to, and partly inspired by the papers of, Weyl and Eddington, Einstein suggested a unified field theory (UFT) based on an asymmetric connection or 'metric'.[14] In addition to getting the pseudo-Riemannian spacetime of standard GTR, he also aimed at some further geometrical structure responsible for electromagnetism. Although complicated in mathematics, Einstein's UFT was relatively simple in its physical ideas: in addition to the metric structure of spacetime representing gravitation, there must be some other structure of spacetime representing electromagnetism. But, in Einstein's opinion, the idea that there were two structures of spacetime independent of each other was intolerable. So one should look for a theory of spacetime in which two structures formed a unified whole (Einstein, 1923a, 1930a). From this basic idea we can see that Einstein's UFT, just like his GTR, bore a distinctly geometrical impress, and was indeed another phase of GP.

In Einstein's works on UFT in (1945) and (1948b), the total unified field was described by a complex Hermitian tensor g_{ik}. This was quite different from the gravitational field in GTR, which was described by a symmetrical tensor $g_{\mu\nu}$. The fields (or potentials) could be split into a symmetric part and an anti-symmetric part:

$$g_{ik} = s_{ik} + \mathrm{i}a_{ik}, \quad \text{with} \quad s_{ik} = s_{ki}, \quad a_{ik} = -a_{ki};$$

here s_{ik} can be identified with the symmetric tensor of the metric or gravitational potentials, and a_{ik} with the anti-symmetric tensor of the electromagnetic field. As the final result of his thirty-year search for a UFT, Einstein generalized his theory of gravitation by abandoning the restriction that the infinitesimal displacement field $\Gamma^\sigma_{\mu\nu}$ must be symmetric in its lower indices. In this way, Einstein obtained an anti-symmetric part of $\Gamma^\sigma_{\mu\nu}$, which was expected to lead to a theory of the electromagnetic field, in addition to a symmetric part, which led to a theory of the pure gravitational field (1954b).

Thus it was esthetically satisfactory to have the gravitational and electromagnetic fields as two components of the same unified field. But the pity was that Einstein's UFT had no support from experimental facts, certainly nothing similar to the support that GTR obtained from the equality of inertial and gravitational masses. The reasons for the lack of support are far deeper than the absence of suitable experiments so far. Mathematically, as Einstein himself acknowledged, 'we do not possess any method at all to derive systematic solutions . . .. For this reason we cannot at present compare the content of a nonlinear field theory with experience' (1954b). Judging from our present

knowledge about the unification of fundamental interactions, the underlying physical ideas of Einstein's UFT, such as the relationship between non-gravitational fields and the geometrical structures of spacetime, and the structural relations between different component fields, and between component fields and the total field, are purely speculative without empirical basis.

Between 1923 and 1949, Einstein expended much effort, but made no major breakthrough in UFT. In the same period, however, electromagnetism became the first completely special relativistic quantum field theory, with no direct geometrical features being recognized. Later developments also showed that weak interactions and at least certain strong interactions also appeared to be describable in terms of such a relativistic quantum field theory without direct geometrical features.[15] The successes of quantum field theories have stimulated a revival of the idea of UFT since the mid-1970s (see section 11.2). Yet in the revived unification scheme, the structural relations between the total field and component fields, and between different component fields, turn out to be much more complicated than the combination of symmetric and anti-symmetric parts. A number of new ideas, such as hierarchical symmetry-breaking and the Higgs mechanism are indispensable for understanding these relationships. In terms of internal tight-fit and coherence, therefore, Einstein's UFT can hardly be taken as a mature theory.

It is worth noting that according to Einstein's view of spacetime, the geometrical structures constructed from the infinitesimal displacement field in his UFT describe the structural properties of the total field rather than the structures of spacetime as an underlying physical reality. Ontologically, the latter were already reduced to the structural relations of the field without remainder. So, what the 'unity' means in the context of Einstein's UFT is just the unity of two kinds of field, rather than the unity between physical fields and spacetime, as was claimed by John Wheeler (1962).

A less conservative approach that attempted to change the pseudo-Riemannian structure of standard GTR was proposed by Elie Cartan (1922). According to Cartan, a model of spacetime that is responsible for a distribution of matter with an intrinsic angular momentum should be represented by a curved manifold with a torsion, which is related to the density of spin. Cartan's basic idea was developed by several authors.[16] It is conceivable that torsion may produce observable effects inside those objects, such as the neutron stars, which have built-in strong magnetic fields and are possibly accompanied by a substantial average value of the density of spin.

In contrast with Weyl, who introduced the variation of dl^2 and linked it to the existence of electromagnetic potentials, Theodor Kaluza (1921) remained in the realm of metrical geometry and sought to include the electromagnetic field by extending the dimensions of the universe. The line element of Kaluza's five-dimensional geometry can be written as $d\sigma^2 = \Gamma_{ik}\,dx^i\,dx^k$ (i, $k = 0, \ldots,$ 4), here Γ_{ik} are the fifteen components of a five-dimensional symmetric tensor, which are related to the four-dimensional $g_{\mu\nu}$ ($= \Gamma_{\mu\nu}$, μ, $\nu = 1, \ldots, 4$) and the electromagnetic potentials A_ν ($= \Gamma_{0\nu}$, $\nu = 1, \ldots, 4$). Then what is the meaning of the fifteenth quantity Γ_{00}? According to Oskar Klein (1927), we can try to relate Γ_{00} with the wave function characterizing matter, so that a formal unity of matter with field can be achieved. In this sense Γ_{ik} can be viewed as a realization of Eddington's geometry of the world structure.

In order to incorporate quantum effects in the five-dimensional theory, Klein followed Louis de Broglie's idea (see section 6.5) and conjectured that the quantum of action might come from the periodicity of motion in the fifth dimension. These motions were not perceptible in ordinary experiments, so we could average over the entire motion. Klein (1926) even claimed that the introduction of a fifth dimension into the physical picture as 'a radical modification of the geometrical basis of the field equations is suggested by quantum theory'.

Although Klein (1927) himself soon realized that he had failed to incorporate quantum phenomena into a spacetime description, Oswald Veblen and Banesh Hoffmann (1930) were not thereby discouraged. They suggested replacing the affine geometry underlying the Kaluza–Klein five-dimensional theory with a four-dimensional projective geometry, and showed that when the restriction on the fundamental projective tensor (which was imposed to reduce the projective theory to the affine theory of Klein's) was dropped, a new set of field equations that included a wave equation of Klein–Gordon type would be obtained. On the basis of this mathematical demonstration, they claimed, over-optimistically, that the use of projective geometry in relativity theory 'seems to make it possible to bring wave mechanics into the relativity scheme'.

Even at the present, the basic ideas of Kaluza–Klein theory are not dead. Many efforts have been made to show that gauge structures representing internal (dynamic) symmetries are the manifestation of the geometrical structures of a higher-dimensional spacetime (there is more discussion on this in section 11.3).

John Wheeler was one of the most active advocates of GP. He published an influential book *Geometrodynamics* in 1962, in which he considered geometry as the primordial entity, and thought that gravitation was only a manifestation of geometry and that everything else could be derived or constructed from geometry. He believed that there could be energy-bearing waves in the geometry of spacetime itself, which presupposed that the entities of geometry were a kind of physical reality. That is to say, geometrical entities were substantialized by him (Misner, Thorne, and Wheeler, 1973).

As to the quantum phenomena, Wheeler's position was rather modest. He had no ambition to derive quantum theory from the geometry of spacetime. Rather, he regarded quantum principles as more fundamental to the make up of physics than geometrodynamics. He supplemented geometrodynamics with quantum principles and thereby reformulated geometrodynamics. This standpoint led to some significant consequences. At first, it led him to quantum geometrodynamics. But in the end, it ironically led him to a denial of GP.

According to quantum geometrodynamics, there are quantum fluctuations at small distances in geometry, which lead to the concept of multiply connected space. Wheeler argued that for these reasons, space of necessity had a foamlike structure. He regarded electricity as lines of force trapped in a multiply connected space, and took the existence of charge in nature as evidence that space in the small distances is multiply connected. In his view, a particle can be regarded as a 'geometrodynamical exciton, and various fields likewise can be interpreted in terms of modes of excitation of multiply connected geometry' (1964b).

However, 'quantum fluctuations' as the underlying element of his geometrical picture of the universe paradoxically also undermined this picture. Quantum fluctuations entail change in connectivity. This is incompatible with the ideas of differential geometry, which presupposes the concept of point neighborhood. With the failure of differential geometry, the geometrical picture of the universe also fails: it cannot provide anything more than a crude approximation to what goes on at the smallest distances. If geometry is not the ultimate foundation of physics, then there must exist an entity – Wheeler calls it 'pregeometry' – that is more primordial than either geometry or particles, and on the foundation of which both are built.

The question then is what the pregeometry is of which both geometry and matter are manifestations. Wheeler's answer was: a primordial and underlying chaos (1973). In fact, Wheeler's understanding of pregeometry, which was based on the quantum principles, went beyond the scope of GP. The evolution of Wheeler's thoughts reveals one of inherent difficulty in incorporating quantum principles in GP, a difficulty of reconciling the discrete with the continuous.

Roger Penrose was another active advocate of GP. The evolution of his ideas is of special interest. At first, Penrose (1967a) proposed a spinor approach to spacetime structure. Considering the fact that the simplest building blocks, out of which the values of all tensor and spinor fields of standard field theories can be constructed, are two-component spinors, he suggested that we could take spacetime primarily as the carrier of such spinor fields and infer the structures of spacetime from the role of the spinor fields. He thereby obtained both the pseudo-Riemannian and spinor structures of spacetime, the latter being needed since there are fermions in nature.

Afterwards, Penrose (1967b, 1975) proposed a more ambitious theory, the twistor theory, to bridge the geometry of spacetime and quantum theory. He noted that there were two kinds of continuum: a real continuum of four dimensions, representing the arena of spacetime, within which the phenomena of the world were presumed to take place; and a complex continuum of quantum mechanics, which gave rise to the concept of probability amplitude and the superposition law, leading to the picture of a complex Hilbert space for the description of quantum phenomena. Usually, the idea of quantum mechanics was simply superposed on the classical picture of four-dimensional spacetime. The direct aim of twistor theory, however, was to provide a framework for physics in which these two continua were merged into one.

Penrose found that his aim was achievable because there was a local isomorphism between the spinor group SU(2,2) and the fifteen-parameter conformal group of Minkowski spacetime $C_+^{\uparrow}(1,3)$. This isomorphism entails that the complex numbers consisting of two spinor parts, which define the structure of twistor space, are intimately tied in with the geometry of spacetime, and also emerge in a different guise as quantum-mechanical probability amplitudes.

In Penrose's theory, however, twistors representing classical massless particles in free motion were basic, and spacetime points not initially present in the theory were taken to be derived objects. This suggested, according to Penrose, that 'ultimately the continuum concept may possibly be eliminated from the basis of physical theory altogether' (1975). But eliminating the spacetime continuum undermines the ontological foundation of the strong version of GP. Here we have found another renegade of the strong version of GP, although his theoretical motivations and intellectual journey were quite different from Wheeler's.

5.4 Topological investigations: singularities, horizons, and black holes

Curiously enough, Penrose's abandonment, at the ontological level, of the strong version of GP is closely connected, directly prompted, or even irresistibly dictated, at the descriptive level, by his bringing GP to a new stage and pushing it through to its very end. More specifically, Penrose, together with Stephen Hawking and others, by their topological investigations[17] of the behavior of intense gravitational fields and the effects of these fields on light and matter, have successfully clarified the nature of singularities in the formulations of GTR and revealed the fine structure of spacetime, and thus have helped to develop a new interpretation of GTR since the end of the 1960s. These achievements, however, have also exposed the internal limits of GTR. Thus quantum principles have been invoked so that the hope of developing a consistent picture of the world can be maintained.

Three concepts are central to the topological interpretation of GTR: singularity, horizon, and black hole.

Singularities (A)

Physicists were long puzzled by singularities appearing in some solutions to the field equations from the time that GTR was formulated. If we look at the Schwarzschild solution

$$ds^2 = (1 - 2Gm/rc^2)c^2 dt^2 - dr^2/(1 - 2Gm/rc^2) - r^2(d\theta^2 + \sin^2\theta d\phi^2) \quad (1)$$

(here m is a point-mass or the mass of a non-spinning and spherically symmetric star that is removed from all other bodies, and G and c are Newton's constant and the velocity of light), it is easy to find that when $r = 2Gm/c^2$ or $r = 0$, the metric is not well defined. The same also happens to the de Sitter solution

$$ds^2 = (1 - \lambda r^2/3) dt^2 - (1 - \lambda r^2/3)^{-1} dr^2 - r^2(d\theta^2 + \sin^2\theta d\phi^2) \quad (2)$$

when $r = (3/\lambda)^{1/2}$ (or when $r = \pi R/2$ in a transformed expression; see section 4.4).

Concerning the nature and interpretation of these metric singularities, physicists had divergent opinions. For Einstein (1918a), the singularity in the de Sitter metric indicated that the de Sitter world was 'a world in which the matter completely concentrated on the surface $r = \pi R/2$.' Weyl (1918a) interpreted the de Sitter singularity as an unreachable horizon, but insisted that 'there must at least be masses at the horizon'. He shared the opinion with Einstein that the world can be empty of mass only up to singularities.

Eddington's interpretation of metric singularities was quite different from Einstein's and Weyl's. For him, a metric singularity 'does not necessarily indicate material particles'. The reason for this is that 'we can introduce or remove such singularities by making transformations of coordinate. It is impossible to know whether to blame the world structure or the inappropriateness of the coordinate-system' (Eddington, 1923).

Eddington's view that metric singularities were only coordinate singularities was elaborated by his student Lemaître. Lemaître (1932) was the first person to give a strict mathematical proof of the coordinate nature of the Schwarzschild singularity. On the basis of the proof, he concluded that 'the singularity of the Schwarzschild field is thus a fictitious singularity similar to the one of the horizon of the center in the original form of the universe of de Sitter'.

Eddington's other interpretation of metric singularities was also influential: they consisted of an impenetrable sphere, a magic circle, on which matter and light aggregate but cannot penetrate. He took the de Sitter singularity (discontinuity located at a finite distance, $r = \pi R/2$, in space) as a physically inaccessible horizon, and said that where

light, like everything else, is reduced to rest at the zone where time stands still, and it can never get round the world. The region beyond the distance $(\pi R/2)$ is altogether shut off from us by this barrier of time.

(Eddington, 1918)

In his influential book *The Mathematical Theory of Relativity* (1923), Eddington further developed his horizon view of metric singularities. Commenting on a generalized solution

$$ds^2 = (1 - 2Gm/rc^2 - \lambda r^2/3)c^2\,dt^2 - dr^2/(1 - 2Gm/rc^2 - \lambda r^2/3)$$
$$- r^2(d\theta^2 + \sin^2\theta\,d\phi^2) \tag{3}$$

(with which the Schwarzschild solution or de Sitter solution can be obtained by taking $\lambda = 0$ or $m = 0$), Eddington said:

At a place where g_{44} vanishes [i.e., singularities occur] there is an impassable barrier, since any change dr corresponds to an infinite distance ds surveyed by measuring-rods. The two positive roots of the cubic g_{44} are approximately $r = 2Gm/c^2$ and $r = (3/\lambda)^{1/2}$. The first root [the Schwarzschild singularity] would represent the boundary of the particle and . . . give it the appearance of impenetrability. The second root [the de Sitter singularity] is at a very great distance and may be described as the horizon of the world.

(1923)

Horizons

The horizon nature of the surface $r = \pi R/2$ in the de Sitter model was immediately accepted. The surface was taken to be a temporal periphery at a finite yet physically inaccessible spatial distance, because there all physical phenomena would cease to have duration and nothing could reach it. The view that $r = 2Gm/c^2$ in the Schwarzschild solution is inaccessible and impenetrable in a finite time was also accepted – because when r approaches $2Gm/c^2$, the light front tends to stand still (or equivalently, is redshifted infinitely) and all events tend to be pushed to infinity – but was not without challenge.

The conceptual foundation for the challenge was laid down by Lemaître's non-singular solution to the Schwarzschild model (1932). Only with this understanding of the spurious character of the Schwarzschild singularity would meaningful investigations about the physical process leading to the formation of horizons be possible and mathematical clarification about the topological nature of horizons justifiable. Physical reasons for the existence of the seemingly impenetrable horizon were suggested in the 1930s by Subrahmanyan Chandrasekhar (1935), Fritz Zwicky (1935, 1939), Robert Oppenheimer and George Volkoff (1939), and Oppenheimer and Hartland Snyder (1939), in the course of investigating stellar evolution, gravitational collapse (implosion), and the reaching of the critical circumference (i.e. the formation of a black hole; see below). Most of these investigations, however, challenged only the inaccessibility, but not the impenetrability, of the Schwarzschild horizon. It was Howard Robertson (1939) who argued in 1939 that Lemaître's non-singular solution permitted a perfectly regular description of the trajectory of any particle or photon crossing the horizon ($r = 2Gm/c^2$) and up to the center of the Schwarzschild sphere ($r = 0$), thus defying the dogma of the impenetrable horizon.

Robertson noticed a puzzling implication of this description: an observer would never see the particle reach $r = 2Gm/c^2$ (as the particle approaches $r = 2Gm/c^2$, the observer receives more and more redshifted light from the particle; that is, the particle crosses the horizon at $t = \infty$), although the particle passes $r = 2Gm/c^2$, and reaches $r = 0$ in a finite proper time. Thus it seems to have two incompatible points of view: from that of an outsider, the time comes to stop and events come to be frozen at the horizon. This seems to suggest that a horizon cannot be viewed as physically real and that GTR breaks down at horizons. Yet a falling particle would fall through the horizon and would not notice any slowing down of clocks, nor see infinite redshifts or any other pathological effects at the horizon. This means that from the viewpoint of the falling particle, a horizon poses no challenge to GTR. It was David Finkelstein (1958) who introduced a new reference frame and made the two points of view reconcilable.

The meaning of horizon was not quite clear for physicists until 1956, when Wolfgang Rindler defined the particle horizon (PH) as a light front that divided, at any cosmic instant, all particles (photons included) into two classes: those already in our view and all others. According to Rindler, PH is different from the event horizon (EH), which is a spherical light front converging on us and separating all actual and virtual photons on every geodesic through us into two classes: those that reach us in a finite time and those that do not (Rindler, 1956). Rindler's definitions are excellent. But the implications of horizons are far more complicated than any definition could have exhausted. It is impossible to clarify the implications of the horizon without addressing the concept of the black hole.

Black holes

Although the term *black hole* was first suggested by John Wheeler as late as the end of 1967, the idea is by no means a novel one. As early as 1783, John Michell, starting from Newton's corpuscular view of light, in a paper read to the Royal Society, suggested the ideas of critical circumference (within which no light can escape) and the possible existence of 'dark bodies' (as Pierre Simon Laplace was to call them twelve years later), by supposing

the particle of light to be attracted in the same way as all other bodies . . .. Hence, if the semi-diameter of a sphere of the same density with the sun were to exceed that of the sun in the proportion of 500 to 1, all light emitted from such a body would be made to return toward it, by its own proper gravity.

(Michell, 1784)

With the wave theory of light replacing the corpuscular view at the beginning of the 19th century, Michell's ideas fell into oblivion. It was too difficult to imagine how gravitation would act upon light waves.

In the context of GTR, Michell's ideas were revived by the Schwarzschild solution, which represented the gravitational field of a single massive center (e.g. a star) in the vacuum and determined the behavior of geometry around the center. The solution predicts that for each star there is a critical circumference, whose value depends only on its mass. It turns out that the critical distance in the solution, $r = 2Gm/c^2$, the so-called Schwarzschild singularity or Schwarzschild radius, is nothing other than Michell's critical circumference of a star, below which no light can escape from its surface. Or, in modern terms, it is an event horizon.

Although the numerical estimations about the critical circumference of a massive star are similar in the two cases, the conceptions underlying the estimation are quite different. (i) According to Michell's Newtonian conception, space and time were absolute, the velocity of light was relative; the Schwarzschild solution presupposes the opposite. (ii) In Michell's view, the corpuscles of light could fly out of the critical circumference a little bit, when they climbed their velocity slowed down by the star's gravity, and finally they were pulled back down to the star. Thus it was possible for an observer near the star to see the star by its slow-moving light. Yet in the case of the Schwarzschild solution, the light emitted from the event horizon must be redshifted infinitely because there the flow of time is dilated infinitely. Thus, for an outsider, no light from the star exists. The light exists inside of the star, but it cannot escape the horizon and must move toward the center. (iii) From the Newtonian point of view, the stability of a star that is smaller than the critical circumference can be maintained because the gravitational squeeze is counter-balanced by its internal pressure. According to the modern view, however, when the nuclear energy source of a star becomes exhausted, no internal pressure can possibly counterbalance the gravitational squeeze, and the star will collapse under its own weight.

The first person to have noticed the above implications of the Schwarzschild solution was Eddington (1923). In his book *The Internal Constitution of the Stars* (1926), Eddington further summarized what would happen in a region within the critical circumference:

Firstly, the force of gravitation would be so great that light would be unable to escape from it, the rays falling back to the star like a stone to the earth. Secondly, the red shift of the spectral lines would be so great that the spectrum would be shifted out of existence. Thirdly, the mass would produce so much curvature of the space-time metric that space would close up round the star, leaving us outside (i.e., nowhere).

(1926)

His conclusion was simply that 'things like that cannot arrive'.

Eddington's view was challenged by his Indian pupil Subrahmanyan Chandrasekhar (1931, 1934), who, on the basis of existence of the degenerate state implied by quantum mechanics, argued that a massive star, once it had exhausted its nuclear source of energy, would collapse; and the collapse must continue indefinitely till the gravitational force became so strong that light could not escape from it. He also argued that this kind of implosion implied that the radius of the star must tend to zero. Eddington (1935) rejected his pupil's argument: 'this was almost a *reductio ad absurdum* of the relativistic degeneracy formula . . .. I think that there should be a law of nature to prevent the star from behaving in this absurd way'.

Eddington's rejection, however, carried no logical power, and investigations of the implications of the gravitational implosion continued. Fritz Zwicky (1935, 1939) showed that when the imploding star had a small mass, it would trigger a supernova explosion and lead to the formation of a neutron star. When the mass of the imploding star was much larger than the two-sun maximum for a neutron star, Robert Oppenheimer and Hartland Snyder (1939) argued that the formed neutron star would not be stable and the implosion would continue, the size of the star would cross the critical circumference (and form a black hole), until it reached a 'point' of zero volume and infinite density. That is, the implosion would end with a singularity (with infinite tidal gravitational force) at the center of the black hole, which would destroy and swallow everything that fell into the black hole. (The collapse of a massive rotating star is a black hole with an external metric, which will eventually become the black hole that is described by Kerr's rotationally symmetric solution to the field equation. See Kerr, 1963.)

In the 1960s, black holes were conceived as holes in space, down which things could fall, but out of which nothing could emerge. Since the mid-1970s, however, it began to be realized that black holes were not holes in space, but rather dynamical objects that had no other physical properties than spin, mass, and charge.

A black hole, that is, a region with $r < 2Gm/c^2$, is a region with enormous density and an intense gravitational field. Thus the investigations of black holes become a convenient way of examining GTR in the context of intense gravitational fields. Since the consistency of GTR can only be clarified when GTR is carried to its logical conclusion, and its last consequences are examined, a clarification of the conceptual situation of black holes is crucial for testing the consistency of GTR. But this would be impossible without a clear understanding of the nature of singularities, which sit at the centers of black holes.

Singularities (B)

A proper understanding of the nature of the singularity cannot be obtained within the neo-Newtonian framework in which a mathematical adoption of a Riemannian spacetime manifold is underlain by a Newtonian vision of space and time. Within this framework, a singularity is defined in the sense of function theory. That is, when a gravitational field is undefinable or becomes singular at a certain spacetime point, it is said to have a singularity. Yet within the framework of GTR (see section 4.2), a spacetime point itself has to be defined by the solution of the field equations because the metric field is responsible for the geometry and is constitutive of spacetime. A point P can only belong to the structure of spacetime if the metric is definable at P. The world can contain no points at which the metric field cannot be defined. Thus the statement that a metric is singular at a spacetime point, in analogy to the singular point of the pre-GTR field or to the theory of the function, within the framework of GTR, is meaningless.

According to Hawking and Penrose (1970), the spacetime singularity has to be defined in terms of geodesic incompleteness, and a singularity-free spacetime is one that is geodesically complete. A curve is incomplete if it is inextensible. An incomplete timelike geodesic implies that there could be a test particle that could emerge from (or vanish into) nothingness before (or after existing for) a finite interval of proper time. A singularity is thus something that is lacking in spacetime. But this lacking something and its relation to other spacetime regions can still be represented by ideal points enmeshed with spacetime, although these ideal points themselves are not world points or the sites for possible events (Schmidt, 1971).

Many people argue that singularities that crush matter and the spacetime geometry are the result of a naive application of GTR to intense fields, that is, to the big bang or to gravitational collapse, although the big bang is supported by the observation of the expansion of the universe and the cosmic microwave radiation, and the collapse is supported by the observation of neutron stars. For example, Einstein rejected the idea of the singularity and held that

one may not assume the validity of the [gravitational field] equations for very high density of field and matter, and one may not conclude that the 'beginning of expansion' must mean a singularity in the mathematical sense.

(1956)

Penrose and Hawking argued differently. With some apparently reasonable assumptions about the global properties of the universe, such as (i) the validity of the field equations, (ii) the non-existence of negative energy densities, (iii) causality, and (iv) the existence of a point P such that all past-directed time-like geodesics through P start converging again within a compact region in the past of P (which is a statement of the idea that the gravitation due to the material in the universe is sufficiently attractive to produce a singularity), and by using a topological method, they succeeded in proving that either collapsing stars or the evolution of the universe will inevitably result in a singularity.

The first person to introduce the topological method into the exploration of the most intimate structure of Schwarzschild spacetime and who was thus able to examine the spacetime inside a black hole was Martin Kruskal (1960). But it was Penrose who introduced global concepts systematically and, together with Hawking, used them to obtain results about singularities that did not depend on any exact symmetries or details of the matter content of the universe, the famous Hawking–Penrose singularity theorems.

Among the above assumptions, (ii) is a very weak and plausible condition, the violation of (iii) would be a total breakdown of our physical reasoning, and (iv) is supported not only by the observations of the microwave background (Hawking and Ellis, 1968), but also by the following plausible reasoning. If a star's gravity is strong enough to form a horizon, to pull outgoing light rays back inward, then after this happens, nothing can prevent the gravity from growing stronger and stronger so that the star will continue imploding inexorably to zero volume and infinite density, whereupon it creates and merges into a singularity, where the curvature of spacetime or the tidal force becomes infinite and spacetime ceases to exist. This means that every black hole must have a singularity inside itself because it always has a horizon. Thus the occurrence of physical singularities in collapsing stars and the expanding universe, which is entailed by GTR, as is demonstrated by the singularity theorems, has posed the most serious conceptual problems to the validity of GTR itself.

Mathematically, two singularities in the Schwarzschild line element ($r = 2Gm/c^2$ and $r = 0$) have different topological implications. The singularity represented by $r = 2Gm/c^2$ (which can be physically interpreted as an event horizon rather than a real singularity) raised the problem of discontinuity at the boundary of two spheres, and suggested a radical revision of our conception of the causal connections of the world. The other one, represented by $r = 0$, is a real singularity and represents the sharp edge of spacetime, beyond which there is no spacetime. Since the Hawking–Penrose singularity theorems have ruled out bouncing models for the big bang cosmology, the classical concept of time must have a beginning at the singularity in the past, and will come to an end for at least part of spacetime when a star has collapsed.

Physically, gravity at the singularity is so strong that there all matter will be destroyed and spacetime be so strongly warped that time itself will stop existing. The infinite gravity at singularities is a clear message that the laws of GTR must break down at the center of black holes (singularities), because no calculations will be possible. And this means that GTR is not a universally valid theory.

In order to save the consistency of GTR, therefore, a mechanism must be found which can halt the imploding crunch. Considering the extremely high density and small distance near singularities, quantum effects must play an important role in the physical processes there. Thus the almost unanimous opinion among physicists is that only a quantum theory of gravity, a marriage of GTR with quantum mechanics, can provide the necessary mechanism for preventing the occurrence of singularities and save the consistency of GTR, although this would also mean that GTR can no longer be regarded as a complete fundamental theory.

The first successful application of quantum mechanics to the problems of GTR, and probably the only one up to now, was Hawking's quantum-mechanical calculation of the thermal radiation of black holes (1975). These days there is much rhetoric about the quantum fluctuations of gravitational fields, and about the probabilistic quantum foam (see Hawking, 1987). Yet the lack of a consistent quantum theory of gravity has forced physicists to appeal to Penrose's cosmic censorship conjecture, which says that a singularity is always clothed by an event horizon, and thus is always hidden from the outside world (Penrose, 1969). Although all attempts to find genuine counter-examples have failed, no proof of the conjecture has ever been provided.

Notes

1. For a historical account of the subject, see Earman (1989).
2. The existence of absolute motions is not necessarily incompatible with relationalism because the spacetime structures with respect to which the absolute motions are defined can still be shown to be ontological constituted by certain substances, such as gravitational fields in the case of GTR.
3. Newton's own views are much more complicated. Quotations can be cited to show that Mach's interpretation can be contradicted. For example, Newton writes, 'space does not subsist absolutely in itself' (1978, p. 99) and 'space is an affection of a thing qua thing, . . . nothing exists or can exist which is not in some way referred to space, . . . space is an amanative effect of the first existing thing, because if any thing is posited, space is posited' (*ibid.*, p. 163). John Earman claims that Newton rejects one form of the substantivalist doctrine, namely, 'the existence of space is not dependent upon the existence of anything else' (1979). Whatever the position Newton actually took, however, Mach's interpretation and criticism of Newton's position were influential at the time when Einstein's views on the foundations of physics were taking shape, and were accepted wholeheartedly by Einstein.
4. Here quotations from Newton can also be cited to contradict Mach's interpretation of Newton's position. For example, Newton says in 'De gravitation' that space 'does not stand under those characteristic affections that denominate a substance, namely actions, such as are thoughts in a mind and motions in a body' (1978, p. 99), and thus apparently rejects the view that space is active or the source of activity.
5. It is worth noting that this practical view of geometry agrees with Newton's opinion, though the underlying views of space are diametrically opposed. Newton said, 'geometry is found in mechanical practice, and is nothing but that part of universal mechanics' (1678, in 1934).
6. Poincaré wrote that when astronomical observations were not in agreement with Euclidean geometry, 'two courses would be open to us: we might either renounce Euclidean geometry, or else modify the laws of optics and suppose that light does not rigorously propagate in a straight line' (1902).
7. Adolf Grünbaum held this position on Einstein's view, which was rejected by Howard Stein. The heated debate between Grünbaum and Stein was documented by Earman, Glymour, and Stachel (1977). Here I intend only to outline my own understanding of Einstein's view about the relationship between (chrono)geometry and measuring instruments.
8. See appendix A1.
9. Even though Einstein certainly did not know the possibility of a supernova as a source of gravitational radiation, which convincingly suggests the mutual transmutability between gravitational and other kinds of fields.
10. This conclusion is supported by Einstein's own statement: 'I don't agree with the idea that the General Theory of Relativity is geometrizing physics of the gravitational field' (1948a).
11. Here, geometrical fields mean gravitational fields because of the close connection between the geometry of spacetime and the gravitational field.
12. For a criticism of Hilbert's erroneous conclusion, which was based on his misunderstanding of the role of the contracted Bianchi identities resulting from the assumed general covariance of his theory, see Stachel (1991).

13. Weyl said, 'it is necessary to regard electromagnetic phenomena, as well as gravitation, as an outcome of the geometry of the universe' (1921).
14. See also Schrödinger (1950).
15. The geometrical aspects of some gauge invariant quantum field theories, which were recognized much later, will be discussed in section 11.3.
16. See Papapetrou (1949), Sciama (1958), Rodichev (1961), Hayashi and Bregman (1973), Hehl *et al.* (1976).
17. Topology is a branch of mathematics that deals with the qualitative ways in which things are connected to each other or to themselves. It cares only about connections, but not shapes, sizes, or curvatures. Thus geometrical figures obtained from smoothly and continuously deforming a figure, but without tearing it, are topologically equivalent. Questions such as 'does spacetime come to an end or have an edge beyond which spacetime ceases to exist', which is central to the problem of the singularity, and 'which regions of spacetime can send signals to each other and which cannot?', which is central to the question of the formation and existence of black holes and also to cosmology, are topological questions.

Part II

The quantum field programme for fundamental interactions

In this part of the book an analysis of the formation of the conceptual foundations of the quantum field programme for fundamental interactions (QFP) will be given, with special concern for the basic ontology and the mechanism for transmitting fundamental interactions posited by QFP. Chapter 6 reconstructs the history of quantum physics up to 1927 along two lines: (i) the quantization of the mechanical motions of atomic systems, and (ii) the quantization of wave fields. It also describes the basic ideas of uncertainty and complementarity, which were suggested by Werner Heisenberg and Niels Bohr to characterize quantum mechanics. Chapter 7 reviews, historically and critically, the positions adopted with respect to the conceptual foundations of quantum field theory (QFT), both by its founders and by later commentators. Its first three sections serve to analyze the ontological shift that occurred in the early history of QFT, namely, a shift from the particle ontology to an ontology of a new kind. Section 7.4 examines the dilemma facing the original ontological commitment of QFT, which was embodied in Dirac's notion of the vacuum. Section 7.5 reconstructs the evolution of the ideas about local coupling, the exchange of virtual quanta, and invariance principles, which were supposed to be obeyed by quantum interactions and thus to impose restrictions on the forms of the interactions. Section 7.6 reviews the recognition of divergences and the formulation of the renormalization programme in the late 1940s and early 1950s. Chapter 8 summarizes the essential features of QFP, its ups and downs, and various attempts to explore alternatives, until its revival, in the form of gauge field theories, in the early 1970s.

6

The rise of quantum theory

The origin of the relativity theories was closely bound up with the development of electromagnetic concepts, a development that approached a coherent field-theoretical formulation, according to which all actions may vary in a continuous manner. In contrast, quantum theory arose out of the development of atomic concepts, a development that was characterized by the acknowledgment of a fundamental limitation to classical physical ideas when applied to atomic phenomena. This restriction was expressed in the so-called quantum postulate, which attributed to any atomic process an essential discontinuity that was symbolized by Planck's quantum of action.

Quantum field theory (QFT) is a later phase of the conceptual developments of quantum theory, and has the old quantum theory and non-relativistic quantum mechanics, essentially the preliminary analyses of the interactions between atoms and radiation, as its predecessors. This chapter will review some features of quantum physics that are relevant to the rise of QFT.

6.1 The quantization of motion

In solving the problem of the equilibrium between matter and radiation, Max Planck (1900) showed that the laws of heat radiation demanded an element of discontinuity in the description of atomic processes. In the statistical behavior of atoms represented, in Planck's description, by linear resonators in their interactions with radiation, only states of vibration should be taken into account whose energy was an integral multiple of a quantum, $h\nu$, where h is Planck's constant and ν is the frequency of the resonator.

Planck himself believed that the discontinuity of energy was only a property of atoms, and was not ready to apply the idea of energy quantization to radiation itself. Furthermore, he had provided these resonators only with very formal properties, among which one essential property of a real atom, namely the capacity to change the frequency of its radiation, was absent.

This was quite different from the ideas of other physicists, such as Wilhelm Wien and Arthur Haas. Wien (1909) viewed the electromagnetic resonators as real atoms, which, in addition to their ability to absorb and emit radiation energy, also possessed other characteristics. For example, they could be ionized by ultraviolet light or by X rays. According to Wien, 'the element of energy, if it has any physical meaning at all, can probably be derived only from a universal property of atoms' (1909).

From careful reading of Wien's paper and J. J. Thomson's book *Electricity and Matter* (1904), concerning the problem of the constitution of the atom, Haas (1910a, b, c) was led to substitute real atoms for the idealized Hertzian oscillators used by Planck, and to relate the nature of the quantum of action with atomic structure. By assuming that the potential energy of the atomic electron, e^2/a (a is the radius of hydrogen atom), can be described by Planck's element of energy $h\nu$, that is

$$|E_{\text{pot}}| = h\nu, \tag{1}$$

and using the classical relation equating centrifugal force with Coulomb attraction, $m\omega^2 a = e^2/a^2$, Haas obtained an equation,

$$h = 2\pi e(am)^{1/2}, \tag{2}$$

where the frequency ν in equation (1) was directly identified with the orbital frequency of the electron, $\omega = 2\pi\nu$.[1] The establishment of the relationship between the quantum of action h and the atomic quantities e, m, and a (or atomic dimension) was undoubtedly Haas's great contribution to the conceptual development of the idea of quantization of atomic motion.

Arnold Sommerfeld did not want to explain h on the basis of an atomic dimension, but rather to view the existence of the atom as the result of the existence of all elementary quanta of action, which, as a new physical fact, could not be derived from other facts or principles. At the first Solvay congress in Brussels in 1911, Sommerfeld claimed that

an electromagnetic or mechanical 'explanation' of h appears to me quite as unjustified and hopeless as a mechanical 'explanation' of Maxwell's equation . . .. If, as can hardly be doubted, physics needs a new fundamental hypothesis which must be added as a new and strange element to our electromagnetic view of the world, then it seems to me that the hypothesis of the quantum of action fulfills this role better than all others.

(Sommerfeld, 1911b)

This was a further development of his quantum-theoretical fundamental hypothesis proposed at the earlier time of the same year (1911a), according to which the interaction between electrons and atoms was definitely and uniquely controlled by Planck's quantum of action. Commenting on Sommerfeld's position, Lorentz said:

Sommerfeld does not deny a connection between the constant h and the atomic dimensions. This can be expressed in two ways: either the constant h is determined by these dimensions (Haas), or the dimensions, which are ascribed to atoms, depend on the magnitude of h. I see no great difference between these views.

(Lorentz, 1911)

Bohr was strongly influenced by Sommerfeld's above ideas. From mid-March until the end of July 1912, Bohr worked in Rutherford's Institute in Manchester, and accepted Rutherford's planetary model of the atom. The acute question faced by the model was: How could a positively charged atomic nucleus be in equilibrium with a negatively charged rotating electron? What prevented this electron from falling into the center of the atom? This question of stability actually constituted the starting point of Bohr's work.

By mid-1912, Bohr had become convinced that the stability required by Rutherford's model was of non-mechanical origin and could only be provided by a quantum hypothesis:

This hypothesis is that for any stable ring (any ring occurring in the natural atoms) there will be a definite ratio between the kinetic energy of an electron in the ring and the time of rotation, [and to this hypothesis] there will be no attempt to give a mechanical foundation (as it seems hopeless).

(Bohr, 1912)

Bohr's rule of quantization of mechanical motion associated with the stationary state ('stable ring') can be considered a rational generalization of Planck's original hypothesis for the possible energy values of a harmonic oscillation: it concerned atomic systems for which the solution of the mechanical equation of motion was simply periodic or multiply periodic, and hence the motion of a particle could be represented as a superposition of discrete harmonic vibrations.

Bohr's great achievement was the synthesis of Rutherford's model of the atom with Planck's quantum hypothesis, in which a set of assumptions was introduced concerning the stationary states of an atom and the frequency of the radiation emitted or absorbed when the atom passed from one such state to another. Together with these assumptions, and by using the results of Wien, of André Debierne, and especially of Johannes Stark[2], Bohr (1913a, b, c) was able to give a simple interpretation of the main laws governing the line spectra of the elements, especially the Balmer formula for the hydrogen spectrum. Bohr's theory of atomic spectra embodied the idea of quantization of atomic motion, and can be viewed as a milestone in the development of the idea.

6.2 The quantization of radiation

When Planck introduced the idea of the quantum to describe the spectral properties of pure radiation, he did so by a procedure of quantization that was applied only to ponderable matter, that is, to his material oscillators. He was unaware, however, of the fact that his proposal implied the need for a new idea of the classical field itself, an idea that the quanta were intrinsic to radiation and ought to be imagined as a kind of particle flying about. His reasoning purported to involve only a modification of the interaction between matter and radiation, since the interaction was full of obscurities, but no modification of free electromagnetic radiation, which was believed to be much better understood.

By contrast, when Einstein proposed the light quantum in (1905a), he dared to challenge the highly successful wave theory of pure radiation. In his 'very revolutionary' paper, Einstein emphasized that

> it should be kept in mind that optical observations refer to waves averaged over time and not to instantaneous values. Despite the complete experimental verification of the theory of diffraction, reflection, refraction, dispersion, and so on, it is conceivable that a theory of light operating with continuous three-dimensional functions will lead to conflicts with experience if it is applied to the phenomena of light generation and conversion.

> *(Ibid.)*

Einstein appeared to have set himself the task of eliminating the profound difference between the essentially discrete atomic theory of matter and the essentially continuous electromagnetic field theory, at least partially, by introducing a corpuscular structure of radiation. What he suggested was that 'the energy of light consists of a finite number of energy quanta, localized at various points of space', and that these quanta 'can be produced or absorbed only as units' (*ibid*.). It is clear that the light-quantum hypothesis is an assertion about a quantum property of free electromagnetic radiation which should be extended to the interaction between light and matter. That indeed was a very revolutionary step.

Einstein observed in (1906a) that the two critical formulae on which Planck based his radiation theory contradicted each other. By using the oscillator model and tacitly assuming that the amplitude and energy of an oscillator were continuously variable, Planck (1900) obtained the connection between the density ρ of radiation and the average energy U of an oscillator,

$$\rho = 8\pi v^2/c^3 U, \qquad (4)$$

from the laws of Maxwell's electrodynamics. However, in deriving the expression

$$U = hv/(e^{hv/KT} - 1) \qquad (5)$$

which was crucially different from the equipartition theorem, $U = KT$, in statistical thermodynamics, Planck assumed discrete energy steps. The dilemma consisted in the simultaneous application of equations (4) and (5), which were derived from contradictory assumptions.

Einstein pointed out that in fact Planck's theory involved a second assumption, in addition to that of the discreteness of energy. That is, when the oscillator's energy was quantized, equation (4) must continue to hold even though the basis for its derivation (i.e., continuity) had been removed. This second assumption constituted a further break with classical physics and demonstrated that 'in his radiation theory Planck (tacitly) introduced a new hypothetical principle into physics, the hypothesis of light quanta' (Einstein, 1906a), even though Planck himself rejected the hypothesis as late as 1913.

It is interesting to note that Paul Ehrenfest (1906) independently discussed the same problem in the same spirit in the same year. He pointed out that it was possible to get the Planck spectrum by exploiting the analogy of the amplitude of a proper vibration of the radiation cavity with the coordinate of a material oscillator and using Rayleigh–Jeans summation of the resonance vibrations of the radiation cavity, if we make the following hypothesis: the amount of field energy residing in a normal mode of frequency can only be integral multiples of hv. The same hypothesis enabled Peter Debye (1910b) to actually derive the Planck spectrum. Referring to the Rayleigh–Jeans density of modes $N \, dv = 8\pi v^2/c^3 \, dv$, Debye directly obtained Planck's radiation formula after assigning an average energy $U = hv/(e^{hv/KT} - 1)$ to each degree of freedom. He thus showed that Planck's law followed from the single assumption that the energy of radiation was quantized and 'no intermediate use of resonators is required' (1910a).

In (1909a), Einstein gave new arguments to support his light-quantum hypothesis, which, unlike the 1905–1906 arguments based on equilibrium statistical analysis (the dependence of entropy on volume), were based on an analysis of the statistical fluctuations of blackbody radiation energy and momentum. The formulae he obtained for energy and momentum fluctuations were:

$$\langle \epsilon^2 \rangle = (\rho hv + c^3 \rho^2/8\pi v^2)V \, dv \qquad (6)$$

and

$$\langle \Delta^2 \rangle = 1/c(\rho h \nu + c^3 \rho^2 / 8\pi \nu^2) A \tau \, d\nu \tag{7}$$

where V is the volume of the radiation cavity, A the area of the mirror placed inside the cavity, and τ the time interval. It appeared as though there were two independent causes producing the fluctuations. The first mechanism (independent light quantum) alone would lead to Wien's law, and the second (classical wave) to the Rayleigh–Jeans law. Neither of them alone would lead to Planck's law, but their sum did.

With such an analysis, Einstein announced at the Salzburg Congress of 19–25 September 1909:

It is my opinion, therefore, that the next phase of the development of theoretical physics will bring us a theory of light that can be interpreted as a kind of fusion of the wave and emission theories . . . that a profound change in our views of the nature and constitution of light is indispensable.

(1909b)

To prove that these two structural properties (wave and quantum structures) need not be incompatible, at the end of his paper Einstein suggested a conception that was later picked up by Louis de Broglie in his (1923a, b)[3] and became the foundation of the later developments of quantum mechanics and QFT. The conception can be summarized as this. The energy of the electromagnetic field was assumed to be localized in singularities surrounded by fields of forces, which were subject to a superposition principle and thus acted as a field of waves similar to that of Maxwell's theory.

The year 1909 saw a subtle change in Einstein's conception of the structure of light. Earlier he had inclined to the idea of entirely replacing fields by particles. That is, particles were the only reality, and the apparent fields should be reduced to direct interactions between particles. In 1909, Einstein still favoured such a corpuscular model of light, which is testified to by his letter to Hendric Lorentz dated 23 May 1909:

I imagine a light quantum as a point surrounded by an extending vector field, which somehow diminishes with distance. The point is a singularity without which the vector field cannot exist . . . . The vector field should be completely determined by the setting of the motions of all singularities, so that the number of parameters that characterize radiation would be finite.

(1909c)

But a subtle change can already be detected in the same letter:

What seems to me important, however, is not the assumption of these singular points, but the writing of such field equations, that will allow solutions, according to which finite amounts of energy could move without dissipation in a definite direction with velocity *c*. One could think that this purpose could be achieved with a minor modification of Maxwell's theory.

(Ibid.)

In his Salzburg paper (1909b), the physical reality of the field was also acknowledged, and the first hint of what was later to be called wave–particle duality appeared.[4]

The notion of duality assumes the reality of both wave and particle aspects of a physical entity. However, a crucial ingredient of a fully fledged particle conception of radiation[5] was still wanting. The light quanta introduced by Einstein in 1905 were in fact only energy quanta. Einstein himself did not explicitly mention the momentum of light quanta until 1916, although his first term in equation (7) might have led him to such an idea. It was Stark (1909) who derived from this term the statement that 'the total electromagnetic momentum emitted by an accelerated electron is different from zero and is given by $h\nu/c$'.

When Einstein returned to the radiation problem in 1916, after having concentrated on gravitation for several years and been directly stimulated by the publication of Lorentz's calculations (1916) of the fluctuations that were dealt with by Einstein in (1909a, b), the quantum theory had taken a new shape. Bohr had opened a domain of applications for the idea of the quantum in his theory of the hydrogen atom and its spectrum. Bohr's work had noticeably influenced Einstein's ideas. This can be seen clearly from Einstein's new ideas of 'the internal energy states of atoms' and of 'transition from the state E_m to the state E_n by absorption (and emission) of radiation with a definite frequency ν'. By using these ideas, and introducing a new concept of transition probability and new coefficients A_{mn}, B_{mn}, and B_{nm} (characterizing, respectively, the spontaneous emission, induced emission, and absorption), Einstein (1916c) rederived Planck's law when Bohr's frequency condition $E_m - E_n = h\nu$ was assumed, thus establishing a bridge between Planck's radiation theory and Bohr's theory of spectra.

What Einstein himself considered more important in this paper than the above results was the implication of his analysis for the directed nature of the radiation process: 'There is no radiation of spherical waves' (1916c). As a consequence, 'the establishment of a truly quantum theory of radiation seems to be almost inevitable' (*ibid.*). Here came the fully fledged particle conception of radiation in which a light quantum carries a momentum $h\boldsymbol{K}$ (here $\boldsymbol{K}$ is the wave vector, $|\boldsymbol{K}| = \nu/c$).

Under the impact of Einstein's idea of light quanta, the conflict between the corpuscular and the undulatory conceptions of radiation became more and more acute. During the years 1921 to 1924, however, it became clear that the light-quantum hypothesis could be applied, not only to the Stefan–Boltzmann law, Wien's displacement law, and Planck's law, but also to other optical phenomena which had been regarded as irrefutable evidence for the undulatory conception of light, such as the Doppler effect (Schrödinger, 1922) and Fraunhofer diffraction (Duane, 1923; Compton, 1923a). But it was Compton's experimental study of X-ray scattering that had put the quantum view on a firm empirical foundation.

In 1923, Compton (1923b) and Debye (1923) independently derived the relativistic kinematics for the scattering of a light quantum off an electron at rest:

$$ h\mathbf{K} = \mathbf{P} + h\mathbf{K}', \quad hc|\mathbf{K}| + mc^2 = hc|\mathbf{K}'| + (c^2\mathbf{P}^2 + m^2c^4)^{1/2}. \tag{8} $$

Here $h\mathbf{K}$, $h\mathbf{K}'$, and $\mathbf{P}$ are initial and final momenta of the light quantum, and the final momentum of the electron, respectively. These equations imply that the wavelength difference $\Delta\lambda$ between the final and the initial light quanta is given by

$$ \Delta\lambda = (h/mc)(1 - \cos\theta), \tag{9} $$

where θ is the light-quantum scattering angle. Compton found this relation to be satisfied within the errors and concluded: 'The experimental support of the theory indicated very convincingly that a radiation quantum carries with it directed momentum as well as energy' (1923b).

Compton's contribution to the acceptance of the quantum view of radiation, for the majority of the physics community, can be compared to Fresnel's a century earlier to the acceptance of the classical wave theory of light. Yet Einstein's response to it was interestingly modest:

The positive result of the Compton experiment proves that radiation behaves as if it consisted of discrete energy projectiles, not only in regard to energy transfer, but also in regard to momentum transfer.

(Einstein, 1924)

More illuminating to Einstein's position, in this context of the triumph of the quantum concept of light he initiated, is his clear statement of the wave–particle duality of light: 'there are therefore now two theories of light, both indispensable, and without any logical connection' (*ibid.*).

Einstein's position can be seen as a reflection of the fact that a decision between the two competing conceptions could not be enforced. The quantum view seemed indispensable for the interpretation of optical processes involving interactions between light and matter, whereas phenomena like interference and diffraction seemed to require the conceptual apparatus of the classical wave theory of light. What is even worse is that the light-quantum hypothesis, supported by incontestable experimental evidence, became physically significant only by the use of its own negation, the wave hypothesis. This is because, as pointed out by Bohr, Hendric Kramers, and John Clarke Slater (1924), the light quantum is defined by a frequency which could be measured only by applying an undulatory notion such as diffraction.

6.3 The birth of matrix mechanics

Research work in atomic physics during the years 1919–1925 was primarily based on Bohr's theory. Bohr (1913a, b, c) expressed his underlying ideas as follows: (i) stationary states fixed by the quantum condition, and (ii) the frequency condition $E_1 - E_2 = h\nu$, which indicates that the frequencies of the spectral lines were each connected with two states. In addition, an important heuristic principle, whose generalized form Bohr himself later called the correspondence principle, was formulated explicitly in 1918:

[The frequencies calculated by the frequency law,] in the limit where the motions in successive stationary states comparatively differ very little from each other, will tend to coincide with the frequencies to be expected on the ordinary theory of radiation from the motion of the system in the stationary states.

(1918)

This principle made it possible to retain the classical description of the motion of electrons in an atom, but allowed at the same time a certain tailoring of the results so as to fit observational data.

However, two difficulties remained. First, the theory was a lamentable hodgepodge of classical theory and the quantum hypothesis, lacking logical consistency. Second, according to the mechanical model of the stationary states, the quantum conditions could easily be connected with the periodic orbits of the electrons in an atom, and the optical frequencies of spectral lines should coincide with the Fourier orbital frequencies of the electron's motions, a result never borne out by experiments, in which the frequencies of observed lines were always connected with the difference of two orbital frequencies.

Under the influence of Einstein's paper (1917b) on transitions, which were defined as quantities referring to two states of an atom, the attention of physicists moved over from the energy of the stationary states to the transition probability between stationary states. It was Kramers (1924) who started seriously to study the dispersion of an atom, and to relate the behavior of

Bohr's model under radiation with the Einstein coefficients A_{mn}, B_{mn}, and B_{nm} seriously.

Max Born (1924) extended Kramers's idea and method, and applied them to the interactions between the radiation field and a radiating electron, and to the case of the interactions between several electrons of an atom. In carrying out this programme, he showed that the transition from classical mechanics to what he called 'quantum mechanics' can be obtained if a certain differential was replaced by a corresponding difference. For large n and small τ, the quantum-theoretical frequency $\nu_{n,n-\tau}$ for a transition from a stationary state $n' = n - \tau$ coincides, according to Bohr's correspondence principle, with the classical frequency $\nu(n, \tau)$, that is, the τth harmonic (overtone) of the fundamental frequency of the classical motion in state n:

$$\nu_{n,n-\tau} = \nu(n, \tau) = \tau\nu(n, 1), \tag{10}$$

where $\nu(n, 1)$ is the classical fundamental frequency and is equal to the derivative of the Hamiltonian with respect to the action: $\nu = \mathrm{d}H/\mathrm{d}J$.[6] Now comparing (classical) $\nu(n, \tau) = \tau \, \mathrm{d}H/\mathrm{d}J = (\tau/h) \, \mathrm{d}H/\mathrm{d}n$ with (quantum) $\nu_{n,n-\tau} = \{H(nh) - H[(n - \tau)h]\}/h$, Born recognized that $\nu_{n,n-\tau}$ could be obtained from $\nu(n, \tau)$ by substituting the difference $H(n) - H(n - \tau)$ for the differential $\tau \, \mathrm{d}H/\mathrm{d}n$.

Kramers and Werner Heisenberg (1925) discussed scattering phenomena where the frequency of the scattered light was different from that of the incident light. Their method was clearly related to that of Born's (1924), and carried out entirely in terms of quantities connected with two states, using multiple Fourier series, and replacing differential quotients by difference quotients. Here the scattered light quantum was different from the incoming quantum because during the scattering the atom made a transition. When they tried to write down formulae for the dispersion in these cases, they were forced not only to speak about the transition probabilities of Einstein, but also about transition amplitudes, and to multiply two amplitudes, say the amplitude going from state m to state n_i with the amplitude going from n_i to k or so, and then to sum over the intermediate states n_i.

These sums of products were already almost products of matrices. It was only a very small step from these to replace the Fourier components of the electronic orbit by the corresponding matrix elements, and it was taken by Heisenberg in his historic paper 'On a quantum theoretical re-interpretation of kinematic and mechanical relations' (1925).

The main ideas in Heisenberg's paper were, first, that in the atomic range, classical mechanics was no longer valid; and second, that what he was looking for was a new mechanics that had to satisfy Bohr's correspondence principle. With respect to the first point, Heisenberg (1925) wrote:

The Einstein–Bohr frequency condition already represents such a complete departure from classical mechanics, or rather from the kinematics underlying this mechanics, that even for the simplest quantum-theoretical problems the validity of classical mechanics just cannot be maintained.

In his search for new kinematics, the kinematical interpretation of the quantity x as a location depending on time had to be rejected. Then what kind of quantities were to be substituted for x in the equation of motion? Heisenberg's idea was to introduce the 'transition quantities' depending on two quantum states n and m. For example, instead of a Fourier expansion of $x(t)$ in the classical case of a periodic motion

$$x(t) = \sum_{\alpha=-\infty}^{\infty} a_\alpha \, e^{i\alpha\omega t}, \tag{11}$$

he wrote a new kind of term $a(n, n - \alpha) \, e^{i\omega(n,n-\alpha)t}$ to replace the term $a_\alpha \, e^{i\alpha\omega t}$. Heisenberg motivated the introduction of $a(n, n - \alpha)$ by saying that the intensities and hence $|a(n, n - \alpha)|^2$ were observable in contrast to the classical function $x(t)$ which was nonobservable (*ibid.*).

An idea which Heisenberg stressed very much in this paper was the postulate 'to establish a quantum-theoretical mechanics based entirely upon relations between observable quantities'. (*ibid.*). This suggestion had been much admired as the root of the success of quantum mechanics. The fact is, however, that on the one hand, Heisenberg regarded the location of an electron as non-observable, but he was wrong because according to the fully developed quantum mechanics the three coordinates x, y, and z of an electron are observable (Van der Waerden, 1967). On the other hand, the wave function of Schrödinger is not observable, but no one would doubt its theoretical meaning (Born, 1949). Thus under the criticism of Einstein and others, Heisenberg gave up his claim in his later years (Heisenberg, 1971).

With respect to the second point, Heisenberg had made three observations. First, an important device for satisfying the requirement of the correspondence principle was Born's replacement of differential quotients by difference quotients. Second, the Hamiltonian form of mechanics,

$$Q_r = \partial H / \partial P_r, \quad P_r = -\partial H / \partial Q_r, \tag{12}$$

could be preserved by replacing all dynamical variables by the corresponding matrix. The representation of physical quantities by matrices may be regarded as the essence of the new quantum mechanics, and its introduction, rather than the claim about observables, may be viewed as Heisenberg's important contribution to the conceptual development of quantum theory. Third, with such a representation, the old quantum conditions $\oint P\,\mathrm{d}Q = nh$ can be transcribed as the commutation relation:

$$PQ - QP = (h/2\pi\mathrm{i})I. \tag{13}$$

As a particular scheme of quantization of atomic motion, matrix mechanics in its original form applied obviously only to closed systems with discrete energy levels, and was not applicable to free particles and collision problems. For these applications of the quantum theory, some new conceptual apparatus was needed.

6.4 The duality of matter, individuality, and quantum statistics

Einstein's idea about the wave–particle duality of radiation was believed by Louis de Broglie to be absolutely general and to be necessarily extended to all of the physical world. The epochal new principle of de Broglie 'that any moving body may be accompanied by a wave and that it is impossible to disjoin motion of body and propagation of wave' was first formulated in his (1923a). In his (1923b), de Broglie indicated that a stream of electrons traversing an aperture whose dimensions were small compared with the wavelength of the electron waves 'should show diffraction phenomena'.

De Broglie's main idea was a beautiful one and exactly complemented Einstein's work on radiation. Where Einstein assigned particle properties to radiation, de Broglie assigned wave properties to matter. The concept of matter waves, whose frequency v and length λ were related to the particle's energy E and momentum P in the following way:

$$E = hv, \quad P = h/\lambda \tag{14}$$

did enable him to give a very remarkable geometrical interpretation of the old quantum condition $\oint P\,\mathrm{d}Q = nh$: since $P = h/\lambda$ implied that $\oint 1/\lambda\,\mathrm{d}Q = 2\pi r/\lambda = n$, the quantum condition was just the condition that the number of wavelengths which cover the orbital circumference be an integral number.

At the very time when Compton's experiments had finally convinced a good many physicists of the reality of light quanta, Einstein joined de Broglie in his proposal that this same duality must hold for ponderable matter as well as radiation. In his 1925 paper on the quantum theory of the ideal gas, Einstein offered new arguments from the analysis of fluctuations in support of de Broglie's idea.

One of Einstein's starting points was Satyendra Bose's work. Bose (1924) derived Planck's law directly from Einstein's hypothesis of light quanta, using only the method of statistical mechanics, and avoiding any reference to classical electrodynamics. He considered the quanta as particles. But, as Ehrenfest (1911) had already realized, independent quanta with their distinct individuality could only lead to Wien's law, but not Planck's law. Thus, if light quanta were to be used in interpreting the Planck distribution, they had to lack the statistical independence (which is normally associated with free particles), show certain kinds of correlation (which can be interpreted either as an indication of the wave characteristic of the phenomena, as physicists usually do, or as an indication that the particles are under the influence of a quite mysterious kind, as Einstein did in (1925a)), and obey a statistics that is different from the Boltzmann statistics of independent particles, that is, obey a statistics of indistinguishable particles, as first developed by Ehrenfest and his collaborators in (1915), and independently rediscovered by Bose in (1924). By using an unusual counting procedure, latter called Bose–Einstein statistics[8], Bose implicitly incorporated these correlations into his theory, and effectively denied the individuality of the light quanta.

On the basis of 'the hypothesis of a far-reaching formal relationship between radiation and gas', Einstein in (1925a) rewrote his mean square energy fluctuation of electromagnetic radiation (equation (6)) as

$$\Delta(\nu)^2 = n(\nu) + n(\nu)^2/Z(\nu),\tag{15}$$

by putting $V\rho\,d\nu = n(\nu)h\nu$, $\langle\epsilon^2\rangle = \Delta(\nu)^2(h\nu)^2$, and Z (the number of states per interval) $= (8\pi\nu^2/c^3)V\,d\nu$. Then he showed that equation (15) held equally well for his quantum gas as long as ν was defined in the latter case by $\nu = E/h = P^2/2mh$, and Bose statistics, instead of Boltzmann statistics, was used. The first term in (15) was the only one that would be present for a gas of independent particles. The second term corresponded to the classical wave interference term in the case of radiation. While the first term was the surprising one for radiation, the problem for the gas case was what to do with the second term, which incorporated indistinguishability effects of particles. Since this second term was associated with waves in the case of radiation, Einstein was led to 'interpret it in a corresponding way for gas, by associating with the gas a radiative phenomenon' (*ibid.*), that is, a wave phenomenon.

But what were these waves? Einstein suggested that a de Broglie-type wave field should be associated with gas, and pursued the implications of the de Broglie waves:

This wave field, whose physical nature is still obscure for the time being, must in principle be demonstrable by the diffraction phenomena corresponding to it. A beam of gas molecules which passes through an aperture must, then, undergo a diffraction analogous to that of a light ray.

(Ibid.)

However, he hastened to add that the effect would be extremely small for manageable apertures.

Stimulated by Einstein's paper, Walter Elsasser (1925) suggested that slow electrons of energy values below 25 eV would be ideally suited for testing '[the] assumption that to every translational motion of a particle one must associate a wave field which determines the kinematics of the particle'. He also pointed out that the existing experimental results of Carl Ramsauer, Clinton Davisson, and Charles Kunsman already seemed to give evidence of the diffraction and interference of matter waves.

6.5 The birth of wave mechanics

Inspired by de Broglie's idea and 'by the first yet far-seeing remarks of Einstein [in (Einstein, 1925)]', Erwin Schrödinger in his 'On Einstein's gas theory' (1926a) applied de Broglie's idea to gas theory. According to Schrödinger, the essential point of Einstein's new theory of a gas was that the so-called Bose–Einstein statistics was to be applied to the motion of gas molecules. In view of the fact that, in addition to Bose's derivation of Planck's law, there was also Debye's derivation, by applying 'natural' statistics to the field oscillators or the degrees of freedom of radiation, Einstein's theory of a gas can also be obtained by applying the natural statistics to the oscillators of a wave field representing molecules. Schrödinger then asserted that 'this means nothing else but taking seriously the de Broglie–Einstein wave theory of moving particles' (1926b).

In (1926c), Schrödinger developed de Broglie's analogy between classical mechanics and geometrical optics, both of which failed to apply to small dimensions, and the analogy between wave mechanics and undulatory optics. He maintained thereby:

We must treat the matter strictly on the wave theory, i.e., we must proceed from the wave equation and not from the fundamental equations of mechanics, in order to form a picture of the micro-structure [of physical processes].

(Ibid.)

He also pointed out that the recent non-classical theories concerning the micro-structure

[bear] a very close relation to the Hamiltonian equation and the theory of its solution,

i.e., to that form of classical mechanics which already points out most clearly the true undulatory character of mechanical processes.

(Ibid.)

With the Hamiltonian principle, Schrödinger replaced the quantum condition by a variation problem. That is, he tried to find such a function ψ that for any arbitrary variation of it the integral of the Hamiltonian density, 'taken over the whole coordinate space, is stationary, ψ being everywhere real, single-valued, finite and continuously differentiable up to second order' (1926b). With the help of this procedure,

the quantum levels are at once defined as the proper values of the wave equation, which carries in itself its natural boundary conditions, [and] no longer takes place in two separated stages: (1) Definition of all paths dynamically possible. (2) Discarding of the greater part of those solutions and the selection of a few by special postulations [quantum conditions].

(Ibid.)

The interpretation of the wave function introduced by Schrödinger is the most difficult question in the history of quantum physics and will be examined in section 7.1.

6.6 Uncertainty and complementarity

The rise of quantum mechanics[9] had raised the difficult question of how to characterize it. It was widely accepted that quantum mechanics 'is character-ized by the acknowledgment of a fundamental limitation in the classical physical ideas when applied to atomic phenomena' (Bohr, 1927). The reason for the limitation, as Heisenberg explained, was believed to lie in the fact that the physical reality possessed by 'the electron and the atom' was radically different from that of classical entities (Heisenberg, 1926b). But how to characterize the difference was a topic for debate even among quantum theorists who accepted the radical difference, which was different from the debate between them and the group of classically oriented physicists, such as Einstein, Schrödinger, and de Broglie.

The general strategy of the first group was, first, to retain, rather than to discard, the concepts of classical physics; and, second, to impose restrictions upon them by prohibiting their full use at one and the same time, or more precisely by separating them into two disjoint categories with the rule that only concepts belonging to the same category could be applied simultaneously to the entities and processes that were supposed to be described by quantum mechanics. The retention of classical concepts was justified, mainly by Bohr, on philosophical grounds. It was argued that the human capacity of conceptualization was inescapably confined within classical intuitive notions. Thus an unambiguous intersubjective communication about empirical data cannot be carried out without these classical notions.

What is interesting is, of course, the concrete form of these restrictions. The conceptual development in this regard started from Born's work on collision (1926), in which a probability interpretation was suggested. Yet the paper was much richer than this. Since in Born's scattering theory off-diagonal matrix elements were calculated by using wave functions related by Fourier transformations, Pauli saw it as entailing that

one can see the world with p-eyes and one can see it with q-eyes, but if one opens both eyes together, one can go astray.

(1926)

This implication of Born's paper was brought into light with greater mathematical clarity by Dirac (1926b). In his work on the theory of transformations, Dirac demonstrated that Born's probability amplitude was in fact a transformation function between different (canonically conjugate) representations. Considering the fact that a pair of canonically conjugate variables are subject to the commutation relations, such as $PQ - QP = (h/2\pi i)I$, Dirac's theory provided Heisenberg with a mathematical framework for conceptualizing the limitation of the simultaneous measurability of the conjugate variables that were involved in Born's probability amplitudes, although Heisenberg had already recognized, as an implication of the commutation relations, the impossibility of measuring canonically conjugate variables in the same experiment. In a letter to Pauli of 28 October 1926, before he saw Dirac's paper, Heisenberg claimed that

the equation $pq - qp = h/2\pi i$ thus corresponds always in the wave representation to the fact that it is impossible to speak of a monochromatic wave at a fixed point in time (or in a very short time interval) .... Analogously, it is impossible to talk of the position of a particle of fixed velocity.

(Quoted by Hendry, 1984)

The next step in characterizing quantum mechanics was Heisenberg's paper on uncertainty relations. In this paper, Heisenberg (1927) tried to set up 'exact definitions of the words position, velocity, energy, etc. [of an electron]'. He claimed that 'it is not possible to interpret quantum mechanics in the[se] customary kinematical terms' because inseparably linked with these classical concepts are contradictions and a 'struggle of opinions concerning discontinuum and continuum theory, particles and waves'. According to Heisenberg, the revision of these intuitive concepts had to be guided by the mathematical formalism of quantum mechanics, the most important part of which was the commutation relations. He stressed that the 'necessity of revising kinematic and mechanical concepts appears to follow directly from the basic equations of quantum mechanics'. Here he referred to the commutation relations again, and drew from them important lessons. Not only had these relations made us aware that we 'have good reason to be suspicious about uncritical application of the words "position" and "momentum"', but also they had imposed restrictions upon these intuitive notions.

To derive concrete forms of restrictions on classical concepts, such as position, momentum, and energy state, Heisenberg started from a kind of operationalist position[10] (according to which the meaning of concepts can only be defined by experimental measurement) and appealed to thought experiments. Among the well-known thought experiments was the γ-ray microscope experiment which illustrated the uncertainty relation between position and momentum,

$$\Delta P \Delta Q \sim h \tag{16}$$

where ΔP and ΔQ are errors in determination of momentum and position. The uncertainty relation (16) means that a precise description of the location in space (or time) of a physical event or process precludes a precise account of the momentum (or energy) exchange accompanying that event or process. This implication of the uncertainty relations has been exploited in various ways by physicists in explaining exotic features they encountered in the exploration of quantum domains, such as quantum fluctuations.

A more rigorous derivation of the uncertainty relations similar to (16) was given on the basis of Dirac's general theory of transformations, which was interpreted in terms of principal axis transformation. A matrix (or an operator) associated with a dynamical quantity (or an observable) is diagonal in a reference system that is along a principal axis. Every experiment performed on a physical system specifies a certain direction that may or may not be along a principal axis. If it is not, then there is a certain probable error or inaccuracy that is denoted by the transformation formulae to principal axes. For example, measuring the energy of a system throws the system into a state where the position has a probability distribution. The distribution is given by the transformation matrix, which can be interpreted as the cosine of the angle of inclination between two principal axes.

Physically, Heisenberg claimed, the uncertainty relations entailed that each experiment divided physical quantities into 'known and unknown (alternatively: more or less precisely known variables)' in a unique way, and the relationship between results from two experiments that effected different divisions into known and unknown could only be a statistical one. By emphasizing the cognitive significance of experimental setups for measuring particles, Heisenberg tried to explain away the peculiar terms in quantum mechanics, i.e., the interference terms, which were usually interpreted as a manifestation of the wave nature of quantum entities.

Heisenberg's characterization of quantum mechanics with the uncertainty relations was criticized by Bohr because of his neglect of the wave–particle duality of matter and light. The duality escaped our attention in the classical domain (where our usual causal spacetime description was adequate) mainly because of the smallness of Planck's constant. But in the atomic domain, Bohr (1928) stressed, Planck's constant linked the measuring apparatus to the system under investigation in a way that 'is completely foreign to the classical theories'. That is, 'any observation of atomic phenomena will involve an interaction (characterized by Planck's constant) with the agency of observation not to be neglected'. Consequently, Bohr claimed that the notion of duality had to be considered seriously, that the notion of an undisturbed system developing in space and time could only be taken as an abstraction, and that 'the classical mode of description must be generalized' (*ibid.*).

On the basis of these considerations, at the International Congress of Physics (held at Como, Italy, on 16 September 1927), Bohr, for the first time, presented his complementary view of quantum physics. He argued that the classical conception, that physical entities can only be described either as continuous waves or as discontinuous particles but not both, had to be revised, as was required by duality. In the atomic domain, Bohr emphasized that the wave and particle modes of light and matter were neither contradictory nor paradoxical, but rather complementary, that is, mutually exclusive, in the extreme cases. (According to Bohr, then, Heisenberg's uncertainty relations were only a particular case of complementarity because the canonically conjugate quantities were not really mutually exclusive.) Yet for a complete description of the atomic phenomena, both wave and particle modes were required.[11]

With the formulation of the uncertainty relations and the complementarity principle in 1927, quantum mechanics had acquired, in addition to mathematical formalism, an interpretive framework and had become a mature physical theory. Although many puzzles and inconsistencies, such as the reduction of the wave packet and the inseparability of the quantum process, have persisted ever since, these have no direct bearings on our discussion of the conceptual developments associated with the extension of quantum mechanics to relativistic quantum field theory, and will not be addressed in this volume.

Notes

1. Equation (1), when applied to the ground state of the hydrogen atom, led to a result that was in agreement with Bohr's condition. Therefore, *a* here can be taken as the first version of the 'Bohr radius' of the hydrogen atom.
2. Wien's work draws a parallel between the various states of Planck's oscillators and those of real atoms and entails the concept of various energy states of one and the same atom; Debierne's hypothesis of internal degrees of freedom of the atom is based on the phenomena of radioactivity; Stark's idea of the origin of spectra is like this: a single valence electron is responsible for all lines of a spectrum, which are radiated during the successive transitions of the electron from a state of almost complete separation to a state of minimum potential energy. See Pais (1986).
3. This conception of Einstein's was inherited by de Broglie even more faithfully in his double-solution theory (de Broglie, 1926, 1927a, b). However, owing to the dominance of Max Born's probability interpretation of the wave function, when de Broglie's new theory appeared, it played no role in the history. (See section 7.1.)
4. Even more, in his talking of particles as singularities of field, Einstein moved, consciously or unconsciously, into a line of thought in which particles (light quanta or electrons) were to be explained as singularities, or other cohesive structures in a non-linear field. This kind of pursuit can be found in Einstein's unified field theory and de Broglie's nonlinear wave mechanics (de Broglie, 1962).
5. Such a concept was named the photon by G. Lewis in his (1926). A photon has, in addition to its energy $E (= h\nu)$, a momentum $\boldsymbol{P} (= h\boldsymbol{K}$, $\boldsymbol{K}$ being the wave vector), which satisfies the dispersion law $E = c|\boldsymbol{P}|$.
6. Since H is a function of $J (= nh)$, Bohr's frequency condition states that $\nu_{n,n-\tau} = \{H(nh) - H[(n - \tau)h]\}/h = \tau\, \mathrm{d}H/\mathrm{d}J = \tau\nu$.
7. The intensity is proportional to Einstein's emission probability, and the latter is proportional to $|a(n, n - \alpha)|^2$.
8. Bose–Einstein statistics applies to certain indistinguishable particles (bosons) whose number in a cell is not limited. For a different kind of indistinguishable particle (fermions) whose number in each cell can only be 0 or 1, a different kind of quantum statistics, so-called Fermi–Dirac statistics, should be applied.
9. Either in the form of matrix mechanics, or in the form of wave mechanics, which are mathematically and descriptively almost equivalent as was shown by Schrödinger in (1926e).
10. The American physicist P. W. Bridgeman published his operationalismn (1927) in the same year.
11. Bohr characterizes quantum mechanics with his complementary principle. In addition to the complementary descriptions in terms of waves and particles, there are spatio-temporal and causal (in the sense of the conservation of energy and momentum) descriptions that are also complementary: 'We have thus either space-time descriptions or descriptions where we can use the laws of conservation of energy and momentum. They are complementary to one another. We cannot use them both at the same time. If we want to use the idea of space-time we must have watches and rods which are outside and independent of the object under consideration; in that sense we have to disregard the interaction between the object and the measuring rod used. We must refrain from determining the amount of momentum that passes into the instrument in order to apply the space-time point of view' (Bohr, 1928).

7

The formation of the conceptual foundations of quantum field theory

Quantum field theory (QFT) can be analyzed in terms of its mathematical structure, its conceptual system for physical descriptions, or its basic ontology. The analysis can be done logically or historically. In this chapter, only the genesis of the conceptual foundations of QFT relevant to its basic ontology will be treated; no discussion of its mathematical structures or its epistemological underpinnings will be given. Some conceptual problems, such as those related to probability and measurement, will be discussed, but only because of their relevance to the basic ontology of QFT, rather than their intrinsic philosophical interest. Here, by basic ontology I mean the irreducible entities that QFT is invented to describe.[1] The often mentioned candidates for the basic ontology of QFT, in fact of the physical world, are the discrete particle and the continuous field. Another possible candidate (the spacetime point) has also been suggested recently (Redhead, 1983). Since the aim of this chapter is to analyze the historical process in which the conceptual foundations of QFT were laid down, rather than the logical structure of QFT which philosophers of the present day treat, no discussion of the last possibility will be given.

The content of this chapter involves the formation and interpretation, in a roughly chronological order, of the concepts of the wave function, quantization, quantization of the field, the vacuum, interactions between fields, and renormalization. The first two topics will be discussed in relation to the quantization of the field, with their role being taken as the starting point of the conceptual development of QFT. As to interactions, which were the origin of field theories (classical theories as well as quantum ones) and the main subject of this volume, in addition to a brief treatment given here, further discussions will be given in part III.

7.1 Interpretations of the wave function

The wave function introduced by Schrödinger in his wave mechanics, as we shall see in section 7.3, was to be treated as a field to be quantized. Thus the interpretation of the wave function is crucial to the discussion of the ontology of QFT, and is therefore the topic of this first section.

A general review of various interpretations of the wave function can be found in Max Jammer's book *The Philosophy of Quantum Mechanics* (1974). We shall leave aside metaphysical speculations and unsuccessful physical attempts (such as the hidden-variable hypothesis, or hydrodynamic and stochastic interpretations) and concentrate on the interpretations that acted as guides in the historical development of quantum physics. The common opinion is that the accepted interpretation of the wave function is the probabilistic interpretation first suggested by Max Born (1926). This has been evaluated as 'the decisive turning point' in the interpretation of quantum mechanics and 'the decisive step to the final elucidation of the physical interpretation' (Ludwig, 1968). This assertion, however, is true only of non-relativistic quantum mechanics, but not of QFT. As far as QFT is concerned, the situation is far more complicated than that imagined by the holders of the above opinion, for a certain realist element can be clearly discerned within the overall probabilistic framework. For the convenience of the following discussion, therefore, it is better to introduce briefly both the realist and probabilistic interpretations, in a historical and comparative way.

As we have already seen in the last chapter, Schrödinger's wave function, satisfying a certain wave equation and natural boundary conditions, originated from de Broglie's idea of the matter wave. In de Broglie's theory, there is associated with the motion of any material particle a system of plane waves, each of which is represented by a wave function $\Psi_{dB} = \exp\left[2\pi i/h(Et - Px)\right]$ (where E denotes the energy and P the momentum of the particle), such that the velocity of the particle is equal to the group velocity of the waves V_g, and the phase velocity of the wave is equal to c^2/V_g. Representing the wave of ordinary light of frequency ν by a wave function $\Psi_1 = \exp\left[2\pi i/h(Et - Px)\right]$ (where E denotes the energy $h\nu$ and P the momentum $h\nu/c$ of the corresponding photon), and comparing Ψ_{dB} with Ψ_1, we see that an ordinary wave of light is simply the de Broglie wave belonging to the associated light quantum. From this fact, two inferences of different ontological implications can be drawn. An inference of realist inclination, originally drawn by de Broglie himself and accepted later on by Schrödinger, asserts that the de Broglie waves, just like the ordinary light waves, are real, three-dimensional, continuous, substantial waves. The other inference is probabilistic in nature and was first drawn by Born and then accepted by many others, which I shall discuss shortly.

Here we should be very careful in dealing with de Broglie's idea of matter waves expressed in terms of wave functions. In order to make the meaning of this idea clearer, and to avoid a confusion which occurs so frequently in the literature, it is necessary to distinguish two basic usages of the term 'matter wave', which is crucial for clarifying the ontology of QFT. We can indicate these two usages by two contrasts: (1) the wave of matter versus the wave of light, and (2) the material (substantial) wave (of matter or light) versus the probabilistic wave (of matter or light). The de Broglie–Schrödinger wave was associated with the motion of material particles in the first usage of the term. That is, it had nothing to do with light quanta. This was the original meaning of de Broglie's phase wave and was not questioned.

What was confusing and controversial in the literature was its second usage. One might think that after the publication of Born's probabilistic interpretation, the controversy had been settled. But this is not true in the context of QFT. Some QFT physicists who generally accepted Born's view nevertheless ascribed a material sense to the de Broglie–Schrödinger wave when they treated the wave function realistically as a matter wave. We shall discuss the cause and implications of this confusion in section 7.3.

Schrödinger was influenced by de Broglie's idea of the phase wave, and generalized from it. He represented an individual atomic system by a wave function and regarded the waves as the bearers of the atomic processes. He connected the wave function 'with some vibration process in the atom' (Schrödinger, 1926b) and regarded the wave 'as a superposition of pure time harmonic vibrations, the frequencies of which agree exactly with the spectroscopic term frequencies of the micro-mechanical system' (1926d). But then the big question was: wave of what?

In his (1926e), Schrödinger attempted to give the wave function an electrodynamic meaning: 'the space density of electricity is given by the real part of $\Psi \, \partial\Psi^-/\partial t'$. But the difficulty here was that this density, when integrated over all space, yielded zero rather than a time-independent finite value. To resolve this inconsistency, Schrödinger in his (1926f) replaced the expression $\Psi \, \partial\Psi^-/\partial t$ for the charge density by the 'weight function in the system's configuration space', $\Psi\Psi^-$, multiplied by the total charge e. This was equivalent to the following interpretation:

The wave-mechanical configuration of the (atomic) system is a superposition of many, strictly speaking of all, point-mechanical configurations kinematically possible. Thus each point-mechanical configuration contributes to the true wave-mechanical configuration with a certain weight, which is given precisely by $\Psi\Psi^-$.

(Schrödinger, 1926f)

That is, the atomic system existed 'simultaneously in all the positions kinematically imaginable, but not "equally strongly" in all' (*ibid.*). Schrödinger stresses that in this interpretation,

the Ψ-function itself cannot and may not be interpreted directly in terms of three-dimensional space – however much the one electron problem tends to mislead us on this point – because it is in general a function in ($3n$-dimensional) configuration space, not real space.

(Ibid.)

Still, he maintained a realist interpretation:

But there is something tangibly real behind the present conception also, namely, the very real electro-dynamically effective fluctuation of electric space-density. The Ψ-function is to do no more and no less than permit of the totality of these fluctuation being mastered and surveyed [by a wave equation].

(Ibid.)

Thus the wave was regarded by him as having a reality of the same type as electromagnetic waves possessed, having a continuous density of energy and momentum.

Relying on the hypothesis that a particle of velocity V can be represented by a wave packet moving with the group velocity V, Schrödinger (1926c) proposed to abandon the particle ontology entirely and maintained that physical reality consisted of waves and waves only. His reasons for rejecting particles as well-defined permanent objects of a determinable identity or individuality were summarized in his (1961) as follows: (1) 'because of the identity of mass and energy, we must consider the particles themselves as Planckian energy quanta'; (2) in dealing with two or more particles of the same type, 'we must affect their identity, otherwise the results will simply be untrue and not agree with experience'; and (3) 'it follows from [the uncertainty principle] that it is never possible to observe the same particle twice with apodeictic certainty'.

A similar but more sophisticated view of the continuous field as the basic ontology of quantum physics was suggested by de Broglie (1926, 1927a, b) in his theory of the double solution. The main difference between his and Schrödinger's view lies in the fact that while Schrödinger viewed physical reality as consisting of waves only, de Broglie accepted the reality of classical particles, in addition to that of the waves, and identified a particle as an energy concentration in a singularity region of the non-linear wave, which was taken to be guided by an extended linear wave Ψ. Hence, according to de Broglie, physical reality consisted of waves and particles, though the latter were taken to be merely a manifestation of the former.

The realist interpretations of the wave function had to face serious difficulties. First, the wave could not be conceived as a real one because, as pointed out by Bohr in his (1927), the phase velocity of a wave was c^2/V_g (V_g being the group velocity), which was in general greater than the velocity of light c. Second, as far as the dimensionality of the configuration space of the wave function Ψ is concerned, one could not, as Lorentz pointed out in his letter to Schrödinger of 27 May 1926, interpret the waves physically when they were associated with processes which were classically described with the help of several particles. This is because the wave then becomes a function of $3n$ (n is the number of particles) position coordinates and requires for its representation a $3n$-dimensional space. Third, Schrödinger's wave function is a complex function. Fourth, the Ψ function is representation dependent. The last and most serious difficulty is related with the so-called spread of the wave packet. Lorentz pointed out in the above-mentioned letter that 'a wave packet can never stay together and remain confined to a small volume in the long run', so it is not suitable for representing a particle, 'a rather permanent individual existence'. Encouraged by his success in getting a non-spreading wave packet in the harmonic oscillator case, Schrödinger (1926g) thought that it was 'only a question of computational skill' to get it in the general case. But Heisenberg (1927) soon realized that it was an impossible task.

Now I turn to Born's probabilistic interpretation, which was consistent with viewing particles as the basic ontology of quantum physics and quickly became a dominant interpretation.

Born was awarded the Nobel Prize of 1954 'for his fundamental work in quantum mechanics and specially for his statistical interpretation of the wave function' (the official declaration of the Royal Swedish Academy of 3 November 1954). Born's idea, however, originated from Einstein's observation (1909a, b) on the relationship of a wave field and light quanta. In his paper on a collision process, in which he proposed for the first time a probabilistic interpretation of the wave function, Born indicated that

[Einstein said that] the waves are present only to show the corpuscular light quanta the way, and he spoke in the sense of a 'ghost field'. This determines the probability that a light quantum, the bearer of energy and momentum, takes a certain path; however, the field itself has no energy and no momentum.

(1926)

Here, the wave of light lost its substantiality and became a kind of probabilistic wave.

We can fill the gap between Einstein's view of 1909 and Born's view of 1926 with several important developments. As is well known, Bohr, Kramers, and John Slater in their 1924 paper suggested that the field of electromagnetic waves determined only the probability that an atom would absorb or emit light energy by quanta in the relevant space. This idea actually belongs to Slater.[2] A few weeks earlier than the submission of the paper, Slater sent a letter to *Nature* in which he proposed that part of the field was 'to guide discrete quanta' and to induce a 'probability that it [the atom] gains or loses energy, much as Einstein has suggested' (Slater, 1924).

In Einstein's 1925 paper on gas theory we can find another link between Einstein's and Born's view of waves: Einstein related de Broglie's wave with his moving gas molecules and made use of the mathematical formalism of particle physics, indicating that the basic ontology of his theory was that of the particle rather than the wave, although there was a wave aspect in the phenomenon under investigation, as was revealed in the mysterious interaction among the molecules. Pascual Jordan's follow-up paper (1925) was dominated by his acceptance of the spirit of wave–particle duality, but the analysis there was also entirely in terms of the particle.

What Born planned to do in his 1926 paper, as regards the interpretation of the wave function, was just to extend 'Einstein's idea', mentioned above, on the relation between the light wave and the light quanta to the relation between the Ψ wave function, which was also regarded by him as a 'ghost field', and ponderable particles. Born's main arguments were based on 'the complete analogy' between these two sets of relations, and can be summarized as follows:

(1) the Ψ waves, as ghost fields, themselves have no energy and no momentum, and hence no ordinary physical meaning;
(2) the Ψ waves, in contrast with de Broglie's and Schrödinger's view mentioned above, do not represent the motion of matter at all; they only determine the possible motion of matter, or the probability of measured results about moving matter;
(3) by probability Born meant 'a definite frequency with which states occur in an assembly of identical, uncoupled atoms';
(4) so the concept of the Ψ wave presupposes the existence of a very large number of independent particles.

It is therefore clear that Born regarded particles as the basic ontology of wave mechanics or quantum mechanics.

Max Jammer summarizes Born's probabilistic interpretation of the Ψ wave as follows:

$|\Psi|^2\,d\tau$ measures the probability density of finding the particle within the elementary volume $d\tau$, the particle being conceived in the classical sense as a point mass possessing at each instant both a definite position and a definite momentum.

(Jammer, 1974)

Whether the particle was conceived by Born in the classical sense is a disputable issue.

Born's interpretation led to Dirac's transformation theory (1926b) and the latter in turn led to Heisenberg's uncertainty relationship (1927), from which Born quickly realized that it was

necessary to abandon not only classical physics but also the naive conception of reality that thought of the particles of atomic physics as if they were exceedingly small grains of sand, [having] at each instant a definite position and velocity.

(Born, 1956)

Born acknowledged that in atomic systems 'this is not the case'. He also agreed with Schrödinger, in regard to quantum statistics, that 'the particles are not to be identified as individuals' (*ibid.*). As regards non-classical, non-individual particles of this kind, Born explained that the reason 'for the retention of the particle idea' lay in the fact that 'they represent invariants of observables' (*ibid.*). But for the same reason, the wave idea should also be retained. Born went even further than this and took 'a probability wave, even in $3n$-dimensional space, as a real thing, certainly as more than a tool for mathematical calculations' (1949). He asked, 'how could we rely on probability predictions if by this notion we do not refer to something real and objective?' (*ibid.*).

All these show that Jammer's misleading exposition not only oversimplified Born's position on the ontology of quantum mechanics, but also neglected the important fact that Born's interpretation, though it took a kind of particle as the ontology of quantum physics, was incompatible with the classical particle ontology, and shared certain features with the wave ontology.

Another misleading exposition of Born's 1926 paper can be found in Heisenberg's famous article 'The development of the interpretation of the quantum theory', in which he asserts that

[Born's] hypothesis contained two important new features in comparison with that of Bohr, Kramers and Slater. The first of them was the assertion that, in considering 'probability waves', we are concerned with processes not in ordinary three-dimensional space, but in an abstract configuration space; the second was the recognition that the probability wave is related to an individual process.

(Heisenberg, 1955)

It is obvious that the second point does not tally with Born's original intention. As we observed above, Born used the concept of probability in the sense of the frequency interpretation. This means that the probability wave contains statistical statements about an assembly of systems, not only about one system. As to the first feature, Heisenberg's observation was superficial and missed the point. In fact, Born's main purpose of proposing a probability interpretation was precisely to restore the strange configurational wave introduced by Schrödinger to the ordinary three-dimensional space. Born himself clearly pointed this out in 'the closing remarks' of his 1926 paper: 'it [the probability interpretation] permits the retention of the conventional ideas of space and time in which events take place in a completely normal matter'.

This achievement, however, was not without cost. As Heisenberg realized clearly in his (1927), it reduced the interpretation of quantum mechanics to the problem of how to draw experimental information from the formulae of quantum mechanics. In this way, one can easily explain the 'reduction of the wave packet', but not the diffraction of electrons in the double-slit experiment. This experiment can be carried out in such a way that only one electron passes the apparatus at a time. In this case, the Ψ wave associated with each electron interfered with itself. The Ψ wave must therefore be some physically real wave and not merely a representation of our knowledge about particles.

As a response to this difficulty faced by Born's interpretation, Heisenberg reinterpreted the probability wave as one related to an individual process mentioned above, and regarded the probability wave as a new kind of physical reality, related to a possibility or *potentia*, 'a certain intermediate layer of reality, halfway between the massive reality of matter and the intellectual reality of the idea' (Heisenberg, 1961). In this view, a closed system represented by a wave function (or by a statistical mixture of such functions) is potential but not actual in character, the 'reduction of wave packets' occurring during a process of measurement is thus a consequence of the transition from the potential to the actual, which is produced by connecting the closed system with the external world (the measuring apparatus) (Heisenberg, 1955).

The introduction of the metaphysical concept of the *potentia* was characteristic of the dilemma faced by the probabilistic interpretation based on a particle ontology, and was a significant concession to the realist interpretation of the wave function, or to the wave ontology. In fact, Heisenberg made this concession as early as 1928, when Jordan, Oskar Klein, and Eugene Wigner proved the equivalence between Schrödinger's wave description of a system of *n* particles and the second-quantization description of particles obeying quantum statistics (Bose–Einstein as well as Fermi–Dirac statistics).[3] For example, Heisenberg thought then that the particle picture and the wave picture were merely two different aspects of one and the same physical reality (cf. Jammer, 1974). He also realized that

> only the waves in configuration space, that is the transformation matrices, are probability waves in the usual interpretation, while the three-dimensional material waves or radiation waves are not. The latter, according to Bohr, Jordan and Wigner, have just as much (and just as little) 'objective reality' as particles; they have no direct connection with probability waves, but have a continuous density of energy and of momentum, like a Maxwell field.
>
> *(Heisenberg, 1955)*

That is to say, Heisenberg accepted that there were material waves, different from the probability waves, having the same ontological status as particles.

What is not so clear here is what was meant by 'material waves'. If they can be interpreted as some substantial waves of ponderable matter, then it is a big concession to the realist interpretation of the wave function, even though the restriction 'three-dimensional' seems to hint that only the de Broglie wave of one particle can be regarded as 'objective reality'. If, on the other hand, 'material waves' was just a synonym for 'radiation waves', then the question is: When the radiation waves (as de Broglie waves of photons), which are different from the probability waves, are regarded as 'objective reality', what about the three-dimensional de Broglie waves of just one electron or other ponderable particle? Are they still probability waves, or objective reality? In fact, Heisenberg and all other physicists in QFT have continuously shifted back and forth between these two senses. The equivocal use of the concept of 'material waves' is actually the origin of all the conceptual confusions about the ontology of QFT. But on the other hand, it is precisely this equivocality that makes possible the development of QFT on the basis of the concept of the material wave field. Any clear-cut and unequivocal use of this concept would cause serious conceptual difficulties. So it seems that the equivocality reflects and marks the deepest difficulty in grasping the ontology of QFT conceptually. This also explains why the equivocality has surprisingly lasted for more than 60 years, and shows no signs of being clarified.

7.2 Quantization

To a great extent, the discussion of the ontology of QFT also depends on one's understanding of the concept of quantization: What is meant by quantization? What is meant by the mathematical procedure of quantization? What is the relation between the physical implications and mathematical procedures of quantization? Some far-reaching confusions in the literature can be traced back to the misunderstanding of this concept. While the physical idea about quantization had already occurred as early as Planck's 1900 theory of black-body radiation, a mathematical procedure of quantization was established only after the emergence of quantum mechanics in 1925 as developed by Heisenberg, Born, Jordan, and, especially, Dirac. Initially, the mathematical procedure had served as a tool for realizing the physical implications of the idea of quantization. But once the mathematical procedure was established, it took on an independent existence.

Then the pitfall was to forget or neglect or cut the intimate connection between the physical and mathematical aspects of quantization, to take the formal procedure as the essence of quantization, as a model within, or basis on, which the discussion of quantization, second quantization, and quantization of the field could be carried out, and hence to give misleading interpretations. In this section I shall pinpoint the proper meaning and the primary misunderstanding of this concept by a brief historical survey.

For Planck, quantization was demanded by the laws of heat radiation and meant only the quantization of the energy of a simple oscillator: in the statistical behavior of electrons in a heated atom executing simple harmonic oscillations, only those vibrational states should be taken into account whose energy was an integral multiple of a quantum. Here the quantization of energy was a property of the electrons in atoms, represented by oscillators in their interactions with radiation.

For Bohr, quantization meant the quantization of the periodic orbits of the electrons in an atom, and the quantized electronic orbits were associated with the discrete stationary states. So the mechanical motions associated with the stationary states were to be chosen by the quantization condition (QC) from the continuous manifold of such motions. According to Bohr, QC should be considered 'a rational generalization of Planck's original result for the possible energy values of an harmonic oscillator', since here the quantization of mechanical motion involved not only that of energy, but also that of momentum and angular momentum in atomic systems.

Mathematically, Bohr's QC for every periodic coordinate Q of the motion can be expressed in the form $J = \oint P \, dQ = nh$, where P is the canonically conjugate momentum corresponding to Q (P may be the angular momentum if Q is an angle) and the integration extends over a period. The most convincing theoretical argument for choosing this integral J was given by Ehrenfest (1916), who showed that if the system was subject to a slow external perturbation, J was invariant and therefore well suited to be equated to a discontinuous 'jumping' quantity nh.

Bohr's conception of quantization, both in its physical aspect (as a quantization of mechanical motion of electrons in an atom) and its mathematical aspect ($J = \oint P \, dQ = nh$), constituted a starting point from which two schools of quantization had grown. The first was the de Broglie–Schrödinger school. According to de Broglie (1923a, b), by using the expression ($P = h/\lambda$, $J = \oint P \, dQ = 2\pi r h/\lambda = nh$, $2\pi r/\lambda = n$), QC could be interpreted as follows. The wave associated with a moving electron must be such that the number of wavelengths that cover the orbital circumference must be a integral number. For Schrödinger, quantization could be reduced to a problem of proper values (1926b, c, e, f), as was indicated by the title of his four seminal communications 'Quantization as a problem of proper values'. If, for any variation of a wave function Ψ (associated with some vibration process in the atom) the integral of the Hamiltonian function (taken over all of coordinate space) is stationary, then the Ψ function can be chosen for a discrete set of energy proper values, and the chosen Ψs themselves then become the proper functions corresponding to the proper values.

Mathematically, Schrödinger's variation problem was far more complicated than the Bohr–Sommerfeld QC: $\oint P \, dQ = nh$. The kinship of its guiding principle with the latter, however, can be found from the similarity of the idea of 'stationary integral' to Ehrenfest's hypothesis of adiabatic invariant. As to its relation with de Broglie's conception, Schrödinger himself pointed out that he obtained his inspiration directly from de Broglie's conclusion that connected with the space distribution of the phase wave there was always a whole number measured along the orbit of the electron. Schrödinger went on, 'the main difference is that de Broglie thinks of progressive waves, while we are led to stationary proper vibrations' (1926b).

The second school was represented by Heisenberg (1925), Born and Jordan (1925), and Dirac (1925). Here the basic physical implication of quantization was the same as in Bohr's conception: the mechanical motion in the atom was quantized. But its guiding principle for getting the mathematical expression for this quantization went beyond Ehrenfest's adiabatic hypothesis. The decisive step was taken by Heisenberg (1925). In addition to the quantization of the whole phase space integral, which represented the mechanical motion in the atom, Heisenberg called attention to the quantal nature of dynamical quantities (or physical observables) in atomic physics. These were transitional quantities depending on two quantized stationary states, and hence should be represented, as Born and Jordan (1925) demonstrated two months later, by matrices. Taking this observation into account, Born and Jordan showed that the Bohr–Sommerfeld QC, $\oint P \, dQ = \oint_0^{1/\nu} PQ \, dt = nh$ can be transformed into $\mathbf{1} = 2\pi i \sum_{-\infty}^{\infty} (\tau \partial/\partial J)(Q_\tau P_{-\tau})$ when the Fourier expansions of P and Q ($P = \sum_{-\infty}^{\infty} P_\tau e^{2\pi i \nu \tau t}$,

$Q = \sum_{-\infty}^{\infty} Q_\tau \, e^{2\pi i \nu \tau t})$ are introduced. Considering Kramers' dispersion relation (1924), which says that the sum $\sum_{-\infty}^{\infty} (\tau \partial/\partial J)(Q_\tau P_{-\tau})$ must correspond to $1/h \sum_{-\infty}^{\infty} [Q(n+\tau, n)P(n, n+\tau) - Q(n, n-\tau)(P(n-\tau, n)]$, QC becomes $\sum_{-\infty}^{\infty} [P(nk)Q(kn) - Q(nk)P(kn)] = h/2\pi i$, or in matrix notation,

$$PQ - QP = (h/2\pi i)\mathbf{1}. \tag{1}$$

By reinterpreting the classical equations of Hamiltonian dynamics, Dirac (1925) obtained the same non-commutative relations for the q numbers.

The establishment of mathematical expressions for the non-commutative relations of dynamical quantities marks the completion of the mathematical description of quantization of mechanical motion in the atom. On the basis of such a description, a certain procedure for quantization was put forward: substituting the canonically conjugate dynamical quantities in the classical Hamiltonian formulation by corresponding q numbers, and subjecting them to the non-commutative relations (1), then the quantum formulation can be obtained.

The mathematical procedure of substitution and subjection quickly became a model of quantization, on the basis of which various interpretations of quantization were given. Therefore, it profoundly influenced the subsequent development of quantum physics, not only in its mathematical description, but also in its physical and ontological interpretation of concepts. What I want to stress here, however, is that we should keep in mind the basis, premises, proper implications, and restrictions of this mathematical procedure.

Heisenberg's decisive starting point was Bohr's frequency condition connecting the radiation frequency (transitional quantity or the so-called observable) with two stationary states. Heisenberg's idea of an 'observable' therefore presupposed the existence of quantized stationary states in the atom. Originally, Heisenberg's, Born and Jordan's, and Dirac's expression of QC was no more nor less than a rewriting of Bohr's QC. So it would be meaningless to consider the procedure of quantization without relating it with the quantized stationary states in the atom.

This restriction, however, was soon removed by subsequent developments initiated by Schrödinger and Dirac. Looking back for a connection between wave mechanics and matrix mechanics, Schrödinger recognized that the essential feature of a matrix was that it represented a linear differential operator acting on a vector (one-column matrix). In this way, Schrödinger (1926e) came to his operator calculus. If a coordinate q is taken as an ordinary variable corresponding to a diagonal matrix representing a dynamical variable (say Q) in its own representation (the Q representation), then the corresponding canonically conjugate momentum p is to be replaced everywhere by the operator $h/i(\partial/\partial q)$. In this way QC (1) becomes a trivial operator identity. Then Schrödinger's idea of quantum mechanics is like this. A definite state of a system is defined by a wave function $\Psi(q)$ in some q representation, q being an observable quality of the system. A dynamical variable (or observable) A can be represented by a linear operator, and $A\Psi(q)$ means a new wave function representing a new state of the system, by operating with A on $\Psi(q)$. Then if this new function is, apart from a constant factor, identical with $\Psi(q)$: $A\Psi(q) = a\Psi(q)$, then $\Psi(q)$ is called an eigenfunction of A, and the constant a an eigenvalue. The whole set of eigenvalues is characteristic of the operator A and represents the possible numerical values of the observable, which may be continuous or discrete. Operators whose eigenvalues are all real numbers are called Hermitian operators. It is clear that all physical quantities have to be represented by Hermitian operators, since the eigenvalues are supposed to represent the possible results of a measurement of a physical quality.

On the basis of this idea of Schrödinger, Dirac (1927a) obtained an insight into the various possible representations of quantum mechanics. He recognized that what the original matrix mechanics took was an energy representation, and what the original wave mechanics took was a coordinate representation. These were just two special cases of possible representations. In addition to them, there were also angular momentum representation, particle number representation, and so on. Dirac pointed out in his own notation that if a dynamical variable G originally took an α representation $\langle \alpha'|G|\alpha \rangle$, then its β representation could be obtained through a unitary transformation: $\langle \beta'|G|\beta \rangle = \int\int \langle \beta'|\alpha \rangle \, \mathrm{d}\alpha \langle \alpha|G|\alpha' \rangle \, \mathrm{d}\alpha' \langle \alpha'|\beta \rangle$, where the transformation function $\langle \alpha'|\beta \rangle$ was precisely the eigenfunction of the operator representing the dynamical variable. For example, the eigenfunction of Schrödinger's wave equation (as an energy eigenequation) was exactly the transformation function from coordinate representation to energy representation. This is the so-called general transformation theory, since α and β can be any set of commuting observables.

The basis of QC (1) was thus widened. First, the physical system under consideration was no longer restricted to atoms with discrete energy levels, since Schrödinger's wave equation made it possible to deal with continuous eigenvalues. The free particles and collision problems therefore also came into the range of quantum mechanics. Second, Dirac's transformation theory entailed that the physical states of a system that could be treated by quantum mechanics were no longer restricted to the energy states of the system (e.g., the quantized stationary states of an atom), which were represented by energy eigenfunctions, and various physical states, represented by eigenfunctions of operators representing observables other than energy, could also be treated.

In view of these generalizations, QC (1) seems to have become a universal condition for quantization. A severe restriction, however, remained on all these generalizations: only those pairs of operators, or matrices, or dynamical variables, that are genuine observables can be subjected to QC (1), by which a true quantization of mechanical motion of a system is obtained. By 'genuine observable' I mean an observable represented by a Hermitian matrix of which a representation can be found in which the matrix is a diagonal one.

One might think this restriction an academic one because the dynamical variables in quantum physics seem to be equal to the observables. But this is actually not true in the case of QFT, or even in the case of many-body problems in quantum mechanics. It is strange to find that all of the forerunners and founders of QFT, such as Jordan (1925b, c), Dirac (1927b, c), Klein (with Jordan, 1927), Wigner (with Jordan, 1928), Heisenberg and Pauli (1929), and Fermi (1932), took a certain set of non-observables (such as the creation and annihilation operators in the particle number representation; see section 7.3), subjected them to a variation of QC (1), and then claimed that a quantized or second quantized theory was obtained. This procedure of quantization, on which physical interpretations of QFT were given, therefore became another source of confusion about the ontology of QFT, as we shall see in detail in the next section.

Another development in the conception of quantization was more pertinent to QFT. Here the physical system under consideration was the continuous field rather than the discrete particle. The originator of this development was Einstein. On the basis of Planck's quantization of energy of material oscillators, Einstein proposed that 'the energy of light consists of a finite number of energy quanta, localized at various points of space', and that 'these quanta can be produced or absorbed only as units' (1905a). This idea was taken over by Ehrenfest who mentioned in his (1906) that the amounts of electromagnetic field energy residing in a normal mode of frequency v, represented by a simple harmonic oscillator, could only be integral multiples of hv. Ehrenfest's idea that the radiation itself could be regarded as a system of simple oscillators, each of which represented the amplitude of a plane wave, was inherited by Debye (1910b), Jordan (1925), and Dirac (1927b, c). Thus the quantization of field energy can be easily obtained by applying Planck's hypothesis to the field oscillators (Debye), or by applying the mathematical method that was used in quantum mechanics for treating the material oscillators (Jordan and Dirac).

Einstein's idea of the quantization of the field, however, was actually deeper than this. In the same paper mentioned above (1905a) he speculated on the possibility that radiation itself was 'composed of independent energy quanta'. Einstein's hypothesis of light quanta was further supported by his contributions of (1909a, b, 1916c, d, 1917b) and by Compton's experiment (1923a, b), which was reviewed in chapter 6. Here I just want to make two remarks:

(i)   Einstein's idea of light quanta involved a new kind of quantization, the quantization of a substantial field rather than the mechanical motion of particles or fields.
(ii) There was no mathematical procedure for the quantization of the field accompanying Einstein's light quantum hypothesis. What was used as a mathematical procedure in the quantization of fields was just an analogue of what was used in the quantization of mechanical motion. Therefore there was a conceptual gap to be bridged, and justification for taking the analogy seriously to be given.

Let us now turn to the discussion of these issues.

7.3 The quantization of fields

As we noticed above, the concept of the quantization of a field had two meanings: (i) the quantization of field energy (or, more generally, of the mechanical motion of the field), a quantization that was similar to the quantization of the mechanical motion of particles dealt with in quantum mechanics; (ii) the quantization of the field as a substantial entity.

If we accept Maxwell's idea that energy must reside in a substance, and also accept Einstein's idea that the substance as the carrier of electromagnetic field energy is not the mechanical ether, but the electromagnetic field itself, which, as an independently existing entity, is ontologically on a par with ponderable matter, then 'the field' cannot be regarded as a synonym for 'the energy of the field', but the former is related to the latter as an owner is to his possessions. In this view, meaning (i) does not entail meaning (ii), and vice versa. The non-entailment claim is supported by the fact that there is no necessary connection between the continuity or discontinuity of energy and that of the carrier of energy, a fact that can be checked easily in the case of particles. The energy of an electron can be either continuous when it is free, or discrete when it is found within an atom.

In the history of QFT, however, these two meanings have often been lumped together simply under the vague heading of 'the quantization of fields', and some unsound inferences have been carelessly drawn from such an equivocality. Since the concept of the quantization of fields is a cornerstone of the conceptual foundations of QFT, a careful historical study of the formation of this concept seems necessary and is attempted in the following.

It is a common opinion that the idea of field quantization can be traced back to Ehrenfest's paper (1906) and Debye's paper (1910b).[4] According to these two authors, if we take the electromagnetic field in empty space as a system of harmonic oscillators, and apply Planck's quantum hypothesis about the energy of material oscillators to the field oscillators, then the Planck spectrum can be derived without using the material oscillators. The justification of these heuristic ideas is no more nor less than Planck's idea itself, that is, the observed blackbody radiation spectrum. It is obvious that the quantization of fields here means only the quantization of field energy, that is, only meaning (i).

Jordan's works occupy a prominent place in the history of QFT. In his paper co-authored with Born (1925), he gives not only a proof of QC (1): $PQ - QP = (h/2\pi i)I$, but also the original idea, which is more pertinent to the present discussion, that the electric and magnetic fields should be regarded as dynamic variables, represented by matrices and subject to QC, so that the classical formula for the emission of radiation by a dipole can be taken over into the quantum theory. In his paper co-authored with Born and Heisenberg (1926), Jordan, like Ehrenfest and Debye, considered the electromagnetic field as a set of harmonic oscillators and subject to QC. In this way he derived the mean square energy fluctuation in a field of blackbody radiation.[5]

Since Einstein derived the justification of the concept of corpuscular light quanta from the laws of fluctuation in a field of waves, the above achievement convinced Jordan that 'the solution of the vexing problem of Einstein's light quanta might be given by applying quantum mechanics to the Maxwell field itself' (Jordan, 1973). Here, two points should be noted. (i) The field to be quantized was the substantial and continuous Maxwell field, and (ii) the corpuscular light quanta can be obtained as a result of quantizing the field. Considering the fact that Einstein's light quanta possessed not only a definite amount of energy, $h\nu$ (as he initially proposed), but also a definite amount of momentum, $P = hK$ (as he realized in his 1916 papers, see section 6.2), and thus were fully fledged particles rather than merely quanta of energy, the quantization of the field here was obviously in the sense of meaning (ii).

To judge whether or not Jordan's conviction was justifiable, we have to make it clear what Jordan had actually done. An examination of his 1925 paper shows that what he had done was simply apply the Heisenberg–Born QC to the wave field. Remember that the Heisenberg–Born QC presupposed (i) the existence of quantized stationary states of the atom, and (ii) that observables were related to two states of the atom (see section 7.2). Then we find that Jordan's application of the Heisenberg–Born QC implies that he tacitly adopted two hypotheses, namely, the generalizations to the Maxwell field of the presuppositions of the Heisenberg–Born QC: (i) the energy states of the Maxwell field were also quantized, and (ii) the field variables as dynamical variables were always related with two states of the field. No explicit justification for the generalization was given, though the close connection between the energy change of an atom and of a field seemed to support such a generalization indirectly. But anyhow, with these two assumptions what Jordan could 'prove' was merely that the energy states of the field were quantized, a result that was exactly what he had presupposed but that had nothing to do with the quantization of substantial Maxwell fields themselves. In fact, the significance of Jordan's paper lies in its mathematical aspect: he provided a mathematical formulation for realizing the physical hypothesis about the quantization of field energy, with the help of the oscillator model.

The situation in 1925 was relatively simple. Although there was de Broglie's speculation on the matter wave, it did not change the situation much. In 1925, when one mentioned a field, it always referred to the substantial field. The situation became much more complicated after Schrödinger's work took the stage in 1926. Schrödinger's waves soon became fundamental entities to be quantized. Even the electromagnetic wave was regarded as a kind of Schrödinger wave associated with the light quanta (see the discussion below on Dirac's paper). Then, as we have already noticed, there were two interpretations of Schrödinger's wave function: one was realist, and the other probabilistic. They each assumed different basic ontologies for QFT. In addition to the confusion about the interpretations of the wave function, there was another kind of confusion about two meanings of the quantization of the field. The discussion of the quantization of the field therefore needs a careful analysis so that the complicated situation can be clarified. This will involve criticisms of prevailing misunderstandings and clarifications of misleading assumptions contained in the original papers by the founders of QFT.

I shall begin with Dirac's paper 'The quantum theory of emission and absorption of radiation' (1927b), which is regarded as the germ out of which QFT developed, owing to 'the invention of second quantization' contained in it (Jost, 1972). There are a number of widespread misunderstandings about this paper. First, what was the field that Dirac proceeded to quantize? Concerning this question, Jordan tells us in his recollections (1973) that when Dirac's paper had just appeared, he and Born took the field to be 'the eigenfunctions of a particle' or 'the Schrödinger field of the single particle'. In fact Jordan maintained this understanding all his life and regarded his 1925 paper and subsequent works on QFT as also having the Schrödinger field as their starting point. Similarly, Gregor Wentzel (1960), in his authoritative article on the history of QFT before 1947, also asserted that the field was taken to be 'a (complex) Schrödinger wave function' or 'the probability amplitude of the "undisturbed states"'.

In contrast to Jordan, who confused the radiation field treated in his (1925; co-authored with Born, 1925; and co-authored with Born and Heisenberg,

1926) with the Schrödinger wave, Dirac distinguished clearly these two kinds of field in the 'Introduction and Summary' of his paper:

Firstly, the light-wave is always real, while the de Broglie wave associated with a light-quantum . . . must be taken to involve an imaginary exponential. A more important difference is that their intensities are to be interpreted in different ways. The number of light-quanta per unit volume associated with a monochromatic light-wave equals the energy per unit volume of the wave divided by the energy $(2\pi h)\nu$ of a single light-quantum. On the other hand a monochromatic de Broglie wave of amplitude a . . . must be interpreted as representing a^2 light-quanta per unit volume.

(1927b)

He also pointed out that the above interpretation of many-body (many light quanta) Schrödinger waves 'is a special case' of the general probabilistic interpretation of wave functions,

according to which, if (ξ'/α') or $\Psi_{\alpha'}(\xi'_k)$ is the eigenfunction in the variables ξ_k of the state α' of an atomic system (or a simple particle), $|\Psi_{\alpha'}(\xi'_k)|^2$ is the probability of each ξ_k having the value ξ'_k.

(Ibid.)

Now two points underlying Dirac's observations are clear. (i) He identified Schrödinger waves with probability waves and related them to the particle ontology. (ii) He distinguished the true and substantial wave from Schrödinger waves, and claimed that 'there is no such (substantial) wave associated with electrons'. What is not so clear here is his position on the relation between light waves (or quanta of energy in light waves) and light quanta. The clarification of his position on this relation is crucial to the understanding of the conceptual situation in the early stage of QFT, so we have to go into the details.

One of Dirac's underlying ideas in this paper was that there was 'a complete harmony between the wave and light quantum descriptions of the interaction'. To show this, he first actually built up the theory from the light quantum point of view, and then showed that the particle formulation can be transformed naturally into a wave form.

The first step was taken with the help of the so-called 'second quantization' procedure. He proceeded from an eigenfunction Ψ for the perturbed assembly of N similar independent systems, presenting Ψ in terms of eigenfunctions Ψ_r for the unperturbed assembly: $\Psi = \Sigma_r a_r \Psi_r$, and then obtained the probable number N_r of the assembly being in the state r: $N_r = |a_r|^2$. By introducing variables $b_r = a_r \exp(-i\omega_r t/h) = \exp(-i\theta_r/h)N_r^{1/2}$ and $b_r^* = a_r^* \exp(i\omega_r t/h) = \exp(i\theta_r/h)(N_r + 1)^{1/2}$, and taking b_r and ihb_r^* to be canonically conjugate 'q numbers' satisfying the QC: $[b_r, ihb_s^*] = ih\delta_{rs}$, he obtained a theory in which the interaction Hamiltonian F could be written in terms of N_r (particle number)[6]:

$$F = \Sigma_r W_r N_r + \Sigma_{rs} V_{rs} N_r^{1/2} (N_s + 1 - \delta_{rs})^{1/2} \exp\left[i(\theta_r - \theta_s)/h\right]. \quad (2)$$

Here the starting point was the many-particle Schrödinger wave and the basic ontology was the light quantum, a kind of permanent particle carrying energy and momentum. No other textual evidence of this claim about Dirac's view of the light quantum is clearer than the following:

> The light-quantum has the peculiarity that it apparently ceases to exist when it is in one of its stationary states, namely, the zero state, in which its momentum, and therefore also its energy, are zero. When a light-quantum is absorbed it can be considered to jump into this zero state, and when one is emitted it can be considered to jump from the zero state to one in which it is physically in evidence, so that it appears to have been created.

(1927b)

All that had been done in the first step had nothing to do with the quantization of the field because, first, there was no real field at all (any realist interpretation of the many-particle Schrödinger wave function encountered serious difficulties that were mentioned in section 7.1), and second, there was no quantization of the wave field or second quantization properly defined.

The reason for the last statement is simple but often neglected by physicists and historians of quantum physics. Remember that a severe restriction to which QC should be subjected is that QC is meaningful only when the dynamical variables involved in it are the observables represented by Hermitian operators. Then Dirac's assumption that one can take the non-Hermitian operators b and ihb^*, representing probability field amplitudes, as canonical q numbers and can subject them to QC: $[b_r, ihb_s^*] = ih\delta_{rs}$ or $[b_r, b_s^*] = \delta_{rs}$, is obviously misleading. In fact, what is misleadingly called 'second quantization' is nothing else than an equivalent formulation of QC for the mechanical motion of particles, with the help of convenient 'creation' and 'annihilation' operators $a^*(b^*)$ and $a(b)$ in the particle number representation. In particular, the commutation relations satisfied by 'creation' and 'annihilation' operators have nothing to do with the quantization of the probability field since no Planck constant is involved ($[b_r, b_s^*] = \delta_{rs}$). They are just algebraic properties of these operators.

Now let me turn to Dirac's second step. Only in this step did he deal with the quantization of fields. The starting point was the classical radiation field. He resolved the radiation field into its Fourier components, considered the energy E_r and phase θ_r of each of the components to form a pair of canonically conjugate dynamical variables describing the field, and subjected them to the standard QC: $[\theta_r, E_s] = i\hbar\delta_{rs}$.[7] Here the radiation field was indeed quantized. The question then is, in what sense, the first one or the second, was the radiation field quantized?

The answer is at hand if we note, first, that the quantization of the field was effected by the quantization of the field oscillators (since each of the harmonic components of the field was effectively a simple harmonic oscillator), and second, that the quantization of the field oscillators, being a kind of quantization of oscillators, can only be understood according to the first meaning, namely, the quantization of the energy of the field or the field oscillators.

In this interpretation, Dirac's assertion that the assumption that the field variables E_r and θ_r (being the q numbers) 'immediately give light-quantum properties to the radiation' is justified only when the quanta of energy of the wave field (here, the basic ontology was that of the field, and the energy quanta were only a way of describing the property of the field) were identified with the light quanta described by Schrödinger probability waves (here the basic ontology was that of the particle; the Schrödinger wave, according to Born, was only a way of assigning the probability to the occurrence of the particles with certain properties). Dirac tacitly assumed this identification by taking the number N_r' of light quanta per stationary state in Schrödinger's scheme of the many-particle problem as the number N_r of quanta of energy in the field component r on page 263 of (1927b), where he rederived Einstein's probability coefficients for emission and absorption from his new theory. In this way, he obtained a Hamiltonian that described the interactions of the atom and electromagnetic waves.

By calling attention to the fact that this Hamiltonian was identical to the Hamiltonian for the interaction of the atom with an assembly of light quanta, Dirac claimed that 'the wave point of view is thus consistent with the light-quantum point of view'. This claim of consistency is so impressive that it has been taken to be a proof, by many physicists, including Dirac himself, that 'the riddle of the particle–wave nature of radiation is solved'. (Jost, 1972; see also Dirac, 1983).

Here, however, we should note that:

(i) The quanta of energy as the crests in a wave system are creatable and destroyable. Thus, as an epiphenomenon, they presuppose the existence of the field as the basic ontology, although its number can be used as a parameter to specify the states of field oscillators and the configuration of the field.

(ii) In Dirac's mind, the light quantum was an independent, permanent particle, carrying energy, momentum, and polarization, and being in a certain state specified by these parameters. Their creation or annihilation was merely the appearance of their jump from or into the zero state. Here the particle was the fundamental substance and the probabilistic wave was merely a mental device for calculation.

(iii) Dirac's identification of energy quanta with light quanta was therefore a big conceptual jump, by which the property of the field (its quantized energy) was identified with the existence of a discrete substance (light quanta), and the continuous substance (the field) was confused with the non-real probabilistic wave. No justification for this jump was given.

If we take these remarks seriously, then Dirac's assertion that quantization of the field can effect a transition from a field description to a light quanta description cannot be taken as sound in its ontological aspect. The reason for this statement is that the transition relies heavily on the identification, and this is an assumption as radical and significant as the transition itself.

Dirac's next paper (1927c) shows that in Dirac's scheme, as far as the quantization is concerned, the field to be quantized referred, contrary to the widely prevailing opinion, always to the classical radiation field rather than the Schrödinger wave field, and the quantization, also contrary to the widely prevailing opinion, referred only to the quantization of the classical field rather than the second quantization of the Schrödinger wave. But the question then is why the misleading opinion was so prevalent that it was even accepted by most of the founders of QFT, such as Jordan, Klein, Wigner, Heisenberg, Pauli, and Fermi. For them, the misleading opinion was the underlying idea and starting point in their fundamental papers. This opinion, therefore, had profoundly influenced the conceptual developments of QFT. The reason for this is twofold. First, it came from the special treatment of field quantization in Dirac's original papers. Second, it came from a deeper conceptual confusion involved in the theory of Fermion fields that was associated with the particles satisfying the Fermi–Dirac statistics.

Let me examine the first reason first. The main steps in Dirac's (1927b, c), where he obtained a quantum theory of the radiation field, were as follows: (A_1) start from the classical radiation field; (A_2) regard the field as a dynamical system, that is, treat the energy and phase of each component of the field, E_r and θ'_r (or equivalently the number of energy quanta N_r and the canonically conjugate phase $\theta_r = h\nu\theta'_r$) as dynamical variables describing the field; (A_3) assume that N_r and θ_r are q numbers satisfying QC, $[\theta_r, N_s] = i\hbar\delta_{rs}$. In his original paper, especially in (1927b), however, Dirac mixed this line of reasoning with another one, which also consisted of three main steps: (B_1) start from the Schrödinger wave of the many-light-quanta system $\Psi = \Sigma_r b_r \Psi_r$; ($B_2$) regard the probability amplitude b_r and its conjugate $i\hbar b_r^*$ as canonical q numbers satisfying QC, $[b_r, i\hbar b_s^*] = i\hbar\delta_{rs}$; ($B_3$) relate b_r and b_r^* with another pair of canonical q numbers, N_r (the number of light quanta) and θ_r (the phase of the Schrödinger wave) satisfying QC $[\theta_r, N_s] = \mathrm{i}h\delta_{rs}$, through

$$b_r = (N_r + 1)^{1/2} \exp{(-\mathrm{i}\theta_r/h)} = \exp{(-\mathrm{i}\theta_r/h)}N_r^{1/2},$$

$$b_r^* = N_r^{1/2} \exp{(\mathrm{i}\theta_r/h)} = \exp{(\mathrm{i}\theta_r/h)}(N_r + 1)^{1/2}. \tag{3}$$

The reason for the mix-up seems to be twofold. First, ontologically, Dirac confused the energy quanta in the field ontology and the light quanta in the particle ontology (1927b, p. 263; 1927c, p. 715). He even showed a little inclination to interpret the real electromagnetic field as the de Broglie wave of light quanta when he considered each component of the electromagnetic field as 'being associated with a certain type of light-quanta' (1927c, p. 714). This interpretation of the electromagnetic field in terms of the particle ontology was, of course, consistent with the view of Einstein of 1909, Slater, and Born and Jordan (see section 6.2), and paved the way for the later particle-ontology-inclined interpretation of QFT, which is now dominant among physicists. Second, technically, Dirac cannot obtain a quantitative relation between the amplitude of the vector potential and the number of energy quanta in each component of the field without relying on the light quantum picture and identifying the number of energy quanta with the number of light quanta.

The mix-up gave people a false impression. Jost's opinion was typical when he summarized Dirac's QFT programme as this: 'quantize the vector potential and substitute it into the classical interaction' (Jost, 1972).

The idea of the quantization of the vector potential is closely linked with the idea of 'second quantization' because in both cases the amplitude a_r of the vector potential or the probability amplitude b_r was assumed to be a q number. The assumption that the b_r was a q number originated from Dirac's paper (1927b, c). Dirac, however, did not make any explicit assumption that the a_r was also a q number, although the role the a_r played in his QFT approach was completely parallel to the role the b_r played in his many-particle approach in his (1927b, c). Then, from 1929 on, Fermi in his extremely influential lectures and papers (1929, 1930, and especially 1932) extended Dirac's idea about the b_rs to the a_rs, and made it acceptable to the quantum physics community.

Conceptually, however, all these ideas were confused. First, the amplitude a_r or b_r cannot be regarded as a q number in its original meaning, because it is not an observable and cannot be represented by a Hermitian operator. In fact, $a^*(b^*)$ or $a(b)$ can be interpreted as the creation or annihilation operator in an oscillator model, where it increases or reduces the excitation of one of the oscillators by one quantum. Thus the amplitudes have no real eigenstates and eigenvalues. That is, they cannot be represented by Hermitian operators.

Second, b_r (or a_r) is not a c number owing to the relation $[b_r, b_s^*] = \delta_{rs}$. But this has nothing to do with quantization.

In order to see this clearly, let me underline the following fact. In a quantum theory of radiation, N_r, the number of the light quanta, and θ_r, the phase of the wave field, as the observables must be q numbers. If this is the case, then the fact that b_r is not a c number follows from the transformation equation (3). But the converse is not true. That N_r and θ_r are q numbers will not follow automatically from the non-commutative relation $[b_r, b_s^*] = \delta_{rs}$. The statement is justified by the following observation: the second equations in both lines of equation (3) cannot be obtained from the fact that b_r and b_r^* are not c numbers but have to be derived from a special assumption that N_r and θ_r are canonical q numbers (Dirac, 1927b). In short, the fact that the b_r and b_r^* (or a_r and a_r^*) are not c numbers is inherited from the fact that N_r and θ_r are q numbers. It is a necessary consequence of, but not a sufficient condition for, the quantum nature of the theory. A classical radiation theory cannot be quantized by only taking the amplitude as a non-c-number. On the other hand, in a quantum theory, the amplitude can always be represented by a quantity which is not a c number.

Thus the idea of 'second quantization' is inadequate for two reasons. First, in the quantum theory, it is not an additional quantization, but a formal trans- formation. And second, it cannot serve as a means for quantizing a classical theory.

Now let me turn to the second reason why the misleading opinion that Dirac's starting point for QFT was the Schrödinger wave is so prevalent. As we have noted above, Dirac, like all other founders of QFT, interpreted the Schrödinger wave as a probability wave. Ontologically, the probability inter- pretation of Schrödinger's wave function is inconsistent with field theory because in the former, the basic ontology is the discrete particle, and in the latter, the basic ontology is the continuous field. In spite of this apparently serious difficulty, physicists still believe that the starting point for QFT must be Schrödinger's wave function, which has to be taken as a kind of classical field and subjected to 'second quantization'. In fact, such a conceptual trick of 'quantal–classical–(second) quantal' has become the accepted conceptual foundation of QFT. The reason for this, I think, is certainly deeper than the confusion of physicists, including Dirac himself, and is closely linked with the peculiarity of the fermion field. For this reason, let us have a closer look at this intriguing point.

What Dirac treated in his theory of radiation was a many-particle problem. According to the newly emerged wave mechanics, this problem was described by the Schrödinger wave function in higher-dimensional configuration space.

Starting from such a configurational wave, Dirac made a transformation of representation and obtained a representation of particle number, in which no configuration space other than ordinary three-dimensional space was needed. This method, which has an inappropriate name, 'second quantization', was based totally on the particle ontology and had nothing to do with field theory. To the many-particle problem Dirac also developed another approach, namely, QFT. His QFT was apparently similar to, but ontologically different from, the method of second quantization, because its starting point was the classical field in ordinary three-dimensional space, rather than the Schrödinger wave in configuration space. It is true that Dirac's QFT also presupposed the existence of discrete particles. But here the particles were considered as the quanta of the continuous field. That is, the particles in QFT were reduced in status to mere epiphenomena: they were manifestations of the underlying substance, the field, and thus could be created and destroyed, and hence were different from the permanently existing field.

It was strange and difficult for physicists to consider the configurational wave as something real. Most physicists detested it and attempted to use it only as a computational device, rather than a representation of physical reality. Born's probability interpretation and Dirac's 'second quantization' can be viewed as good examples of the attempt to leave configurational space for ordinary space. Dirac's QFT provided another example of attacking the many-particle problem without involving configuration space, which was one of the central tasks of the time.

The next step taken by Jordan and Wigner (1928) was to generalize Dirac's idea for the case of bosons (light quanta) to the case of fermions. For Jordan, the QFT approach was particularly attractive because this was essentially what he introduced into quantum physics in his paper (with Born and Heisenberg, 1926): applying the quantum hypothesis to the (Maxwell) field itself. One of the main tasks for the Jordan–Wigner generalization was a quantization of a classical field whose quanta were fermions. Here we come to the crucial point pertinent to our discussion. There was simply no such a classical field at all in nature. Of course, there were also no such classical particles as fermions. But the issue here, in the framework of QFT, was how to derive fermions from some assumed ontology. Faced with such a situation, the only thing Jordan and Wigner could choose as the candidate for the starting point of their QFT was the three-dimensional Schrödinger field of a single fermion.

Proceeding from such a 'classical field', a QFT formulation of the many-fermion problem can be obtained by a procedure formally parallel to Dirac's in the case of bosons. The only difference lay in the definition of QC, which was necessary to satisfy the Pauli exclusion principle.[8] In this formulation, the creation and annihilation operators (i.e., a, a^* or b, b^*) were introduced to increase or decrease the number of particles in a certain quantum state. The amplitudes of the field were expressed as linear combinations of these operators. This was a direct generalization of the quantization of the electromagnetic field as decomposed into oscillator amplitudes. The operator of an oscillator amplitude either created or destroyed a quantum of the oscillator. Thus the fermions were considered as the quanta of a field just as light quanta were the quanta of the electromagnetic field. Such quanta could be created or destroyed and thus were not eternal or permanent.

Looking back on his work co-authored with Jordan, Wigner said in an interview, 'just as we get photons by quantizing the electromagnetic fields, so we should be able to get material particles by quantizing the Schrödinger field' (Wigner, 1963) A big question can be raised about this underlying idea. It is thinkable that the photons are excited from the electromagnetic field, since the latter is substantial (namely, with energy and momentum). But what about the Schrödinger field? Is it also a substantial field, or merely a probability field? Leon Rosenfeld tells us, 'in some sense or other, Jordan himself took the wave function, the probability amplitude, physically more seriously than most people' (Rosenfeld, 1963). But even Jordan himself did not provide a consistent exposition in justifying his use of the probability amplitude as a classical field, which was usually to be conceived as substantial and possessing energy and momentum. In my opinion, the replacement of the classical field by the three-dimensional Schrödinger field of a single fermion initiated a radical change in the conception of the world.

First, the realist interpretation has replaced the probability interpretation of the Schrödinger wave function of a single fermion. That is, it began to be considered as a substantial wave rather than merely a probability (or guiding) wave. The reason for the replacement is simple: otherwise, the wave would be tied up with only a single particle, and have nothing to do with the many-particle problem. An even more serious reason for this is that a non-substantial probability field could not confer substantiality on its quanta, so the particles, as the quanta of the field, could not obtain their substantiality from the field. Then the substantiality of the particles would have no source and becomes mysterious. A definite fact in the history of 20th century field theories is that in QFT the Schrödinger wave of a single fermion has always been regarded as a substantial (material) wave field since the Jordan–Wigner paper was published in 1928. After the replacement, the Schrödinger equation, originally describing the motion of a single fermion, became a field equation. Only then could the procedure parallel to the quantization of the electromagnetic field be justly adopted, and the many-fermion problem solved by using the creation and destruction operators, that is, a field theory of fermions could be developed.

Second, the field ontology has replaced the particle ontology. The material particle (fermion) is no longer to be regarded as an eternal existence or independent being, but rather an excitation of the (fermion) field, a quantum of the field.

Judging from the orthodox interpretation of quantum mechanics, the realist interpretation of Schrödinger waves in QFT cannot be justified as a logically consistent development of quantum mechanics. It is at best a makeshift, demanded by a formal analogy between matter (fermions) and light (bosons). The justification, however, actually lies in the subsequent success it has brought about. Thus the replacement of a probability interpretation by a realist one, or, going back one step further, the replacement of the classical field by the Schrödinger wave marked the introduction of a new kind of substantial field, and the commencement of a new stage of theoretical development.

Once physicists got used to taking the Schrödinger wave of a single fermion as a new kind of substantial wave, as the starting point of the field theory for fermions, it was natural but misleading to assume that the starting point of the field theory for bosons was also a kind of Schrödinger wave (namely the Schrödinger wave of a single boson), rather than the classical (Maxwell) field, so that both cases could be put into a single conceptual framework of 'second quantization'.

The quantization of a field converts the field variables into operators which satisfy certain algebraic relations, $[b_r, b_s^*] = \delta_{rs}$, and so on[9], so the field amplitudes, expressed as linear combinations of these operators, also become operators and satisfy algebraic relations of the same kind. In Dirac's theory (1927b, c) and in Jordan and Klein's (1927) and Jordan and Wigner's (1928) theory, the operators b_r and b_r^* merely serve to cause transitions of a particle from one quantum state to another. However, once these operators are defined, they permit a natural generalization to the case in which particles can actually be created and annihilated, namely, when the total number of particles is no longer a constant of the motion (see section 7.4). This is exactly the situation that is observed experimentally and theorized in the quantized relativistic theories of interacting fields. For example, the light quanta can be emitted and absorbed in the electromagnetic field; so can other bosons, e.g., π mesons. In the case of electrons, we know that electron–positron pairs can be created and annihilated. All these processes can be conveniently and precisely described by the creation and annihilation operators of field theories. These operators also serve to describe concisely and accurately the fundamental interaction between fermions and bosons, which consists of the product of two fermion creation and/or destruction operators and one boson operator. This is interpreted as a change of state of a fermion, destroyed in one state and created in another, accompanied by either the emission or absorption of a boson. More about interaction will be given in section 7.5 and part III.

As a field, the quantum field, like the classical field, must be a continuous plenum. What is treated in QFT, however, is the so-called local field. That is, the field variables are defined (and describe the physical conditions) only at a point of spacetime. The local field, as we have already noted above, is not a c number, but rather an operator, the local field operator. When QFT was first invented, the operator fields had clear and immediate physical interpretations, in terms of the emission and absorption (creation and annihilation) of physical particles. After Dirac introduced his idea of the vacuum (1930a), however, the operator fields became abstract dynamical variables, with the aid of which one constructs the physical states. They in themselves do not have any elementary physical interpretation. More about Dirac's vacuum will be given in section 7.4.

Dirac claimed to have proved, with the help of his QFT (in which particles are treated as the field excitations), that there is a 'complete harmony' or 'complete reconciliation' between the wave and the light-quanta points of view. The claim has been considered valid ever since by the scientific community, including most physicists, historians, and philosophers of physics. For example, Jost asserts that 'the riddle of the particle–wave nature of radiation, which has so strongly motivated theoretical physics since 1900, is solved' by Dirac's 1927 papers (Jost, 1972). This claim, however, was challenged by Michael Redhead (1983), whose arguments can be summarized like this. The representation in which the field amplitude is diagonal can be called the field representation, and likewise the representation in which the number operators are all diagonal can be termed the particle representation. However, in this latter representation, the field amplitude is not sharp because the number operators do not commute with the creation and annihilation operators. Thus one cannot simultaneously ascribe sharp values to both the particle number and the field amplitude, and hence the wave–particle duality is manifested again, albeit in a different form.

Dirac's claim is also challenged in the above discussion, by distinguishing the energy quanta and light quanta, the quantization of motion of the field and that of the substantial field. The latter challenge can be met by arguing that the field quanta excited by the field operators are not the energy quanta, or quantized energy, but rather the quantized carriers of the energy, momentum, charge, and all other properties of the field. In this case, the distinction between the field quanta and Einstein's light quanta (or other bosons or fermions) disappears, and the field system can be described in terms of particles or the field quanta. This is true. But this was not Dirac's idea. What was excited by the field operator in Dirac's 1927 paper was the quantized energy of the field (the energy quanta), but not the quantized carriers of the field energy. Ontologically, therefore, the identification of the quantized energy (momentum, charge, etc.) of the field with the quantized carrier of the field energy (momentum, charge, etc.) is still a big question.

It should be mentioned in passing that the energy and momentum of the electromagnetic field do not in general form a four-vector. They are parts of a four-tensor, behaving quite differently from the corresponding quantities of a particle. Thus, in terms of the Lorentz transformation of its energy and momentum, the field, in general, does not have the properties of a particle. As Fermi showed, however, the field can be separated into a transverse part (light wave) and a longitudinal part (static Coulomb field). The former, in its behavior under a Lorentz transformation, has a certain similarity to a particle since its energy and momentum form a four-vector. For the latter, this is not the case. In Fermi's quantum theory (1932), only the former was subjected to quantization, giving rise to the existence of the light quanta, which, in some other aspects, also behaved like particles; the latter remained unquantized.

It is true that quantizing a field makes manifest the particle aspect of the field. But QFT is certainly more complicated and richer than what is apparent from its particle interpretation. The existence of so-called zero point fluctuations hints that even in the vacuum (i.e., in the state where there are no particles or quanta of the field present) the field exists. The fluctuating vacuum field actually gives rise to observable effects. For example, we may think of the spontaneous emission of radiation as forced emission taking place under the influence of the vacuum fluctuations. Another effect is the fluctuations in the position of a charge in space, which gives rise to a part of the self-energy of the particle (Weisskopf, 1939; see also section 7.6). The decrease in the binding energy of a bound electron due to these fluctuations in position accounts for most of the level shift in the Lamb effect (Welton, 1948; Weisskopf, 1949). In the case of the fermion fields, there are also fluctuations of charge and current density in some finite volume of spacetime when the number of particles is definite.

Dirac did not solve the riddle of the wave–particle duality of the field, since he failed to find a bridge connecting the continuous field and the discrete particle. In fact, he merely reduced the field to the particles by his formulation of quantizing the field and his peculiar interpretation of the formulation. The most peculiar point of his interpretation lay in his conception of the light quanta in the zero state, which showed that for him the light quantum was a kind of real particle with permanent existence (he took the creation and destruction of the light quanta to be merely the appearance of the real processes of jumping from and to the zero state). In short, for Dirac, the quantization of the field meant that the field was composed of the discrete and permanent particles or quanta. This was the basis on which the particle-ontology-based interpretation was developed.

The crucial step in solving the riddle, in fact, was taken by Jordan. Jordan's mathematical formulation was the same as Dirac's. What made them different were the ontological commitments underlying the interpretations of the formulation. Dirac started from the real continuous electromagnetic field, but reduced it to a collection of discrete particles. In contrast, the starting point of Jordan's work was the Schrödinger wave. This is true not only of his 1928 paper dealing with the fermion field, but also of his 1927 paper, where the electromagnetic field was also considered as the Schrödinger wave function of a single light quantum.

We have already noticed above that Jordan took the physical reality of the Schrödinger wave very seriously. This position was closely linked to his complementary understanding of the nature of quantum-physical reality, and of wave–particle duality. According to Jordan,

Quantum physical reality is simpler in a marked way than the system of ideas through which classical theories sought to represent it. In the classical system of representation, wavelike and corpuscular radiation are two fundamentally different things; in reality, however, there is instead only one single type of radiation, and both classical representations give only a partially correct picture of it.

(1928)

That is to say, both the wavelike and corpuscular aspects of the radiation, as different special manifestations of the same basic substance, were taken to be real. Here is the bridge connecting the continuous field and the discrete particle: radiation as the quantum physical reality is both waves and particles.

In classical theory, particles, owing to their permanent existence, were fundamentally different from the energy quanta of the field, which were capable of being created and absorbed. In quantum theory, according to Jordan, considering that the reality was both the wave and particle, the particles were also capable of being created or absorbed, just like the quanta of the wave field; and the fields were also capable of exhibiting their discrete existence. In this way, the riddle of wave–particle duality seems to have been solved by Jordan, rather than by Dirac.

It is interesting to note that the basic ontology on which the riddle is solved is something novel, different from both the classical particle and field. It cannot be taken as the classical particle because the quanta sprung from it possess no permanent existence and individuality. It also cannot be taken as the classical field, because, as the quantized field, it has lost its continuity. A new ontology, which Redhead (1983) calls ephemeral, has emerged, from which QFT has subsequently developed.

7.4 The vacuum

The operator fields and field quantization, especially Jordan–Wigner quantization, had their direct physical interpretation only in terms of the vacuum state. In order to see this clearly, let us examine the physicists' conception of the vacuum before and after the introduction of field quantization in the late 1920s.

After Einstein had put an end to the concept of the ether, the field-free and matter-free vacuum was considered to be truly empty space. The situation, however, had changed with the introduction of quantum mechanics. From then onwards, the vacuum became populated again. In quantum mechanics, the uncertainty relation for the number N of light quanta and the phase θ of the field amplitude, $\Delta N \Delta \theta \geq 1$, means that if N has a given value zero, then the field will show certain fluctuations about its average value, which is equal to zero.[10]

The next step in populating the vacuum was taken by Dirac. In his relativistic theory of the electron (1928a, b), Dirac met with a serious difficulty, namely the existence of states of negative kinetic energy. As a possible solution to the difficulty, he proposed a new conception of the vacuum: 'all of the states of negative energy have already been occupied', and all of the states of positive energy are not occupied (Dirac, 1930a). Then the transition of an electron from the positive energy state to the negative energy state could not happen owing to Pauli's exclusion principle. This vacuum state was not an empty state, but rather a filled sea of negative energy electrons. The sea as a universal background was unobservable, yet a hole in the negative energy sea was observable, behaving like a particle of positive energy and positive charge. Dirac tried first to identify the hole with a proton. This interpretation soon proved untenable. In May 1931, Dirac accepted the criticism of Oppenheimer (1930b) and Weyl (1931) and took the hole as a new kind of particle with a positive charge and the mass of the electron (Dirac, 1931).

It is easy to see that Dirac's infinite sea of unobservable negative energy electrons of 1930–31 was analogous to his infinite set of unobservable zero energy photons of 1927. Such a vacuum consisting of unobservable particles allowed for a kind of creation and annihilation of particles. Given enough energy, a negative energy electron can be lifted up into a positive energy state, corresponding to the creation of a positron (the hole in the negative energy sea) and an ordinary electron. And of course the reverse annihilation process can also occur.

One might say, as Steven Weinberg (1977) did, that all these processes can be accounted for by the concept of the transition between observable and unobservable states, without introducing the ideas of QFT. This is true. But on the other hand, it is precisely the concepts of the vacuum and of the transition that have provided QFT with an ontological basis. In fact, the introduction of the negative energy sea, as a solution to the difficulty posed by Dirac's relativistic theory of the electron, implied that no consistent relativistic theory of a single electron would be possible without involving an infinite-particle system, and that a QFT description was needed for the relativistic problems (more discussion follows below). Besides, the concept of transition between observable and unobservable states did indeed provide a prototype of the idea of excitation in QFT, and hence gave a direct physical interpretation to the field operators.

It was noticed that the concepts of creation and destruction predated quantum mechanics and could be traced back to the turn of the century, when these concepts were based on the ether model of matter and light (Bromberg, 1976). In fact, Dirac's idea of the vacuum as a kind of substratum shared some of its characteristic features with the ether model, which explains why he returned to the idea of an ether in his later years,[11] although in his original treatment of creation and destruction processes he did not explicitly appeal to an ether model.

One of the striking features of Dirac's vacuum, similar to that of the ether, is that it behaves like a polarizable medium. An external electromagnetic field distorts the single electron wave function of the negative energy sea, and thereby produces a charge-current distribution acting to oppose the inducing field. As a consequence, the charges of particles would appear to be reduced. That is, the vacuum could be polarized by the electromagnetic field. The fluctuating densities of charge and current (which occur even in the electro-free vacuum state) can be seen as the electron–positron field counterpart of the fluctuations in the electromagnetic field.

In sum, the situation after the introduction of the filled vacuum is this. Suppose we begin with an electron–positron field Ψ. It will create an accompanying electromagnetic field which reacts on the initial field Ψ and alters it. Similarly, an electromagnetic field will excite the electron–positron field Ψ, and the associated electric current acts and alters the initial electromagnetic field. So the electromagnetic field and the electron–positron field are intimately connected, neither of them having a physical meaning independent of the other. What we have then is a coupled system consisting of electromagnetic field and electron–positron field, and the description of one physical particle is not to be written down a priori but emerges only after solving a complicated dynamical problem.

The idea that the vacuum was actually the scene of wild activities (fluctuations), in which infinite negative energy electrons existed, was mitigated later on by eliminating the notion of the actual presence of these electrons. It was recognized by Oppenheimer and Wendell Furry (1934) that the filled-vacuum assumption could be abandoned, without any fundamental change of Dirac's equations, by interchanging consistently the roles of creation and destruction of those operators that act on the negative states. In this way electrons and positrons entered into the formalism symmetrically, as two alternative states of a single particle, and the infinite charge density and negative energy density of the vacuum disappeared. With this formalism the vacuum became again a physically reasonable state with no particles in evidence. After all, in a relativistic theory the vacuum must have vanishing energy and momentum, otherwise the Lorentz invariance of the theory cannot be maintained.

The same method of exchanging the creation and destruction operators for negative states was used in the same year by Pauli and Weisskopf (1934) in their work on the quantization of the Klein–Gordon relativistic wave equations for scalar particles. The quantum theory of the scalar field contained all the advantages of the hole theory (particles and antiparticles, and pair creation and annihilation processes, etc.) without introducing a vacuum full of particles.

In sum, the above pair of papers showed that QFT could naturally incorporate the idea of antimatter without introducing any unobservable particles of negative energy. It could thus describe, satisfactorily, the creation and annihilation of particles and antiparticles, which were now seen as co-equal quanta of the field.

For most physicists, these developments, especially the publication of Wentzel's influential book *Quantum Theory of Fields* (1943), had settled the matter: the picture of an infinite sea of negative energy electrons had been seen as a historical curiosity and forgotten (Schwinger, 1973b; Weinberg, 1977; Weisskopf, 1983). Yet there are still some others who maintain that Dirac's idea of the vacuum is of revolutionary meaning, and that its essence of regarding the vacuum as not empty but substance-filled remains in our present conception of the vacuum.[12] They argue that the conception of the substance-filled vacuum is strongly supported by the fact that the fluctuations of matter density in the vacuum remain even after the removal of the negative energy electrons, as an additional property of the vacuum besides the electromagnetic vacuum fluctuations.

Here we run into the most profound ontological dilemma in QFT. On the one hand, according to special relativity, the vacuum must be a Lorentz invariant state of zero energy, zero momentum, zero angular momentum, zero charge, zero whatever, that is, a state of nothingness. Considering that energy and momentum have been thought to be the essential properties of substance in modern physics and modern metaphysics, the vacuum definitely cannot be regarded as a substance. On the other hand, the fluctuations existing in the vacuum strongly indicated that the vacuum must be something substantial, certainly not empty. A possible way out of the dilemma might be to redefine 'substance' and deprive energy and momentum of being the defining properties of a substance. But this would be too *ad hoc*, and could not find support from other instances. Another possibility is to take the vacuum as a kind of pre-substance, an underlying substratum having a potential substantiality. It can be excited to become substance by energy and momentum, and become physical reality if some other properties are also injected into it.

In any case, no one would deny that the vacuum is the state in which no real particle exists. The real particles come into existence only when we disturb the vacuum, when we excite it with energy and other properties. What shall we say about the local field operator whose direct physical meaning was thought to be the excitation of a physical particle? First of all, the localized excitation described by a local field operator $O(x)$ acting on the vacuum means the injection of energy, momentum, and other special properties into the vacuum at a spacetime point. It also means, owing to the uncertainty relations, that arbitrary amounts of energy and momentum are available for various physical processes. Evidently, the physical realization of these properties symbolized by $O(x)|\text{vac}\rangle$ will not only be a single particle state, but must be a superposition of all appropriate multi-particle states. For example, $\Psi_{\text{el}}(x)|\text{vac}\rangle = a|1 \text{ electron}\rangle + \Sigma a'|1 \text{ electron} + 1 \text{ photon}\rangle + \Sigma a''|1 \text{ electron} + 1 \text{ positron} + 1 \text{ electron}\rangle + \ldots,$ where $\Psi_{\text{el}}(x)$ is the so-called dressed field operator, and a^2 is the relative probability for the formation of a single bare particle state by the excitation $\Psi_{\text{el}}(x)$, and so on.

As a result, the field operators no longer refer to the physical particles and become abstract dynamical variables, with the aid of which one constructs the physical state. Then how can we pass from the underlying dynamical variables (local field operators), with which the theory begins, to the observable particles? This is a task that is fulfilled temporarily only with the aid of the renormalization procedure. There will be more discussion on renormalization in sections 7.6 and 8.8.

7.5 Interaction

The mechanism of interaction according to which particles act on one another across space is so fundamental a question in physics that the whole field theory programme can be viewed as a response to it. In contrast to Newton's formulation of the law of gravity, in which material bodies could act instantaneously on each other at any distance without intermediary, Faraday and Maxwell developed, in the realm of electromagnetic phenomena, the idea of the field of force, involving the intervention of a medium through which the force between the interacting bodies is transmitted by successive and continuous propagation. But such an idea of the force field ran into two fatal difficulties, one physical, the other conceptual. Physically, one could not find a convincing basis for the idea of a medium in the form of a mechanical model without having consequences that are contradicted by optical evidence. Conceptually, the idea leads to an infinite regress: each force field needs a medium for its propagation between the particles it couples, and the constituting particles of this medium have to be coupled by some other field, which requires some other medium for its propagation, and so on and so forth.

Hertz attempted to circumvent this conceptual difficulty by describing the coupling between particles as a convection process effected by an unknown 'middle term':

The motion of the first body determines a force, and this force then determines the motion of the second body. In this way force can with equal justice be regarded as being always a cause of motion, and at the same time a consequence of motion. Strictly speaking, it is a middle term conceived only between two motions.

(1894)

Hertz's idea is important in the history of field theories because, first, it extricates interaction from a universal medium, and, second, it allows a conveyor of interaction to exist. But the big question remains what the mysterious 'middle term' is. Decisive progress in answering this question was made by Einstein, who thought the electromagnetic field discovered by Hertz was the primary entity and regarded it as the agent for transmitting interaction between particles. It is worth noting that Einstein's idea of the electromagnetic field was more general than Hertz's idea of a 'middle term', because it could exist independently, while the existence of Hertz's 'middle term' depended on the existence of the interacting particles.

Dirac, in developing QFT, followed Hertz's and Einstein's ideas, summarized the view of interaction in relativistic classical electrodynamics as 'the idea of each particle emitting waves traveling outward with a finite velocity and influencing the other particle in passing over them', and claimed that 'we must find a way of taking over this new (relativistic) information into the quantum theory' (Dirac, 1932).

In fact, the classical idea, that charged particles were connected with each other via the electromagnetic field emitted and absorbed by the interacting particles, was indeed taken over, as a conceptual model, into the conception of interaction in QFT. The difference between the classical and quantum views lies in two points. First, according to the quantum view, the interaction is transmitted by the discrete quanta of the electromagnetic field, which 'continuously' pass from one particle to the other,[13] but not by a continuous electromagnetic field as such. So the classical distinction between the discrete particle and the continuous force field is dimmed in quantum theory. Second, in the quantum view, the interacting particles must be considered the quanta of a fermion field, just as photons are the quanta of the electromagnetic field. All such quanta, as distinct from classical permanent particles, can be created and destroyed. So this is a field theory not only in the sense that the agent for transmitting interaction between particles is a field, but also in the sense that the interacting particles themselves should be viewed as a manifestation of the field, the fermion field. Such a fermion field can transmit interaction via its quanta just like the electromagnetic field via photon. Thus the distinction between matter and force field vanishes from the scene, and is to be replaced by a universal particle–field duality affecting equally each of the constituent entities.

Quantum electrodynamics, based on the works of Jordan (with Born, 1925, with Born and Heisenberg, 1926), Dirac (1927b, c; 1928a, b; 1930a; 1931), Jordan and Wigner (1928), and Heisenberg and Pauli (1929; 1930)[14], was a starting point from which the quantum theories of interactions have developed along two lines. One led to Fermi's theory of beta decay, the prototype of the quantum theory of weak interactions, and to Yukawa's theory of mesons, the prototype of the quantum theory of strong interactions. The other led to gauge theories. The rest of this section is devoted to the first development, while the second line will be dealt with in part III.

The early development of the quantum theory of interactions was closely intertwined with nuclear physics. In fact, nuclear beta decay and the nuclear force that stabilizes the nucleus were two topics which provided the main stimulus to the development of novel ideas of interaction.

Before James Chadwick's discovery of the neutron in 1932, which is justifiably viewed as a watershed in the history of nuclear physics, the accepted theory of nuclear structure was the electron–proton model of the nucleus, according to which the nucleus contains, in addition to protons, electrons, so that the extra charges in the nucleus can be compensated and beta decay can be accounted for. The model suffered a serious spin-statistics difficulty from Franco Rasetti's measurement of the Raman spectrum of N_2 (1929), which implies that the nitrogen nucleus consisting of 21 spin-1/2 particles (14 protons and 7 electrons) obeys Bose statistics. For, as Walter Heitler and Gerhard Herzberg pointed out (1929), Rasetti's analysis means that if both the electron–proton model and Wigner's rule[15] were correct, then the nuclear electron together with its spin would lose its ability to determine the statistics of the nucleus.

As a response to this difficulty, the Russian physicists V. Ambarzumian and D. Iwanenko (1930) suggested that the electrons inside nuclei 'lose their individuality' just as photons do in atoms, and that beta emission is analogous to the emission of photons by atoms. The novelty of this idea lay in the possibility that electrons can be created and annihilated, and hence the total number of electrons is not necessarily a constant.[16] It should be noted that the idea is novel because it is different from Dirac's idea of the creation and destruction of an electron–positron pair based on his hole theory, which are simply quantum jumps of an electron between a state of negative energy and a state with positive energy, with the total number of electrons being conserved. It is also different from Jordan and Wigner's work (1928), where the creation and destruction operators, in dealing with the non-relativistic many-particle problem, are only the mathematical device for describing the quantum jumps between different quantum states, without changing the total number of electrons. This novel idea had anticipated by three years an important feature of Fermi's theory of beta decay. In fact, the idea was explicitly used by Fermi (1933, 1934) as the first of the three main assumptions on which his beta decay theory was based.

Chadwick's discovery of the neutron (1932) gave a death blow to the electron–proton model and opened the way to the neutron–proton model of nuclear structure. The first physicist who treated the neutron as a spin-1/2 elementary particle and proposed the neutron–proton model was again Iwanenko (1932a, b). At nearly the same time, Heisenberg (1932a, b; 1933) proposed the same model, referring to Iwanenko, and noted (1932a) that the neutron as an elementary particle would resolve the 'spin and statistics' difficulty. He also suggested that the neutron and proton may be treated as two internal states of the same heavy particle, described by a configurational wave function where an intrinsic coordinate ρ of the heavy particle must be included. The intrinsic coordinate ρ can assume only two values: $+1$ for the neutron state and -1 for the proton state. The introduction of the ρ coordinate indicated the non-relativistic nature of Heisenberg's model, which had nothing to do with QFT. It did, however, anticipate by six years the concept of isospin, an internal (non-spatio-temporal) degree of freedom, which was going to be introduced by Nicholas Kemmer in (1938) and would play an important role in subsequent developments, both in the theory of nuclear force and in gauge theory.[17]

Heisenberg, however, paradoxically tried to incorporate, within the above model, a concept of the neutron that is pictured as a tightly bound compound of proton and electron, in which the electron loses most of its properties, notably its spin and fermion character. He believed that, as a complex particle, the neutron would decay inside the nucleus into a proton and an electron. And this would explain beta decay. More important in Heisenberg's idea of the compound model of neutron was his belief that the neutron would be bound to the proton by exchanging its constituent electron, analogous to the exchange of the electron in the chemical bonding of the H_2^+ ion. Thus the nuclear force was conceived as consisting of the exchange of the 'pseudo'-electron.

The basic idea of Heisenberg's exchange force was taken over by Ettore Majorana (1933), though the chemical bonding analogy was discarded. Majorana started from the observed properties of nuclei, especially the saturation of nuclear forces and the particular stability of the α particle, and realized that he could achieve the saturation by an exchange force different from Heisenberg's. He rejected Heisenberg's picture (in which neutron and proton are located at two different positions and exchange their charges but not their spins), eliminated Heisenberg's ρ coordinate, and involved the exchange of spins as well as charges. In this way, Majorana found that the force was saturated at the α particle, while Heisenberg's model would have the force wrongly saturated at the deuteron.

Another important development obtained its stimulus more directly from the study of beta decay. The discovery by Chadwick (1914) of β rays with a continuous energy spectrum allowed but two possible interpretations. One advocated by Bohr (1930) said that the conservation of energy is valid only statistically for the interaction that gives rise to the beta radioactivity. The other was proposed by Pauli:

There could exist in the nuclei electrically neutral particles that I wish to call neutrons,[18] which have spin 1/2 and obey the exclusion principle . . .. The continuous β-spectrum would become understandable by the assumption that in β decay a neutron is emitted together with the electron, in such a way that the sum of the energies of neutron and electron is constant (1930), [or] equal to the energy which corresponds to the upper limit of the beta spectrum.

(1933)

In Pauli's original proposal, the neutrino was thought to be constituent of the nucleus, pre-existing in the nucleus prior to the emission process. The idea that the neutrino could be created at the moment of its emission with the electron was clearly proposed for the first time by Francis Perring (1933). Perring's idea, together with the idea of Iwanenko and Ambarzumian, was absorbed by Fermi in forming his theory of beta decay, which as a synthesis of prior development is in many ways still the standard theory of interaction.

First, the idea that the process of interaction is not a process of transition, but that of creation or destruction of particles, was clearly expressed in Fermi's theory:

The electrons do not exist as such in the nucleus before the β emission occurs, but acquire their existence at the very moment when they are emitted; in the same manner as a quantum of light, emitted by an atom in a quantum jump ... then, the total number of the electrons and of the neutrinos (like the total number of light quanta in the theory of radiation) will not be constant, since there might be processes of creation or destruction of those light particles.

(1933)

Second, it was assumed that Dirac and Jordan's method of quantizing the probability amplitude could be used to deal with the change in the number of field quanta (electrons, neutrinos):

the probability amplitudes ψ of the electrons and ϕ of the neutrinos, and their complex conjugate ψ^* and ϕ^*, are considered as non-commutative operators acting on functions of the occupation numbers of the quantum states of the electrons and neutrinos.

(Ibid.)

Fermi's interpretation of Dirac–Jordan quantization and the application of it to field theories soon became the orthodoxy in QFT, and his formulation in field theory notation became the standard formulation, leaving the physical interpretation of the quantization of the probability amplitudes open. This is why Fermi's theory of beta decay was viewed by Weisskopf (1972) as 'the first example of modern field theory'.

Third, Fermi's formulation of the four-line interactions had dominated the theories of weak interactions until the early 1970s, and as a good approximation is still used in the field of low energy weak interactions:

the transformation of a neutron into a proton is necessarily connected with the creation of an electron which is observed as a β particle, and of a neutrino; whereas the inverse transformation of a proton into a neutron is connected with the disappearance of an electron and a neutrino.

(Ibid.)

On the whole, Fermi's theory of weak interactions was based on the analogy with quantum electrodynamics. Not only were the creation and destruction of electrons and neutrinos assumed by analogy with those of the photon, but also the interaction term in the Hamiltonian function was taken in a form analogous to the Coulomb term in quantum electrodynamics (Fermi, 1934). Such an analogy has also been assumed by Hideki Yukawa, Chen Ning Yang and Robert Mills, and by many other physicists who wish to extend the existing theory to a new domain.

However important Fermi's theory of beta decay may be, he did not touch the nuclear force and the quanta of the nuclear field at all. As to the heavy particles (proton and neutron) themselves, in contrast to the light particles (electron and neutrino) that were handled by the method of 'second quantization', they were treated, by Fermi as by Heisenberg, as two internal quantum states of the heavy particles, and described by 'the usual representation in configuration space, where the intrinsic coordinate ρ of the heavy particles must be included' (ibid.).

It is obvious that the heavy particles here were viewed as non-relativistic particles rather than the field quanta. So there would be no creation or destruction process for the heavy particles, but only the transition between two states. In this respect, Fermi's theory seems to be only a transitional one rather than a mature and consistent formulation. For this reason, either it tended to be transformed into a consistent QFT formulation by treating the protons and neutrons as field quanta as well, or one returned in a consistent way to the old idea of the transition, regarding the creation or destruction of the electrons and neutrons as the result or appearance of the quantum jumps. The latter route was taken by the so-called Fermi-field model. According to this model, the simultaneous creation of the electron and neutrino was considered a jump of a light particle from a neutrino state of negative energy to an electron state of positive energy.[19] Here we can see clearly the profound influence of Dirac's idea of the filled vacuum on the subsequent development of QFT.

Fermi's theory of beta decay implies also the possibility of deducing the exchange force between neutrons and protons from the exchange of an electron–neutrino pair. This possibility was exploited by the so-called 'beta theory of nuclear force' or 'the pair theory of nuclear force', developed by Igor Tamm (1934) and Iwanenko (1934). In this theory a proton and a neutron can interact by virtually emitting and re-absorbing an electron–neutrino pair. That is to say, the nuclear force is related to the beta process. But as a second-order effect, this interaction would be very weak. Thus even the proposers of this model correctly doubted that the origin of the nuclear force would lie in beta decay and its reverse.

The pair theory of nuclear force was short-lived but historically important. The reason for this assertion is twofold. First, it proved for the first time in the history of physics that interactions could be mediated by quanta with limited (i.e., non-zero) mass. Second, the theory made it clear that the short-range nature of the nuclear interaction was directly determined by the fact that the quanta conveying the nuclear force possessed non-zero mass. It also gave some clue to the general method of estimating the mass of the quanta of the nuclear force field. For example, Tamm pointed out in (1934) that the exchange energy depended on a decreasing function $I(r)$ of the interaction range r, which was equal to 1 when $r << h/mc$, where m is the mass of the electron conveying the interaction.

Theoretically, knowing that 'the interaction of the neutron and proton is at a distance of the order $r \sim 10^{-13}$ cm', Tamm should have been able to infer that the mass of the particle conveying the interaction would be about 200 m_e (m_e is the mass of the electron) as Yukawa would do later. Technically, such a hindsight is meaningless because the parameter m here was already presupposed to be the mass of the electron, and to replace the electron by another kind of particle conveying the nuclear force would be a big step to take. However, the pair exchange theory of nuclear force had indeed in a sense paved the way for such a crucial step.

The turning point in the development of the theory of nuclear force, or of general interactions, was marked by Yukawa's work 'On the interaction of elementary particles' (1935) or Yukawa's meson theory for short. Yukawa began his research in 1933 with an unsuccessful attempt to explain the nuclear force by the exchange of an electron between the proton and neutron, motivated by Heisenberg's model of the nucleus and the composite neutron. The notorious difficulty inherent in this and all other similar attempts was the non-conservation of spin and statistics. Fortunately, Yoshio Nishina gave Yukawa a good suggestion, namely, that the exchange of a boson could overcome the difficulty.[20]

Soon afterwards, the papers by Fermi, Tamm, and Iwanenko were published, and the idea of the exchange force, of the creation and destruction of particles as the field quanta, and of the nuclear force transmitted by the field quanta with non-zero mass were all available to Yukawa. What was also available to him was the recognition that the nuclear force could not be directly related to the β process. Then, in early October of 1934, Yukawa conceived an ingenious idea that a boson with a mass of about 200 m_e strongly interacting with heavy particles could be the quantum of the nuclear force field. This idea was the basis of his meson theory, and is the basis of the whole modern theory of fundamental interactions.

The novelty of Yukawa's idea lies in the following. First, the nuclear forces are mediated by a field of force, a boson field, or single quanta rather than by pairs. Second, the nuclear force is a different kind of interaction from that of the beta process because 'the interaction between the neutron and the proton will be much larger than in the case of Fermi' (Yukawa, 1935). This idea was the origin of the difference between the strong interactions and weak interactions, and became one of the basic tenets in QFT. Third, the nuclear field, or the 'U field', as a new physical object, is not reducible to known fields. This point differentiated Yukawa's work on the nuclear force from all previous work, and was a step beyond the inhibition of western physical thinking in the 1920s, according to which one should not multiply entities 'unnecessarily'. Since Yukawa's work, the dominant idea has been that each fundamental interaction in nature should be characterized by a mediating boson field.

It is interesting to note that Yukawa also suggested that the boson might be β unstable, and thus the nuclear β decay would be explained by a two-step process mediated by a virtual boson. In this way, Yukawa's boson played the roles of both nuclear force quantum and beta decay intermediary in its virtual states. Although Yukawa's scheme had to be separated into two,[21] such a unifying character of the theory was indeed attractive. Since the late 1950s, Yukawa's three-line formulation of the weak interaction (one line represents the intermediate boson coupling with two fermion lines) has become fashionable again. It is true that in later theories the boson mediating the weak interactions is different in kind from the boson mediating the strong interactions. Yet the basic idea in both cases of formulating the fundamental interactions is the same as what Yukawa had suggested, which in itself was modeled on the electromagnetic interaction. Such a formal similarity seems to suggest, for the majority of the physics community, that there is a deeper unity among the fundamental interactions.

The ideas underlying the concept of interactions in QFT, before the non-Abelian gauge theories came onto the stage, can be summarized as follows. The interactions between field quanta (fermions or bosons) are mediated by the field quanta of another kind, and all kinds of quanta are creatable and destroyable. Since the mediating fields, as local fields, have infinite degrees of freedom, divergences in calculating the interactions are unavoidable. This is a fatal flaw inherent in QFT, deeply rooted in the basic idea of interaction. Thus QFT cannot be considered a consistent theory without resolving this serious difficulty. Here we come up against the problem of renormalization, which plays a crucial role in the conceptual structure of QFT, and will be treated in sections 7.6, 8.8, and 11.4.

7.6 Renormalization

Classical roots for infinities

The divergence difficulty in early QFT is closely related to the infinities appearing in the electromagnetic self-energy of the electron. However, it is not a new problem peculiar to QFT because it had already occurred in classical electrodynamics. According to J. J. Thomson (1881), the energy contained in the field of an spherical charge of radius a is proportional to $e^2/2a$. Thus when the radius of a Lorentzian electron goes to zero, the energy diverges linearly. But if the electron is given a finite radius, then the repulsive Coulomb force within the sphere of the electron makes the configuration unstable. Poincaré's response (1906) to the paradox was the suggestion that there might exist a non-electromagnetic cohesive force inside the electron to balance the Coulomb force, so that the electron would not be unstable. Two elements of the model have exercised great influence upon later generations. First, there is the notion that the mass of the electron has, at least partly, a non-electromagnetic origin. Second, the non-electromagnetic compensative interaction, when combined with the electromagnetic interaction, would lead to the observable mass of the electron. For example, Ernest Stueckelberg (1938), Fritz Bopp (1940), Abraham Pais (1945), Shoichi Sakata (1947), and many others obtained their inspiration from Poincaré's ideas in their study of the problem of the electron's self-energy.

As first pointed out by Fermi in 1922 (cf. Rohrlich, 1973) the equilibrium of the Poincaré electron is not stable against deformations, and this observation elicited another kind of response to the difficulty, first stated by Yakov Frenkel in (1925). Frenkel argued, within the classical framework, that

the inner equilibrium of an extended electron becomes . . . an insoluble puzzle from the point of view of electrodynamics. I hold this puzzle (and the questions related to it) to be a scholastic problem. It has come about by an uncritical application to the elementary parts of matter (electrons) of a principle of division, which when applied to composite systems (atoms, etc.) just led to these very 'smallest particles'. The electrons are not only indivisible physically, but also geometrically. They have no extension in space at all. Inner forces between the elements of an electron do not exist because such elements are not available. The electromagnetic interpretation of the mass is thus eliminated; along with that all those difficulties disappear which are connected with the determination of the exact equation of motion of an electron on the basis of the Lorentz theory.

(Frenkel, 1925)

Frenkel's idea of the point-electron quickly became accepted by physicists and became a conceptual basis of the ideas of local excitations and local interactions in QFT. The idea of looking for the structure of the electron was given up because, as Dirac suggested in (1938), 'the electron is too simple a thing for the question of the laws governing its structure to arise'. It is clear, therefore, that what is hidden in the locality assumption is an acknowledgement of our ignorance of the structure of the electron and that of other elementary entities described by QFT. The justification given for the point model, and for the consequent locality assumption, is that they constitute good approximate representations at the energies available in present experiments, energies that are too low to explore the inner structure of the particles.

By adopting the point model, Frenkel, Dirac, and many other physicists eliminated the 'self-interaction' between the parts of an electron, and thus the stability problem, but they could not eliminate the 'self-interaction' between the point-electron and the electromagnetic field it produces without abandoning Maxwell's theory. The problem Frenkel left open became more acute when QFT came into being.

Persistence of infinities in QFT

Oppenheimer (1930a) noticed that when the coupling between the charged particles and the radiation field was considered in detail, the higher terms of the perturbation treatment always contained infinities, though the first term of the lowest order would cause no trouble. In fact, the difficulty had already appeared in Heisenberg and Pauli's paper (1929) on a general formulation of a relativistic theory of quantized fields. In calculating the self-energy of an electron without considering the negative energy electrons, they found that the divergence difficulty of a point-electron occurring in Lorentz's theory remained in QFT: the integral over the momenta of virtual photons diverges quadratically. The Heisenberg–Pauli method was used by Oppenheimer to investigate the energy levels of an atom, which resulted in a dilemma. To obtain a finite energy for an atom, all self-energy terms must be dropped. But then the theory would not be relativistically invariant. If, on the other hand, the self-energy terms were retained, then the theory would lead to an absurd prediction: the spectral lines would be infinitely displaced from the values computed by Bohr's frequency condition, and 'the difference in the energy for two different states is not in general finite'. Here, all the infinite terms come from the interaction of the electron with its own field. So the conclusion Oppenheimer drew was that the divergence difficulty made QFT out of tune with relativity. The calculation of the self-energy of an isolated point-electron by Ivar Waller also led to a divergent result. Thus it was clear enough by 1930 that the self-energy divergence difficulty still remained in QFT.

Failed attempts in modifying classical theories to be quantized

Dirac, Wentzel and some other physicists thought that this difficulty might be inherited from the classical point-electron theory, which already had the self-energy difficulty. Wentzel (1933a, b; 1934) introduced, in a classical framework, a 'limiting process' into the definition of the Lorentz force. Starting with Dirac's multi-time theory (Dirac, 1932), according to which an individual time coordinate t was assigned to each charged particle, Wentzel proposed that the 'field point' should approach the 'particle point' in the time-like direction. In this way, the self-energy difficulty can be avoided in a classical theory of a point particle interacting with the Maxwell field. Here, the so-called 'approach' is realized by introducing a density factor $\rho_0(k)$, which should be 1 for a point electron. With the aid of $\rho_0(k)$, the expression for the electrostatic self-energy can be written as $U = e^2/\pi \int_0^\infty \rho_0^2(k)\,dk$. Assuming that $\rho_0(k) = 1$ directly, the electrostatic energy would diverge linearly. If one first writes $\rho_0(k)$ as $\cos(k\lambda)$, and assumes that $\lambda = 0$ only after integration, then $\rho_0(k)$ will still be 1 and what one has described is still a point-particle, but then U is obviously equal to zero.

In this formal way, one can avoid the self-energy divergence raised by the Coulomb field. This approach can be easily extended to the relativistic case, but can do nothing with the characteristic divergence of the self-energy associated with the transverse field, which is peculiar to QFT. To deal with this peculiar divergence, Dirac in his (1939) tried to modify the commutation relations of the field by using Wentzel's limiting process. He found, however, in his (1942) that to achieve this goal, it was necessary to introduce a negative energy photon and negative probability, the so-called indefinite metric. Thus, the above-mentioned classical scheme had to be abandoned completely.

The self-energy difficulty in QFT

The problem of self-energy was first studied most thoroughly by Weisskopf (1934, 1939), who had demonstrated that in the framework of QFT based on Dirac's picture of the vacuum, the self-energy of an electron with infinitesimal radius was only logarithmically infinite to the first approximation of the expansion in power of e^2/hc.

The main steps of Weisskopf's analysis are as follows. Compared with classical theories, he found that the quantum theory of a single electron had put the self-energy problem in a different situation for three reasons. First, the radius of the electron must be assumed to be zero since, if there is only one electron present, 'the probability of finding a charge density simultaneously at two different points is zero for every finite distance between the points'. Thus the energy of the electrostatic field is linearly infinite as $W_{st} = e^2/a$. Second, the electron has a spin which produces a magnetic field and an alternating electric field. The energies of the electric and magnetic fields of the spin are bound to be equal. On the other hand, the charge dependent part of the self-energy of an electron, to a first approximation, is given by $(\pi/8)\int(E^2 - H^2)\,dr$. So if the self-energy is expressed in terms of the field energies, the electric and magnetic parts have opposite signs. This means that the contributions of the electric and the magnetic fields of the spin cancel one another. Third, the additional energy which diverges quadratically, arising from the existence of field strength fluctuations in empty space, is given by $W_{fl} = e^2/\pi hcm \lim_{p=\infty} p^2$.

Weisskopf further analyzed the new situation created by Dirac's vacuum picture in detail. He noticed that, first, the Pauli exclusion principle implies a 'repulsive force' between two electrons of equal spin. This 'repulsive force' prevents the two particles from being found closer together than approximately one de Broglie wavelength. As a consequence of this, one will find at the position of the electron a 'hole' in the distribution of the vacuum electrons, which compensates the charge of the electron completely. But at the same time, one can also find around the electron a cloud of higher charge density coming from the displaced electrons. Thus there is a broadening of the charge of the electron over a region of the order h/mc. Weisskopf proved that 'this broadening of the charge distribution is just sufficient to reduce the electrostatic self-energy (from an originally linear) to a logarithmically divergent expression', $W_{st} = mc^2(e^2/\pi hc)\log(p + p_0)/mc$ [here $p_0 = (m^2c^2 + p^2)^{1/2}$].

Second, the exclusion principle also implies that the vacuum electrons, which are found in the neighborhood of the original electron, fluctuate with a phase opposite to the phase of the fluctuations of the original electron. This phase relation, applied to the circular fluctuation of the spin, decreases its total electric field by means of interference, but does not change the magnetic field of the spin, because the latter is produced by circular currents and is not dependent on the phase of the circular motion. So the total electric field energy is reduced by interference if an electron is added to the vacuum. That is, the electric field energy of an electron in QFT is negative. Weisskopf's exact calculation gave $U_{el} = -U_{mag}$. Compared with the corresponding result in one-electron theory ($W_{sp} = U_{el} - U_{mag} = 0$), we find that the contribution of the spin to the self-energy in QFT does not vanish: $W_{sp} = U_{el} - U_{mag} = -2U_{mag} = -1/m(e^2/\pi hc) \lim_{p=\infty} [pp_0 - (m^2 c^2/2) \log(p + p_0)/mc]$.

Third, the energy W_{fl} in QFT is not different from the corresponding energy in one-electron theory. Thus the final result of summing up three parts of the self-energy is this. Since the quadratically divergent term in W_{sp} is balanced by W_{fl}, the total value is the sum of the remainder of W_{sp} and W_{st}: $W = W_{st} + W_{sp} + W_{fl} = [(3/2\pi)(e^2/hc)]mc^2 \lim_{p=\infty} \log(p + p_0)/mc$ + finite term. Moreover, Weisskopf had also proved that the divergence of the self-energy is logarithmic in every order of approximation.

Weisskopf's contribution lies in his finding that the electromagnetic behavior of the electron in QFT is no longer completely point-like, but is extended over a finite region. It is this extension, and the consequent compensation between the fluctuation energy and the quadratically divergent term in the spin energy, that keeps the divergence of the self-energy within the logarithmic divergence. This result was one of the starting points of the theory of mass renormalization in the 1940s and played an important role in the further development of the renormalization theory. It is worth noting that Weisskopf's analysis was totally based on Dirac's idea of the vacuum. This fact convincingly justifies the claim that Dirac's idea of the vacuum has indeed provided an ontological basis for QFT, even though some physicists, including Weisskopf himself, reject the importance of Dirac's idea.

Infinite polarization of the Dirac vacuum

However, Dirac's idea of the vacuum was criticized by Oppenheimer (1930b) as soon as it was published for its leading to a new kind of divergence connected with the charge of the electron. Even though we can neglect the observable effect of the distribution of the vacuum electrons in the case of free state, Oppenheimer argued, we cannot avoid the meaningless 'observable effect' of the infinite electrostatic field distribution arising from the vacuum electrons when the external electromagnetic field appears. The only way out seems to be to introduce 'an infinite density of positive electricity to compensate the negative electrons' (*ibid.*).

In a later work with Wendell Furry on the same subject, Oppenheimer made this move, on the basis of a complete symmetry between electron and positron. However, the infinity difficulty connected with charge remained, although in this case it was of another kind. In studying the influence of the virtual pairs of electrons and positrons produced by the external electromagnetic field in a Dirac vacuum upon the energy and charge in the vacuum, Oppenheimer and

Furry (1934) found that the ratio between the energy of the virtual pairs and the energy of the external field was $(-\alpha K)$, here $K = (2/\pi)\int_0^\infty p^2 (1 + p^2)^{-5/2}\,\mathrm{d}p + (4/3\pi)\int_0^\infty p^4(1 + p^2)^{-5/2}\,\mathrm{d}p$. The effective charge would also be affected by the factor $(-\alpha K)$. The second integral in K diverges logarithmically for large values of p. This divergence was thought to have its roots in 'a genuine limitation of the present theory' and 'tends to appear in all problems in which extremely small lengths are involved' (*ibid.*).

The original work on this new kind of divergence, however, had already been done by Dirac himself during the Solvay Conference of October 1933. Dirac first introduced the density matrix (by using the Hartree–Fock approximation to assign each electron an eigenfunction). Then he studied the influence of an external electrostatic potential upon the vacuum charge density in the first order of the perturbative calculation, and showed that the relation between the charge density ρ (producing the external potential) and the charge density $\delta\rho$ (produced by the external potential in the vacuum) was $\delta\rho = 4\pi(e^2/hc)\,[A\rho + B(h/mc)^2\Delta\rho]$; here B is a finite constant, and A is logarithmically infinite. By a suitable cut-off in the integral leading to A, Dirac found that

the electric charges normally observed on electrons, protons or other charged particles are not the real charges actually carried by these particles and occurring in fundamental equations, but are slightly smaller by a factor of 1/137.

(Dirac, 1933)

The distinction proposed by Dirac between the observable quantities and the parameters appearing in the fundamental equations was undoubtedly of great significance in the development of renormalization ideas. However, the dependence on the cutoff appearing in the same work showed that this work did not satisfy relativity.

In (1934), Dirac carried out his study on the characteristic of the charge density in the vacuum. Since the densities would be apparently infinite, 'the problem now presents itself of finding some natural way of removing the infinities, . . . so as to leave a finite remainder, which we could then assume to be the electric and current density' (*ibid.*). This problem required him to make a detailed investigation of the singularities in the density matrix near the light-cone. The result was that the density matrix R could be divided naturally into two parts, $R = R_a + R_b$, where R_a contained all the singularities and was also completely fixed for any given field. Thus, any alteration one might make in the distribution of electrons and positrons would correspond to an alteration only in R_b, but not in R_a; and only those electric and current densities that arose from the distribution of the electrons and positrons and corresponded to

R_b were physically meaningful. And this suggested a way of removing the infinities.

Here the removal of infinities amounted to nothing but a subtraction of R_a from R. The justification for this removal is that R_a has nothing to do with the alteration in the electric and current distributions. Such an idea of subtraction, though totally unreasonable, had been the fundamental means in QFT for dealing with the divergence difficulties in the 1930s and 1940s, and was not discarded until the theory of renormalization had achieved important progress in the late 1940s. The subtraction procedure proposed by Dirac was generalized by Heisenberg (1934), who argued that all the divergent terms occurring in the expressions should be automatically subtracted without using the cutoff technique. This appeared similar to the later renormalization procedure, but without the latter's physical ground.

Peierls's deeper understanding of charge renormalization

In the same year, Rudolf Peierls also discovered a logarithmic infinity contained in the expression of the vacuum polarization, which had no direct connection with the self-energy divergence. As to the reason of the divergence, Peierls said,

the divergence is connected with this fact that there are an infinite number of states for the electrons and the external field can cause a transition between any of them.

(1934)

This in turn is connected with the point-coupling which entails an unrealistic contribution from virtual quanta with very high energy and momentum. By cutting off the integral at a momentum corresponding to a wavelength equal to the classical electron radius, Peierls found that the polarization would be of the order of 1%. This must correspond to a certain dielectric constant, which reduces the field of any 'external' charge or current by a constant factor. Thus we should assume, Peierls claimed, that all the charges we are dealing with are 'in reality' greater than what we observe, an assumption that is consistent with that of Dirac.

The existence of a meaningless divergence indicates that the theory was imperfect and needed further improving. Peierls, however, said that

one does not know whether the necessary changes in the theory will be only of a formal nature, just mathematical changes to avoid the use of infinity quantities, or whether the fundamental concepts underlying the equations will have to be modified essentially.

(Ibid.)

These words show that Peierls, like Furry and Oppenheimer, took a pessimistic and skeptical attitude to QFT, because of its divergence difficulty. The interesting thing to notice is that like Dirac, Furry, and Oppenheimer, although Peierls asked for a radical reform of QFT, the method he proposed to get a finite result was only to introduce a cut-off in the momentum integral, at the price of violating relativity, without changing the basic structure of QFT.

Kramers's suggestion of renormalization

The conjecture had already been expressed in 1936 that the infinite contributions of the high-momentum photons were all connected with the infinite self-mass, with the infinite intrinsic charge, and with non-measurable vacuum quantities, such as a constant dielectric coefficient of the vacuum.[22] Thus it seemed that a systematic theory, that is, a theory of renormalization, could be developed in which these infinities were circumvented. In this direction, Kramers (1938a) clearly suggested that the infinite contributions must be separated from those of real significance and subtracted; then some observable effects, such as the 'Pasternack effect',[23] could be calculated. He (1938b) also suggested that electromagnetic mass must be included in a theory that refers to the physical electron by eliminating the proper field of the electron. Although the elimination is impossible in a relativistic QFT, the recognition that a theory should have a 'proper-field'-free character had substantial influence on later developments. For example, it became a guiding principle of Julian Schwinger in forming his theory of renormalization (Schwinger, 1983).

Studies of scattering

In the late 1930s there appeared another kind of infinity characteristic of the scattering process.[24] At first, it was found that the non-relativistic radiation correction to the scattering cross-section of order α ($= e^2/hc$) diverged logarithmically (Braunbeck and Weinmann, 1938). The result was thought by Pauli and Markus Fierz (1938) to be evidence that the expansion in powers of α could not be valid. Instead, they performed a non-relativistic calculation without involving the expansion in powers of α. They employed a contact transformation method, which would be used widely in the renormalization procedure later on, to separate the infinite electromagnetic mass and to keep the remainder so that only the character of the scattering process would be represented. The result was absurd: the effect of the high frequencies was to make the total cross section vanish. Pauli and Fierz saw in this result another illustration of the inadequacies of QFT.

To investigate the extent to which the inclusion of relativistic effects would modify Pauli and Fierz's conclusion, Sidney Dancoff (1939) treated this problem relativistically. He divided the relativistic effects into three parts: (i) those representing a relativistic modification of terms occurring in the non-relativistic theory; (ii) those involving pairs in initial and final wave functions, with the electrons or positrons being scattered by the scattering potential; and (iii) those involving pair production and annihilation by the scattering potential. His calculation showed that every term in parts (i) and (ii) diverged logarithmically, but the combination of terms in each part gave a finite result. The situation was more complicated in part (iii): there were ten terms altogether. Among them, six terms were finite; but the sum of the corrections to the scattering cross section given by the remaining four terms was logarithmically divergent.

All the processes responsible for these corrections involve the transitions of the negative energy electrons to states of positive energy, influenced by the radiative field produced by the incident electrons. It is interesting to notice that Dancoff mentioned in a footnote that

Dr. R. Serber has pointed out that corrections to the scattering cross section of order α resulting from the Coulomb interaction with the virtual pairs in the field of the scattering potential should properly be considered here.

(Dancoff, 1939)

But, after considering only one term of this type of the interaction, which would consist of three terms, Dancoff drew the conclusion: 'the divergent term corresponding to this interaction follows directly from formulae for the polarization of the vacuum and may be removed by a suitable renormalization of charge density' (*ibid.*). Sin-itiro Tomonaga (1966) commented that if Dancoff had not been so careless, then 'the history of the renormalization theory would have been completely different'. The reason for this claim will be given shortly.

Compensation

By the end of the 1930s, the various kinds of divergence arising from the field reactions (the interactions between the particles and their own electromagnetic field) had made many physicists lose confidence in QFT. In this atmosphere, a new attempt at introducing a 'compensative field' was put forward. Fritz Bopp (1940) was the first physicist to propose a classical field theory with higher-order derivatives, in which the operator $\square$ appearing in D'Alembert's equation was substituted by the operator $(1 - \square/k_0^2)\square$. The field satisfying a fourth-order differential equation could be reduced to two fields which satisfy second-order equations. That is, the potential A_μ of the new field could be written as $A_\mu = A'_\mu - A''_\mu$, where $A'_\mu = (1 - \Box/k_0^2)A_\mu$, and $A''_\mu = (-\Box/k_0^2)A_\mu$. Obviously, the potential A'_μ satisfies the Maxwell equation and A''_μ satisfies the equation of a vector meson with mass kh/c. Then the scalar potential ϕ of a point charge in the new field is $\phi = \phi' - \phi'' = e/r(1 - e^{-kr})$. When $r \gg 1/k$, $\phi = e/r$, that is, the usual electromagnetic scalar potential; when $r = 0$, $\phi = ek$, and this means that $1/k$ serves as the effective radius of an electron. The point-electron would be stable under the action of the new field if the vector-meson field provided a negative 'mechanical' stress which is necessary for compensating the Coulomb repulsive force. In this case the self-energy of an electron is a finite quantity.

Bopp's theory satisfies relativity and can be generalized to the quantum case. The divergence of the longitudinal field self-energy could be cancelled, but the theory was still unable to deal with the divergence of the transverse field self-energy, which was characteristic of the difficulty in QFT. In spite of this, Bopp's idea of compensation inherited from Poincaré had a direct and tremendous influence on Pais (1945), Sakata (1947), Tomonaga (1948; with Koba, 1947, 1948) and others. Moreover, Feynman's relativistic invariant cutoff technique (1948a, b) and the Pauli and Villars regularization scheme (1949) are equivalent to Bopp's theory in the sense of introducing an auxiliary field for removing the singularities.

Directly influenced by Bopp, Pais (1945) published a new scheme to overcome the self-energy divergence difficulty: each elementary particle is the source of a set of fields in such a way that various infinite contributions to the self-energy, to which these fields give rise, cancel each other so as to make the final outcome finite. With regard to the concrete case of the electron, Pais assumed that the electron was the source of a neutral, short-range vector field, in addition to the electromagnetic field. This 'subtractive' vector field coupled directly with the electron. If the convergence relation $e = f$ was satisfied (e is the electric charge, and f the charge of the electron in its coupling with the vector field) then the self-energy of the electron was finite to any order of approximation. However, Pais acknowledged that such a kind of subtractive field was incompatible with a stable distribution of the vacuum electrons, as required by the hole theory. Thus the scheme had to be discarded within the framework of QFT.

Sakata (1947) independently obtained the same result Pais had obtained. He named the auxiliary field the cohesive force field or C-meson field. Sakata's solution to the self-energy divergence, however, was in fact an illusion. As Toichiro Kinoshita (1950) showed, the necessary relation between the two coupling constants will no longer cancel the divergences when the discussion is extended beyond the lowest order of approximation. Nevertheless, the C-meson hypothesis served fruitfully as one of the catalysts that led to the introduction of the self-consistent subtraction method by Tomonaga.

Inspired by Sakata's achievement of 'solving' the self-energy divergence with the C-meson field as a compensative field, Tomonaga tried to extend Sakata's idea to the scattering problem. It was not successful at the beginning (Tomonaga and Koba, 1947) since in their calculation Tomonaga and Koba had repeated Dancoff's error of missing some terms. But soon after that, by using a new and much more efficient method of calculation (see below), they found, on comparing the various terms appearing in both Dancoff's calculation and their own previous calculation, the two terms that had been overlooked. There were only two missing terms, but they were crucial to the final conclusion. After they corrected this error, the infinities appearing in the scattering process of an electron owing to the electromagnetic and cohesive force field cancelled completely, except for the divergence of vacuum polarization type, which can be struck off at once by a redefinition of the charge. In this way, Tomonaga had achieved a finite correction to the scattering of electrons by combining two different ideas: the renormalization of charge and the compensation mechanism of the C-meson field.

Here a comment seems to be in order. The preoccupation of the proposers and followers of the compensation idea, unlike that of the renormalization idea to be discussed shortly, was not with analyzing and carefully applying the known relativistic theory of the coupled electron and electromagnetic fields, but with changing it. They introduced the fields of unknown particles in such a way as to cancel the divergences produced by the known interactions, and hence went beyond the existing theory. Thus this is a different approach to QFT from the renormalization approach, which is a device within the existing theory, although the two are easily confused with one another, owing to their similar aim of eliminating divergences.

The Bethe–Lewis scheme for mass renormalization

The mature idea of renormalization began with the work of Hans Bethe (1947) and H. A. Lewis (1948). Soon after the rediscovery of the level shift of hydrogen by Lamb and Retherford (1947) with the microwave method, Bethe accepted a suggestion by Schwinger and Weisskopf and tried to explain the shift by the interaction of the electron with the radiation field. The shift came out infinite in all existing theories and had to be ignored. Bethe, however, pointed out that

it was possible to identify the most strongly (in non-relativistic calculation linearly, in relativistic calculation logarithmically) divergent term in the level shift with an electromagnetic mass effect. This effect should properly be regarded as already included in the observed mass of the electron, and must therefore be subtracted from the theoretical expression. The result then diverges only logarithmically instead of linearly in non-relativistic theory, and would be convergent in relativistic theory, where the strongest divergence is just logarithmic infinity. This would set an effective upper limit of the order of *mc* to the frequencies of light which effectively contribute to the shift of the level of a bound electron.

(Bethe, 1947)

After explaining the existence of such an upper limit, Bethe cut off the integral, which diverges logarithmically in the non-relativistic calculation, with the limit, and obtained a finite result which was in excellent agreement with the experimental value (1040 megacycles).

Bethe was the first to clearly separate the effect of the electromagnetic mass from the quantum process. He incorporated it into the effect arising from the observed mass instead of simply ignoring it. In addition to the renormalization of the electron mass itself, which was discussed more or less by his predecessors, Bethe's mass renormalization procedure could also be applied to all kinds of quantum processes. Instead of the unreasonable subtraction of infinities simply because they are unobservable, which would leave open the question of whether the theory producing the infinities is valid, their 'incorporation' had paved the way for a further physical explanation for the divergences.

Inspired by Bethe's success, Lewis (1948) applied the mass renormalization procedure to the relativistic calculation of the radiative correction to the scattering cross section. In doing so, Lewis made the nature of the renormalization idea clearer by claiming that the crucial assumption underlying the whole renormalization programme was that

the electromagnetic mass of the electron is a small effect and that its apparent divergence arises from a failure of present day quantum electrodynamics above certain frequencies.

(Lewis, 1948)

It is clear that only when a physical parameter (which when calculated in perturbation in QFT may turn out to be divergent) is actually finite and small can its separation and amalgamation into the 'bare' parameters be regarded as mathematically justifiable. The failure of QFT at ultra-relativistic energies, as indicated by the divergences in perturbation theory, implied that the region in which the existing framework of QFT was valid should be separated from the region in which it was not valid and in which new physics would become manifest. It is impossible to determine where the boundary is, and one does not know what theory can be used to calculate the small effects which are not calculable in QFT. However, this separation of knowable from unknowable, which is realized mathematically by the introduction of a cut-off, can be schematized by using phenomenological parameters, which must include these small effects.

Such a renormalized perturbative theory, in which the formulae of QFT are mixed with an unexplained observed mass value, is sometimes called an incomplete theory, or a reasonable working hypothesis. But it was theoretically more reasonable than all preceding theories, such as Pais's or Sakata's, in which two infinities with opposite signs had combined with each other to give a finite quantity. These theories could only be regarded as an early intelligent exploration for solving the divergence problem, but were unconvincing in physics and mathematics. In contrast to those theorists, Lewis combined the electromagnetic mass, which was assumed small, with a finite 'mechanical' mass to obtain an observed mass, and at the same time subtracted the effect of the electromagnetic mass from the interaction. In this way, the renormalization procedure had obtained a solid physical ground. So the above proposal should be regarded as a major event in the formation of the renormalization programme.

Lewis also justified Bethe's guess that the electromagnetic mass would give rise to divergent terms in the calculations of a quantum process. In his relativistic calculation, Lewis found that the electromagnetic mass did result in an expression which differed only in the multiplicative constant from the divergent radiative correction terms found by Dancoff. The only numerical difference came from Dancoff's omission of certain electrostatic transitions and would disappear once the terms neglected by Dancoff were taken into consideration again. Thus Lewis suggested quite strongly that those divergent terms found by Dancoff could be identified with manifestations of the electromagnetic mass of the electron. So if the empirical mass, including the electromagnetic mass, is used in the Hamiltonian, one must omit the transitions resulting from the electromagnetic mass effects. Using the empirical mass and omitting the effects resulting from the electromagnetic mass are the essential contents of the Bethe–Lewis mass renormalization procedure.

The canonical transformation

In the Bethe–Lewis scheme, the crucial point was the unambiguous separation of a finite part of a formally infinite term. Lewis acknowledged that the method of canonical transformation on the Hamiltonian developed by Schwinger was better suited to this purpose. Schwinger (1948a) claimed that the separation Lewis had mentioned could be realized by

transforming the Hamiltonian of current hole theory electrodynamics to exhibit explicitly the logarithmically divergent self-energy of a free electron, which arises from the virtual emission and absorption of light quanta.

(Ibid.)

He thought that the new Hamiltonian was superior to the original one in essentially three ways:

it involves the experimental electron mass, rather than the unobservable mechanical mass; an electron now interacts with the radiation field only in the presence of an external field . . .; the interaction energy of an electron with an external field is now subject to a finite radiative correction.

(Ibid.)

For the logarithmically divergent term produced by the polarization of the vacuum, Schwinger, like his predecessors, thought that

such a term is equivalent to altering the value of the charge by a constant factor, only the final value being properly identified with the experimental charge. [Thus] all divergences [are] contained in the renormalization factors.[25]

(Ibid.)

Schwinger did not publish his canonical transformation until the Pocono conference of 30 March to 1 April 1948. Although the same method had already been developed by Tomonaga during the period of 1943 to 1946 (see below), it was not available in the United States owing to communication conditions at the time. With this method, Schwinger (1948b; 1949a, b) investigated systematically the divergences in QFT from the viewpoint of the symmetry between the electromagnetic field and the positron–electron field. According to him, the elementary phenomena in which divergences occur are the polarization of the vacuum and the self-energy of the electron. These two phenomena are quite analogous and essentially describe the interaction of each field with the vacuum fluctuations of the other field: the vacuum fluctuations of the positron–electron field generated by the electromagnetic field are the virtual creation and annihilation of electron–positron pairs; and the fluctuations of the electromagnetic field generated by the electron are the virtual emission and absorption of photons. Schwinger concluded:

The effect of these fluctuation interactions is simply to alter the fundamental constants *e* and *m*, although by logarithmically divergent factors, and all the physically significant divergences of the present theory are contained in the charge and mass renormalization factors.

(Ibid.)

On the basis of such an integrated and thorough understanding of various divergence phenomena, and equipped with advanced mathematical tools, such as the canonical transformation, variational principle, and functional integral, Schwinger was able to deal successfully with a series of difficult problems, such as the level shift and anomalous magnetic moment. His results tally with the experiments very well. This fact had powerfully manifested the tremendous force of renormalized perturbative theory and encouraged physicists to make further and thorough studies of renormalization.

Tomonaga's contributions

Although Tomonaga obtained a proper understanding of renormalization only after the publication of the works by Bethe and Lewis, and after his gradual abandonment of the naive idea of the compensative field, he indeed took a parallel and independent part in some technical developments that were crucial to the formation of the renormalization programme. As early as (1943), Tomonaga published in Japanese a relativistically invariant formulation of QFT,[26] in which a relativistic generalization was made of non-relativistic concepts, such as canonical commutation relations and the Schrödinger equation. In doing so, he applied a unitary transformation to all fields, providing them with the equations of motion of non-interacting fields, whereas in the transformed Schrödinger equation (of the state function), only the interaction terms remained. That is to say, he made use of the interaction representation. Then he observed that when a non-interacting field was considered, there was no difficulty in exhibiting commutation relations for arbitrary spacetime points, which were four-dimensional in character. As for the transformed Schrödinger equation, he replaced the time by 'a spacelike surface' to make it relativistically covariant.

In dealing with divergences, however, Tomonaga did not make use of his powerful techniques. Rather, he appealed to nothing but the idea of compensation until the publication of Bethe's work. The latter attracted his attention strongly

because it may indicate a possible path to overcome the fundamental difficulty of the quantum field theory, and that for the first time in close connection with reliable experimental data.

(Tomonaga and Koba, 1948)

Thereupon he proposed a formalism, which he called the 'self-consistent subtraction method', to express Bethe's fundamental assumption in a more closed and plausible form, in which the separation of terms to be subtracted was made by a canonical transformation. This formalism was then applied to the scattering problem to isolate the electromagnetic mass terms, and to obtain a finite result with the aid of a C-meson field (Ito, Koba, and Tomonaga, 1948).

In (Tomonaga and Koba, 1948) the C-meson theory was still regarded as aimed at a substantialization of the subtraction operation for obtaining a finite self-energy and finite scattering cross section. But the problem was that this seemed to be an *ad hoc* device. If the aim was only to deal with the divergence difficulties in QFT, then the introduction of the C-meson field was far less convenient than the renormalization procedure, and what was worse, it could not overcome the divergence difficulty generated by the vacuum polarization. So the C-meson theory or the more general idea of compensation was gradually replaced in Tomonaga's and other Japanese physicists' work by the renormalization procedure.

Feynman's regularization

We mentioned above that one of the necessary assumptions of the renormalization procedure is that the divergent quantities apparently and improperly calculated in QFT are in fact finite and small quantities. Only in this case can one reasonably separate them from the expression obtained in QFT and merge them into the bare quantities, and substitute the sum by the observed values. To render the separation and amalgamation of divergent quantities mathematically tenable, therefore, a set of rules for regularization had to be established that made it possible to calculate physical quantities in a relativistically and gauge invariant manner. Such a set of rules was proposed by Richard Feynman (1948b), based on his discussion of the corresponding classical case (1948a).

Feynman named the set of rules a relativistic cutoff. With a finite cutoff this artifice transforms essentially purely formal manipulations of divergent quantities, i.e., the redefinition of parameters, into quasi-respectable mathematical operations. The main points of his ideas are as follows. For the vacuum fluctuation of a radiative field, we substitute the old density function $\delta(w^2 - k^2)$ by the new one, $g(w^2 - k^2) = \int_0^\infty [\delta(w^2 - k^2) - \delta(w^2 - k^2 - \lambda^2)]G(\lambda)\,d\lambda$. Here, $G(\lambda)$ is such a smooth function that $\int_0^\infty G(\lambda)\,d\lambda = 1$. In the momentum space representation of the propagator of a photon, this means substituting $1/k^2$ by $\int_0^\infty [1/k^2 - 1/(k^2 - \lambda^2)]G(\lambda)\,d\lambda$. Mathematically, this is equivalent to multiplying $1/k^2$ by a convergence factor $c(k^2) = \int_0^\infty -\lambda^2(k^2 - \lambda^2)^{-1}G(\lambda)\,d\lambda$.

For the vacuum fluctuation of the electron field, Feynman tried to use a similar method, that is, to introduce the convergence factor $c(p^2 - m^2)$ and $c[(p + q)^2 - m^2)]$ respectively for two internal lines of the electron in the expression $J_{\mu\nu} = -ie^2/\pi \int_0^\infty \mathrm{sp}[(\gamma \cdot p + \gamma \cdot q - m)^{-1}\gamma_\mu(\gamma \cdot p - m)^{-1}\gamma_\nu]\,d^4p$, which characterizes the vacuum polarization. Here, sp means spur, and $(\gamma \cdot p + \gamma \cdot q - m)^{-1}$ and $(\gamma \cdot p - m)^{-1}$ are the momentum space representations of the internal lines of the electrons with momentum $(p + q)$ and p respectively

(Feynman, 1949b). This approach would make $J_{\mu\nu}$ converge, but only at the price of violating current conservation and gauge invariance. Therefore, it must be given up. However, Feynman realized that the introduction of the convergence factor implied, physically, that the contribution of the internal line of a particle with mass m would be cancelled by the contribution of a new particle with mass $(m^2 - \lambda^2)^{1/2}$. Considering the fact that any internal line in the vacuum polarization diagram would complete a closed loop, we should not introduce different convergence factors for two internal lines. Rather, a contribution from a new particle with mass $(m^2 - \lambda^2)^{1/2}$ for the closed loop should be introduced so that a convergent result could be obtained. Writing $J_{\mu\nu}$ as $J_{\mu\nu}(m^2)$ and substituting for it $J_{\mu\nu}^p = \int_0^\infty [J_{\mu\nu}(m^2) - J_{\mu\nu}(m^2 + \lambda^2)]G(\lambda)\,\mathrm{d}\lambda$, the result of the calculation must be convergent, except for some terms depending on λ.

The dependence on λ seems to violate relativity, just as any cutoff introduced before would do. Moreover, in the limit as λ goes to infinity, the calculation still results in divergent expressions. In fact, this is not the case here. The quantity apparently dependent on λ, even if it is divergent, can be absorbed after renormalization, so that the cutoff introduced in this way would not conflict with relativity. In particular, if after the redefinition of mass and charge, other processes are insensitive to the value of the cutoff, then a renormalized theory can be defined by letting the cutoff go to infinity. A theory is called renormalizable if a finite number of parameters are sufficient to define it as a renormalized one.

Physically, Feynman's relativistic cutoff is equivalent to the introduction of an auxiliary field (and its associated particle) to cancel the infinite contributions due to the ('real') particles of the original field.[27] Feynman's approach was different from realistic theories of regularization or compensation. In the latter, auxiliary particles with finite masses and positive energies are assumed to be observable in principle and are described by field operators that enter the Hamiltonian explicitly. Feynman's theory of a cutoff is formalistic in the sense that the auxiliary masses are used merely as mathematical parameters which finally tend to infinity and are non-observable in principle. Representative of the 'realistic' approach are the papers of Sakata (1947, 1950), Umezawa (with Yukawa and Yamada, 1948; with Kawabe, 1949a, b), and other Japanese physicists, as well as those of Rayski (1948). Among the 'formalists' we find, in addition to Feynman, Rivier and Stueckelberg (1948), and Pauli and Villars (1949).

Feynman's other contribution to the renormalization programme is related to the simplification of the calculation of physical processes. He provided not only neat forms for causal propagators, but also a set of diagram rules, so that each factor in an S-matrix element describing a physical process corresponds, in a one to one manner, to a line or a vertex on the plane (Feynman, 1949b). The diagram rules are a convenient and powerful tool which enabled Feynman to express and analyze various processes described in QFT, and to clearly embody the ideas of Tomonaga, Bethe, Lewis, and Schwinger about canonical transformation, separation of the divergences, and renormalization. All these provided prerequisites for Freeman Dyson to propose a self-consistent and complete renormalization programme through further analysis and combination of the contributions of the diagrams.

Dyson's renormalization programme

Dyson synthesized the ideas and techniques of renormalization by Bethe, Lewis, Tomonaga, Schwinger, and Feynman. He made use of Feynman's diagram, analyzed in detail various divergences appearing in the calculations of S-matrix elements, proposed a renormalization programme for handling these divergences systematically in his (1949a), and made some supplements to this programme and perfected it in (1949b).

Dyson's programme was entirely based on the Feynman diagram analysis. First, he demonstrated that the contributions of disconnected graphs could be omitted. Second, he defined the primitive divergent graphs to which all the divergent graphs could be reduced. Then, by analyzing the topology of the diagram, he obtained the convergence condition $k = 3E_e/2 + E_p - 4 \leqslant 1$, for the graphs; here E_e and E_p are the numbers of external electron and photon lines in the diagram respectively. With the convergence condition, Dyson classified all possible primitive divergences into three types, namely, electron self-energy, vacuum polarization, and scattering of a single electron in an electromagnetic field. In a further analysis, Dyson introduced ideas about the skeleton of the diagram, reducible and irreducible graphs, proper and improper graphs, etc. With these ideas, he suggested that in calculating an S-matrix element, we should (i) draw related irreducible graphs; (ii) substitute the observed mass into the Hamiltonian; (iii) substitute the lowest- order propagators S_f, D_f and Γ_μ, which would not result in divergences, by new propagators $S'_f = Z_2 S_f$, $D'_f = Z_3 D_f$ and a new vertex $\Gamma'_\mu = Z_1^{-1}\Gamma_\mu$;[28] and (iv) substitute the bare wave function ψ, ψ^-, and A_μ, which would not bring about divergences, by new ones $\psi' = Z_2^{1/2}\psi$, $\psi'^- = Z_2^{1/2}\psi^-$ and $A'_\mu = Z_3^{1/2}A_\mu$, where $Z_{1,2,3}$ are 'divergent factors'. Having done this, Dyson claimed, all three types of primitive divergence arising from the two types of vacuum fluctuation, that is, all the radiative corrections, have been taken into consideration, and will result in a 'divergent factor' $(Z_1^{-1}Z_2 Z_3^{1/2})^n$ after mass renormalization. Considering then that in the *n*th order process, there must appear a factor e_0^n (where e_0 is the so-called bare charge appearing in the underlying equations), and assuming the observed charge $e = (Z_1^{-1} Z_2 Z_3^{1/2})e_0$, then there would no longer appear any divergence difficulty.

Obviously, this procedure can be used without limit to an arbitrary order in the fine-structure constant $\alpha = e^2/hc$. One cannot help asking why Dyson called such an elegant scheme a programme rather than a theory. His answer was that he had not given a general proof for the convergence of higher-order radiative corrections. That is to say, although his procedure guarantees that the radiative corrections obtained in an arbitrary-order approximation would be finite, it is not guaranteed that the summation of them would still be finite.

Summary

The formation of the renormalization programme consists of three main steps. First, reveal the logarithmic nature of various divergent terms. Second, reduce all divergent terms to two primitive types that arise from self-energy and vacuum polarization. Third, find an unambiguous and consistent way to treat the divergent quantities, by (i) employing the canonical transformation to get the covariant formulation of QFT (the interaction representation) and to separate the divergent terms; (ii) mass and charge renormalization; (iii) a consistent use of the empirical values of mass and charge. With the aid of the renormalization programme, finite radiative corrections can be obtained and difficult problems can be treated. Thereby QFT becomes a powerful calculating means to predict various physical processes, and one of the most predictive branches in contemporary physics.

Dirac's criticism of renormalization

Ironically, in spite of all its tremendous successes, the idea of renormalization was attacked fiercely by Dirac (1963, 1968, 1983), one of its active pioneers. In Dirac's opinion, all the successes that had been achieved by using renormalization had neither a sound mathematical foundation nor a convincing physical picture. Mathematically, renormalization demands, contrary to the typical custom in mathematics, neglecting the infinities instead of infinitesimals. So it is an artificial or illogical process (Dirac, 1968). Physically, Dirac argued that the presence of the infinities indicates that

there is something basically wrong with our theory of the interaction of the electromagnetic field with electrons. By basically wrong I mean that the mechanics is wrong, or the interaction force is wrong.

(Dirac, 1983)

Yet the renormalization procedure has nothing to do with correcting what is wrong in the theory, or with searching for new relativistic equations and new kinds of interaction. Instead, it is merely a rule of thumb that gives experimental results. So it would mislead physics into taking a wrong road (Dirac, 1963). Dirac strongly condemned renormalization and stressed that 'this is quite nonsense physically, and one should be prepared to abandon it completely' (Dirac, 1983).

Tomonaga's attitude was not so negative. But he also acknowledged that

it is true that our method (of renormalization) by no means gives the real solution of the fundamental difficulty of quantum electrodynamics but an unambiguous and consistent manner to treat the field reaction problem without touching the fundamental difficulty.

(Tomonaga, 1966)

Justifications for the renormalization programme

To be sure, there are some deep insights in Dirac's criticism of renormalization. But on the whole, his criticism seems unfair because it has missed the conceptual ground of renormalization. In fact, the renormalization programme can be justified on two levels.

Physically, the infinities at higher orders, which are all connected with the infinite contributions of the high-momentum virtual photons (and electron–positron pairs), indicate that the formalism of QFT contains unrealistic contributions from the interactions with high-momentum virtual photons (and pairs). Although the physical reality of the virtual quanta processes is testified by experiments,[29] the infinite-momentum virtual quanta are obviously unrealistic. Unfortunately, this unrealistic element is deeply rooted in the foundation of QFT, in the concepts of the operator field and of localized excitations produced by the local operator field. Various Green's functions are the correlation functions among such localized excitations, and the study of their spacetime behavior is the only instrument for the identification of the physical particles and their interactions. In this context, renormalization can be properly understood as an aspect of the transfer of attention from the initial hypothetical world of localized excitations and interactions to the observable world of the physical particles. Its main aim, as Dyson put it,

is not so much a modification of the present theory which will make all infinite quantities finite, but rather a turning-round of the theory so that the finite quantities shall become primary.

(1949b)

Philosophically, renormalization can be viewed as a superstructure on the basis of atomism that is deeply rooted in the western mode of thinking. The various models developed within the framework of QFT for describing the subatomic world remain atomistic in nature: the particles described by the fields that appear in the Lagrangians are to be regarded as elementary constituents of the world. However, the atomism, or the notion of atomicity, adopted by unrenormalized theories has a halfway character. As was clearly pointed out by Schwinger (1973), the reason for this is that the unrenormalized operator field theories contain an implicit speculation about the inner structure of physical particles that is sensitive to details of dynamic process at high energy. Mathematically, this assumption has manifested itself in the divergent integrals.[30] Metaphysically, the assumption implies that there exist more elementary constituents of the physical particles described by the fields appearing in a Lagrangian, and this contradicts the status of the particles, or that of the fields, as basic building blocks of the world.

The fundamental significance of the renormalization procedure in this regard can be described in the following way. By removing any reference to an inaccessible, very high energy domain and the related structure assumption, the renormalization procedure reinforces the atomistic commitment of QFT. This is because the particles or the fields appearing in the renormalized theories do act as the basic building blocks of the world. But note that atomicity here no longer refers to the exact point model. To the extent that one removes the reference to the inaccessible very high energy domain, which arises conceptually in the exact point model by virtue of the uncertainty principle, renormalization blurs any point-like character.

This spatially extended yet structureless quasi-point model, adopted or produced by renormalized QFT, can be justified in two ways. On the one hand, it is supported by its empirical success. It can also be justified philosophically by arguing that, so long as the experimental energy is not high enough to detect the inner structure of the particles, and thus all the statements about their small distance structure are essentially conjectures, the quasi-point model is not only an effective approximation for experimental purposes, but also a necessary phase of cognition we must go through.

Compared with the true theory that Dirac (1983) and Tomonaga (1966) once yearned for, and compared with the 'theory of everything' that superstring theorists are still searching for, the quasi-point model seems to be a mathematical device needed in a transition period. A disciple of Dirac would argue that the quasi-point model should be discarded once the structure of the elementary particles is known. This is true. But those committed to atomism would argue that in physics (as presently formulated and practiced) the structural analysis of objects at any level is always based on (seemingly) structureless objects (genuine or quasi-point-like) at the next level. For example, the analysis of deep inelastic lepton–hadron scattering would be impossible if there were no analysis of elastic scattering of the point-like partons as a basis. Thus the adoption of the quasi-point model seems to be unavoidable in QFT. This is probably what Feynman (1973) meant when he stated that 'we will start by supposing that there are [quasi-point particles] because otherwise we would have no field theory at all'.

The dialectics of atomism in this context can be summarized as follows. On the one hand, the structureless character of particles as we know them at any level is not absolute but contingent and context-dependent, justified only by relatively low energy experimental probing. When the energy available in experiments becomes high enough, some of the inner structure of the particles sooner or later will be revealed, and the notion of absolute indivisibility will turn out to be an illusion. On the other hand, with the revealing of the structure of particles at one level, there emerge at the same time, as a precondition of the revealing, (seemingly) structureless objects at the next level. And thus the original pattern of 'structured objects being expressed in terms of (seemingly) structureless objects' remains, and will remain as long as QFT remains the mode of representation.[31] Since the point model field theory would not make any sense without renormalization, Dirac's prediction that renormalization 'is something that will not be reserved in future' (1963) sounds groundless. Rather, according to the adherents of atomism, renormalization will always be active, together with active local field theories.

Nagging problems

On the other hand, a thoroughgoing atomism entails an infinite regress, which is philosophically not so attractive. To avoid this unpleasant consequence, a different programme, the S-matrix theory (based on the idea of the bootstrap, which is a variant of holism in contemporary particle physics) was proposed at the beginning of the 1960s.[32] The bootstrap philosophy goes well beyond the field-theoretical framework and even tries to be an anti-field-theory programme. The S-matrix programme provides a basis for dual models, on the basis of which string theories and, when supersymmetry is incorporated, superstring theories have been developed.[33] One of the remarkable features of superstring theories is that some of them promised to have a calculational scheme without being hindered by infinities. If this turns out to be true and the framework of superstring theories can be developed into a consistent one, then Dirac's criticism of renormalization will finally turn out to be right and insightful.

Moreover, other nagging problems remain. First, as is well known, the renormalization procedure works in practice only in a perturbative framework. It is contingent on the smallness of e^2/hc (about 1/137).[34] But the trouble is that the effective coupling constant at very short distances becomes larger than unity, and in this case no perturbation method can be used. Will there be a theory that can be renormalized by a non-perturbative method? Or will a future unification of electrodynamics and general relativity heal the disease of divergences because of the fact that the dangerous distances are smaller than the Schwarzschild radius of the electron? Or will a unification of electro-dynamics with strong interactions bring a solution to the problem? Whatever, the unification of electromagnetic with weak interactions has not served this purpose since it remains within the framework of perturbative renormalization. There will be more discussion on this topic in sections 8.2, 8.3, 8.8, and 11.2.

Second, a much deeper problem is related to gauge invariance. Quantum electrodynamics is successful in its renormalization mainly because of its gauge invariance, which entails the zero mass of its gauge quanta, the photons. As soon as bosons of finite mass are brought into theories for describing other fundamental interactions, the gauge invariance is spoilt, and these theories are not renormalizable in general. Further development of the renormalization programme is closely connected with gauge theories, which will be reviewed in part III.

Notes

1. For more detailed discussion on ontology, see section 1.4.
2. See Rosenfeld (1973) and van der Waerden (1967).
3. For the equivalence, see section 7.3.
4. See Wentzel (1960) and Weisskopf (1980).
5. For the ascription of the ideas mentioned here to Jordan, see van der Waerden (1967) and Jordan's recollection (1973).
6. In his papers (1927b, c), Dirac's 'h' is actually equal to $h/2\pi$. I follow Dirac's usage in the discussion of his (1927b, c).
7. As early as August 1926, Dirac had already pointed out: 'It would appear to be possible to build up an electromagnetic theory in which the potentials of the field at a specified point X_0, Y_0, Z_0, t_0 in space-time are represented by matrices of constant elements that are functions of X_0, Y_0, Z_0, t_0' (1926b).
8. In the case of bosons, QC is $[\theta_r, N_s] = i h \delta_{rs}$, and hence $b_r b_s^* - b_s^* b_r = \delta_{rs}$. This is consistent with configuration space wave mechanics with a symmetric wave function. In the case of fermions, QC has to be modified in such a way that $\theta_r N_s + N_s \theta_r = i h \delta_{rs}$, and hence $b_r b_s^* + b_s^* b_r = \delta_{rs}$. This is consistent with configuration space wave mechanics with an anti-symmetric wave function, where the Pauli exclusion principle is satisfied (see Jordan and Wigner, 1928).
9. These relations themselves cannot be regarded as QC, as I have indicated above.
10. The zero-point fluctuations of the field have no direct connection with the zero-point energy, which is of purely formal character, and can be removed by a redefinition known as 'normal ordering' (see Heitler, 1936).

11. For Dirac's idea of the substratum, see his preface to the *Principles of Quantum Mechanics* (1930b); for his later belief in a non-mechanical ether, see his (1951, 1952, 1973a).

12. For example, C. N. Yang holds this position. As to the view that Dirac's concept of the filled vacuum was rejected by subsequent developments after the work by Oppenheimer and Furry and by Pauli and Weisskopf, he commented to me: 'This is superficial without catching the true spirit of Dirac's idea' (private conversation on 14 May 1985). See also Lee and Wick (1974).

13. According to Leon Rosenfeld (1968), this idea also has its classical origin in Hertz, whose idea of a 'middle term', he stressed, can be interpreted as 'another kind of hidden particles' by which the interaction is conveyed.

14. Starting from a relativistic classical Lagrangian field theory, Heisenberg and Pauli in (1929, 1930) developed a general theory of the quantized field, with the aid of canonical quantization, and thereby prepared the tools for Fermi's and Yukawa's theory of interactions.

15. According to Wigner's rule (Wigner and Witmer, 1928), a composite system consisting of an odd number of spin-1/2 particles must obey Fermi statistics, and an even number Bose statistics.

16. The novelty of the idea was noted by Yakov Dorfman (1930) in the same year.

17. See Yang and Mills (1954a, b).

18. After Chadwick's discovery of the neutron, Pauli's neutron was renamed by Fermi and others as the neutrino.

19. This model was adopted by Hideki Yukawa and Shoichi Sakata (1935a, b).

20. See Hayakawa (1980).

21. The so-called 'two-meson hypothesis' was proposed to replace Yukawa's original scheme: two kinds of meson exist in nature, possessing different masses; the heavy meson, π, is responsible for nuclear force, and the light meson, μ, is a decay product of the heavy one and is observed to interact weakly with the electron and neutrino. See Sakata and Inoue (1943), Tanikawa (1943), and Marshak and Bethe (1947).

22. See Euler (1936).

23. The observed level splitting between the $2S_{1/2}$ and $2P_{1/2}$ states of hydrogen was discussed by Simon Pasternack (1938). The effect was confirmed again by Willis Lamb and Robert Retherford (1947), and has been known as the Lamb shift since then.

24. At that time, physicists were greatly concerned with the 'infrared catastrophe' in the scattering. But this is irrelevant to the discussion of renormalization, and so this topic will not be included in the following discussion.

25. It should be mentioned that Schwinger might have been the first physicist to fully appreciate that charge renormalization is a property of the electromagnetic field alone, and results in a fractional reduction of charge. As a comparison, see, for example, Pais's letter to Tomonaga in which he wrote: 'It seems one of the most puzzling problems how to "renormalize" the charge of the electron and of the proton in such a way as to make the experimental value for these quantities equal to each other' (13 April 1948, quoted by Schwinger, 1983).

26. The English translation was published in (1946).

27. It should be stressed that the appearance of the compensation idea here is purely formal since the dependence on λ, as we shall see soon, is transient and will disappear in the final expressions.

28. This procedure works only when divergent subgraphs do not overlap, but fails in handling overlapping divergences which in quantum electrodynamics occur only within proper self-energy graphs. To isolate divergent graphs, Dyson (unpublished) defined a mathematical procedure. Abdus Salam (1951a, b) extended Dyson's procedure to general rules and applied these rules to renormalization for scalar electrodynamics. John Ward (1951) resolved the overlapping divergence difficulty in the electron self-energy function with the help of Ward's identity (1950), which is one of the consequences of gauge invariance and provides a simple relation between the derivative of the electron self-energy function and the vertex function that can be renormalized straightforwardly without the overlapping problem. Robert Mills and Chen Ning Yang (1966) generalized Ward's approach and

resolved the overlapping divergence difficulty in the photon self-energy function. I am grateful to Dr. C. N. Yang for calling my attention to this subtle subject.

29. For example, the electron–positron emission from an excited oxygen nucleus testifies to the physical reality of the virtual photon process.

30. Mary Hesse (1961) asserts that the root of the divergence difficulties can be traced to the basic notion of atomicity. If the notion of atomicity refers to the point model or local excitations and local couplings, the assertion is correct. However, if it refers to, as it usually does, the structureless elementary constituents, then the root of the divergence difficulties is the structure assumption implicitly adopted by unrenormalized operator field theories, rather than the notion of atomicity.

31. Thus the idea expressed by the bootstrap hypothesis in S-matrix theory that everything is divisable is incompatible with the atomistic paradigm adopted by QFT.

32. Chew and Frautschi (1961a, b, c). See also section 8.5.

33. For reviews, see Jacob (1974), Scherk (1975) and Green (1985).

34. Thus, strictly speaking, Yukawa's meson theory has never been a field theory of nuclear force, though it was a step in the right direction towards it, because it cannot be renormalized (owing to the large coupling constant in the strong interaction), and hence no one can calculate anything from it.

8

The quantum field programme (QFP)

The study of the interactions between electrically charged particles and electromagnetic fields within the framework of QFT is called quantum electrodynamics (QED). QED, and in particular its renormalized perturbative formulation, was modeled by various theories to describe other interactions, and thus became the starting point for a new research programme, the quantum field programme (QFP). The programme has been implemented by a series of theories, whose developments are strongly constrained by some of its characteristic features, which have been inherited from QED. For this reason, I shall start this review of the sinuous evolution of QFP with an outline of these features.

8.1 Essential features

QED is a theoretical system consisting of local field operators that obey equations of motion, certain canonical commutation and anticommutation relations (for bosons and fermions, respectively), and a Hilbert space of state vectors that is obtained by the successive application of the field operators to the vacuum state, which, as a Lorentz invariant state devoid of any physical properties, is assumed to be unique. Let us look in greater detail at three assumptions that underlie the system.

First is the locality assumption. According to Dirac (1948), 'a local dynamical variable is a quantity which describes physical conditions at one point of space-time. Examples are field quantities and derivatives of field quantities,' and 'a dynamical system in quantum theory will be defined as localizable if a representation for the wave function can be set up in which all the dynamical variables are localizable'. In QED, this assumption, as a legacy of the point model of particles and its description of interactions among them, takes the form that field operators on a spacelike surface commute (bosons) or anticommute (fermions) with each other, which guarantees that measurements of field quantities at relatively spacelike positions can be made independently of one another.

Second is the operator field assumption. When Jordan in 1925 and Dirac in 1927 extended the methods of quantum mechanics to electromagnetism, the electromagnetic field components were promoted from classical commuting variables to quantum-mechanical operators. The same procedure could also be applied to fields describing fermions (Jordan and Klein, 1927; Jordan and Wigner, 1928). These local field operators have a direct physical interpretation in terms of the creation and annihilation of the quanta associated with the particles, which are realized as a localized excitation or deexcitation of the field. According to the uncertainty principle, a strictly localized excitation implies that arbitrary amounts of energy and momentum are available for the creation of particles. Thus the result of applying a field operator to the vacuum state is not a state containing a single particle, but rather a superposition of states containing arbitrary numbers of particles that is constrained only by the conservation of the relevant quantum numbers.[1] Physically, this means that unlike corpuscularism, the ontology underlying the mechanism of interaction in QED is essentially the field rather than the particle. Mathematically, this means that an operator field is defined by the totality of its matrix elements. It should be clear now that an overwhelming proportion of these matrix elements refer to energies and momenta that are far outside experimental experience.

Third is the plenum assumption of the bare vacuum. There are many arguments against the plenum assumption, the strongest among them being based on covariance reasoning: since the vacuum must be a Lorentz invariant state of zero energy, zero momentum, zero angular momentum, zero charge, zero whatever, it must be a state of nothingness (Weisskopf, 1983). However, when certain phenomena supposed to be caused by the vacuum fluctuations were analysed, the very same physicists who objected to the plenum assumption tacitly took the vacuum as something substantial, namely, as a polarizable medium, or assumed it to be an underlying substratum or the scene of wild activity. In other words, they actually did adopt the plenum assumption.

In sum, considering the fact that in QED both interacting particles and the agent for transmitting the interaction are the quanta of fermion fields and the electromagnetic field respectively, that is, they are the manifestation of a field ontology, QED should be taken as a field theory. Yet the local fields in QED are not fields in the usual sense. As locally quantized fields, they have to a great extent lost their continuity. And this novel aspect of quantum fields turns out to have a profound impact on the formulation of the mechanism of interaction in QFP.

Since interactions in QED are realized by local coupling among field quanta and the transmission of what are known as discrete virtual quanta (rather than a continuous field), this formulation is thus deeply rooted in the concept of localized excitation of operator fields, via the concept of local coupling. Thus in QED calculations of interactions, one has to consider virtual processes involving arbitrarily high energy. However, except for the consequences imposed by such general constraints as unitarity, there exists essentially no empirical evidence for believing the correctness of the theory at these energies. Mathematically, the inclusion of these virtual processes at arbitrarily high energy results in infinite quantities that are obviously undefinable. Thus the divergence difficulties are intrinsic, rather than external, to the very nature of QED. They are constitutive within the canonical formulation of QFT. In this sense the occurrence of the divergences clearly pointed to a deep inconsistency in the conceptual structure of unrenormalized QED. Or put in another way, any quantum field theory can be considered a consistent theory only when it is renormalizable. And this is how, as we shall see, the further development of QFP is constrained.

8.2 Failed attempts

After some spectacular successes of the renormalized perturbative formulation of QED in the late 1940s, QED was taken as a prototype on which were modeled theories of nuclear weak and strong interactions. But attempts to apply the same methods to the nuclear interactions failed. These failures explain why physicists' interest in QFT declined in the 1950s.

Fermi's theory of weak interactions, originally proposed in 1933 to describe beta decay, was modified in the 1950s to include parity violation, and re-formulated in the form of the massive intermediate vector-meson (W-meson) theory,[2] which thereby shared the same theoretical structure as QED. While QED was thought to be renormalizable, Kamefuchi pointed out in (1951) that the Fermi four-fermion direct interactions were non-renormalizable. Later, it turned out that even the W-meson theory was also not renormalizable. Here are the reasons for this defect.

For a theory to be renormalizable, the number of types of primitive divergent graph has to remain finite as we go to higher orders, so that the infinities can be absorbed in a limited number of parameters, such as masses and coupling constants, which are arbitrary parameters and can be determined by experiment. In Fermi's original formulation (four-line coupling), however, the dimensional coupling constant G_w ($\sim m^2$) implied that higher powers of G_w were associated with worse infinities and more numerous types of divergent integral.

Thus an unlimited number of arbitrary parameters were required to absorb the infinities. In the W-meson theory, the coupling constant is dimensionless, but the large-momentum asymptotic form of the massive vector-meson propagator $q^\mu q^\nu / m^2 q^2$ contains the dimensional coefficient m^{-2}. It is non-renormalizable because the factors of m^{-2} in higher orders of perturbation theory have to be compensated by more and more divergent integrals.

The situation for the meson theory of strong nuclear force, in particular in the form of the pseudoscalar coupling of pions to nucleons, was more complicated. Formally, the pseudoscalar coupling was renormalizable. Yet its renormalizability was not realizable because the coupling constant of the renormalizable version was too large to allow the use of perturbation theory, which is the only framework within which the renormalization procedure works. More specifically, the process of renormalization is practically possible only in a perturbative approach in which it involves a delicate subtraction of one 'infinite' term from another at each step in the perturbative calculation. The perturbative approach itself depends crucially on the fact that the coupling constant is relatively small. Then it is immediately obvious that perturbative techniques cannot work for the meson theory, whose corresponding coupling constant is greater than one ($g^2/hc \approx 15$). This failure of the QED-type QFT paved the way for the popularity of the dispersion relation approach, and for the adoption of Chew's S-matrix theory approach, in which the whole framework of QFT was rejected by a considerable number of theorists (See section 8.5).

But even within the domain of electromagnetism, the failure of QFT was also visible. David Feldman observed in (1949) that the electromagnetic interactions of the vector-meson were non-renormalizable. In 1951, Peterman and Stueckelberg noted that the interaction of a magnetic moment with the electromagnetic field (a Pauli term of the form $f\psi\sigma_{\mu\nu}\psi F_{\mu\nu}$) was not renormalizable. Later, Heitler (1961) and others noted that the mass differences of particles (such as the pions and the nucleons) that were identical except for their electric charge could not be calculated using renormalization theory. It is not difficult to establish that if the mass differences were of electromagnetic origin, then the divergent electromagnetic self-energy would lead to infinite mass differences. This difficulty clearly indicated that renormalization theory could not fulfill Pauli's hope that it would provide a general theory to account for the mass ratios of the 'elementary particles'.

In addition to these failures, the renormalized formulation of QFT was also criticized as being too narrow a framework to accommodate the representations of such important phenomena as CP-violating weak interactions and gravitational interactions. But the gravest defect of renormalization theory became apparent since the late 1960s when it was recognized that it was in direct and irreconcilable conflict with the chiral and trace anomalies that occur in high orders of QFT, ironically as a consequence of the demand of renormalization (see section 8.7).

These failures of QFT created a sense of crisis among physicists, and required a clarification of what would be the adequate attitude toward the renormalizability of a quantum field theory.

8.3 Various attitudes toward renormalizability

A fundamental question facing physicists, since the late 1940s, has been whether all interactions in nature are renormalizable, and thus whether only renormalizable theories are acceptable.

Dyson was aware that the answer to the question was negative and reported so to the Oldstone conference (Schweber 1986). This position was supported by detailed, explicit, negative examples (see section 8.2) that appeared immediately after the publication of Dyson's classic papers on the renormalizability of QED.

For other physicists, such as Bethe, who had elevated renormalizability from a property of QED to a regulative principle guiding theory selection, the answer was affirmative (see Schweber, Bethe, and Hoffmann, 1955). They justified their position in terms of predictive power. Their argument was that since the aim of fundamental physics is to formulate theories that possess considerable predictive power, 'fundamental laws' must contain only a finite number of parameters. Only renormalizable theories are consistent with this requirement. While the divergences of non-renormalizable theories could possibly be eliminated by absorbing them into appropriately specified parameters, an infinite number of parameters would be required and such theories would initially be defined with an infinite number of parameters appearing in the Lagrangian. According to their principle of renormalizability, the interaction Lagrangian of a charged spin-1/2 particle interacting with the electromagnetic field cannot contain a Pauli moment. By the same reasoning, Fermi's theory of weak interactions lost its status as a fundamental theory. Another application of the renormalization constraint was the rejection of the pseudovector coupling of pions to nucleons in the strong interactions.

However, the internal consistency of renormalization theory itself was challenged by Dyson, Källen, Landau, and others. In (1953), Källen claimed to be able to show that, starting with the assumption that all renormalization constants are finite, at least one of the renormalization constants in QED must be infinite. For several years this contradictory result was accepted by most physicists as evidence of the inconsistency of QED. However, as was later pointed out by some critics (e.g. Gasiorowicz *et al.*, 1959), his results depended upon some notoriously treacherous arguments involving interchanges of the orders of integration and summation over an infinite number of states, and were thus inconclusive. Källen himself later acknowledged this ambiguity (1966).

A more serious argument challenging the consistency of renormalization theory was expressed in terms of the breakdown of perturbation theory. As is well known, Dyson's renormalization theory was only formulated within the framework of perturbation theory. The output of perturbative renormalization theory is a set of well-defined formal power series for the Green's functions of a field theory. However, it was soon realized that these series (and in particular the one for the S-matrix) were most likely divergent. Thus theorists were thrown into a state of confusion and could not give an answer to the question, in what sense does the perturbative series of a field theory define a solution? Interestingly enough, the first theorist to be disillusioned by perturbative renormalization theory was Dyson himself. In 1952, Dyson gave an ingenious argument that suggested that after renormalization all the power series expansions were divergent. The subsequent discussion by Hurst (1952), Thirring (1953), Peterman (1953a, b), Jaffe (1965), and other axiomatic and constructive field theorists added further weight to the assertion that the perturbative series of most renormalized field theories diverge, even though there is still no complete proof in most cases.

A divergent perturbative series for a Green's function may still be asymptotic to a solution of the theory. In the mid-1970s the existence of solutions for some field-theoretical models was established by constructive field theorists, and these indicated a posteriori that the solution is uniquely determined by its perturbative expansion (Wightman, 1976). Yet these solutions were exhibited only for field-theoretical models in spacetime continua of two or three dimensions. As far as the more realistic four-dimensional QED was concerned, in 1952 Hurst had already suggested that the excellent agreement of QED with experiments indicated that the pertubative series might be an asymptotic expansion.

However, the investigations of the high energy behavior of QED by Källen, Landau, and especially Gell-Mann and Low (1954) showed that the perturbative approach in QED unavoidably breaks down, ironically, as a consequence of the necessity of charge renormalization. Landau and his collaborators argued further that remaining within the perturbative framework would lead either to no interaction (zero renormalized charge),[3] or to the occurrence of ghost states rendering the theory apparently inconsistent (Landau, 1955; Landau and Pomeranchuck, 1955). Both results demonstrated the inapplicability of perturbative theory in renormalized QED.

After the discovery of asymptotic freedom in a wide class of non-Abelian gauge theories, especially in quantum chromodynamics (QCD), the hope was expressed that perturbative QCD would get rid of the Landau ghost and would thus eliminate most doubts as to the consistency of QFT. However, this expectation did not last long. It was soon realized that the ghost which disappeared at high energy reappeared at low energy (Collins, 1984). Thus field theorists were reminded, forcefully and persistently, of the limits of applicability of perturbative theory. As a result, the consistency problem of QFT in general, and of its perturbative formulation in particular, is still in a state of uncertainty.

The attitude of theoretical physicists towards the issue of renormalizability differed sharply. For most practicing physicists, consistency is just a pedantic problem. As pragmatists, they are only guided by their scientific experiences and have little interest in speculating about the ultimate consistency of a theory.

The position adopted by Landau and Chew was more radical and drastic. What they rejected was not merely particular forms of interactions and perturbative versions of QFT, but the general framework of QFT itself (see section 8.5). For them, the very concept of a local field operator and the postulation of any detailed mechanism for interactions in a microscopic space-time region were totally unacceptable, because these were too speculative to be observable, even in principle. Their position was supported by the presence of divergences in QFT and by the lack of a proof of the consistency of renormalization theory, even though Landau's arguments for the inconsistency of renormalized QED could not claim to be conclusive.

Schwinger's view of renormalization is of particular interest, not merely because he was one of the founders of renormalization theory, but principally because he gave penetrating analyses of the foundations of the renormalization programme, and was one of its most incisive critics. According to Schwinger, the unrenormalized description, which adopts local field operators as its conceptual basis, contains speculative assumptions about the dynamic structure of the physical particles that are sensitive to details at high energy. However, we have no reason to believe that the theory is correct in that domain. In accordance with Kramers' precept that QFT should have a structure-independent character, which Schwinger accepted as a guiding principle, the renormalization procedure that he elaborated removed any reference to very high energy processes and the related small distance and inner structure assumptions. He thus helped shift the focus from the hypothetical world of localized excitations and interactions to the observed world of physical particles.

However, Schwinger found that it is unacceptable to proceed in this tortuous manner of first introducing physically extraneous structural assumptions, only to delete them at the end in order to obtain physically meaningful results. This criticism constitutes a rejection of the philosophy of renormalization. But renormalization is essential and unavoidable in a local operator field theory if the latter is to make any sense. To bring his criticism to its logical conclusion, Schwinger (1970, 1973a, b) introduced numerically valued (non-operator) sources and numerical fields to replace the local field operators. These sources symbolize the interventions that constitute measurements of the physical system. Furthermore, all the matrix elements of the associated fields, the operator field equations, and the commutation relations can be expressed in terms of the sources. In addition, it has shown that an action principle can give succinct expression to the whole formalism.

According to Schwinger, his source theory takes finite quantities as primary, and thus is free of divergences. This theory is also sufficiently malleable to be able to incorporate new experimental results, and to extrapolate them in a reasonable manner. Most importantly, it can do so without falling into the trap of having to extend the theory to arbitrarily high energies, which constitute unexplored domains where new, unknown physics is sure to be encountered.

Thus, from Schwinger's perspective, the ultimate fate of renormalization is to be eliminated and excluded from any description of nature. He tried to implement this by abandoning the concept of a local operator field, thus drastically altering the foundations of QFT. The radical character of Schwinger's approach, the foundations of which were laid in his 1951 paper and elaborated in the 1960s and 1970s, was not recognized until the mid-1970s when the renormalizability principle began to be challenged. By that time new insights into renormalization and renormalizability had been gleaned from studies using renormalization group methods, resulting in a new understanding of renormalization and QFT, and also in novel attitudes towards scientific theories in general. Thus a renewed interest in non-renormalizable theories appeared, and the 'effective field theory' approach began to gain its popularity (see section 11.4). In this changed conceptual context, some perspicacious theorists began to realize that Schwinger's ideas were essential in the radical shift of outlook in fundamental physics (Weinberg, 1979).

8.4 The axiomatic approach

Although QFT began to fade away in the 1950s, it did not die. There were physicisits who took a positive attitude toward QFT. The most positive attitude was taken by the axiomatic field theorists,[4] who struggled to clarify the mathematical basis of QFT, with the hope of removing its most glaring inconsistencies.

In the spirit of Hilbert's tradition, the axiomatic field theorists tried to settle the question of the internal consistency of QFT by axiomatization, and took this as the only way to give clear answers to conceptual problems.

While Hilbert tried to legitimize the use of mathematical entities with a proof of the consistency of a formal system consisting of these entities, the axiomatic field theorists went about it the other way round. They tried to prove the internal consistency of QFT by constructing non-trivial examples whose existence was a consequence of the axioms alone. Without radically altering the foundations of QFT, they tried to overcome the apparent difficulties with its consistency step by step. Although many important problems remained, nowhere did they find any indication that QFT contained basic inconsistencies.

For the axiomatic field theorists, the local fields are not entirely local. They are treated as operator valued, and are said to exist only in a distributional sense. That is, the local fields can only be understood as operator-valued distributions that are defined with infinitely differentiable test functions of fast decrease at infinity, or with test functions having compact support. Thus, for the axiomatic field theorists, it is meaningless to ask for the value of a field at a spacetime point P. What is important in this approach is the smeared value of the field in as small a neighborhood of P as we like by letting it act on test functions whose supports are contained in the chosen neighborhood. Essentially, this is a mathematical expression of the physical idea of modifying the exact point model. Considering the fact that the source of divergence in QFT lies in the point model, and that the essence of renormalization is the absorption of the infinities, which is equivalent to blurring the exact point model, the concept of smearing in the axiomatic field theory seems to suggest a deep link between the basic motives of the axiomatic field theory and the theory of renormalization.

For axiomatic field theorists, an unrenormalized theory is certainly inconsistent, because of the presence of ultraviolet divergences and by virtue of the infinities that stem from the infinite volume of spacetime. The latter has to be disentangled from the former and excludes the Fock representation as a candidate for the Weyl form of the canonical commutation relations. The occurrence of these two kinds of infinity makes it impossible to define a Hamiltonian operator, and the whole scheme for the canonical quantization of QFT collapses.

The consistency problem in renormalized theories is very different from that of unrenormalized ones. The ultraviolet divergences are supposed to be circumventable by the renormalization procedure. Some of the remaining difficulties, such as how to define local fields and their equivalence class, and how to specify asymptotic conditions and the associated reduction formula, the axiomatic field theorists claim, can be analyzed in a rigorous fashion with the help of distribution theory and normed algebra (Wightman, 1989).

Of course, a theory defined in this way may still be non-renormalizable in the sense of perturbation theory. Since the mid-1970s, however, there have been major efforts using the approach of constructive field theory to understand the structure of non-renormalizable theories and to establish the conditions under which a non-renormalizable theory can make sense. One of the striking results of this enterprise is that the solutions of some non-renormalizable theories have only a finite number of arbitrary parameters. This is contrary to their description in terms of the perturbative series. It has been speculated that the necessity for an infinite number of parameters to be renormalized in perturbation theory may come from an illegitimate power series expansion (Wightman, 1986).

It is true that in these efforts the axiomatic and constructive field theorists have exhibited an openness and a flexible frame of mind. Yet future developments in understanding the foundations and proving the consistency of the renormalization theory may involve changes in assumptions that have not yet been challenged, and that have not been captured by any axiomatization of the present theory. In any case, the failure to construct a soluble four-dimensional field theory, despite intensive efforts for nearly four decades, indicates that the axiomatic and constructive field theorists have considerable difficulty in solving the consistency problem of QFT in Hilbert's sense. This has also dampened their initial optimism somewhat.

An interesting but unexplored topic is the role played by the development of axiomatic field theory, in the period 1955–56, in the making of S-matrix theory. According to Lehmann, Symanzik, and Zimmermann (LSZ) (1955, 1957), and Wightman (1956), a quantum field theory, or its S-matrix elements, can be formulated directly in terms of Green's functions (or retarded functions or vacuum expectation values of products of fields) alone. The LSZ reduction formula for Green's function, as Wightman (1989) once pointed out, 'was the starting point of innumerable dispersion theory calculations', and the LSZ reduction formulae for the $2 \rightarrow n$-particle reactions in terms of retarded functions 'were, for $n = 2$, the starting point for proofs of dispersion relations'. It is worth noting that Nambu, before making important contributions to dispersion relations (1957), published two papers (1955, 1956) in the period 1955–56, on the structure of Green's function in QFT. These papers were in the spirit of the axiomatic field theory and paved the way for double dispersion relations, which was a crucial step for the conceptual development of the S-matrix theory (see section 8.5).

8.5 The S-matrix theory

As mentioned above, the sense of crisis in QFT went very deep in the 1950s. In addition to the failures in formulating renormalizable theories for weak and strong interactions, Landau and his collaborators (1954a, b, c, d; 1955) pointed out in the mid-1950s that local field theories seemed to have no solution except for non-interacting particles. They argued that renormalizable perturbation theory worked only when infinite terms were removed from power series, and that in general there was no non-trivial, non-perturbative, and unitary solution for any QFT in four dimensions. While formal power-series solutions were generated and satisfied unitarity to each order of the expansion, the series itself could be shown not to converge. They also claimed to have proved that QED had to have zero renormalized charge.[5] Landau took these conceptual difficulties as a proof of the inadequacy of QFT as a research framework that was based on the concepts of quantized local field operators, microspacetime, microcausality, and the Lagrangian approach. He advocated that the inconsistency of QFT seemed to have its root in these unobservables. The arguments that Landau developed against QFT, though not taken seriously by most physicists at the time, did have strong philosophical appeal for some radical physicists.

Another conceptual difficulty in QFT was related to its atomistic thinking. One of the basic assumptions of QFT was that the fundamental fields described by the field equations were associated with the 'elementary particles', which in turn were considered to be the basic building blocks of the entire universe. In the early 1930s it was already realized that despite the existence of beta decay, the neutron should not be considered a composite of a proton and an electron. Thus the division of matter into elementary building blocks could be regarded only as an approximate model, very successful in non-relativistic situations but inapplicable in high-energy physics. It was impossible to tell whether the proton or neutron was the more elementary particle. Each had to be regarded as equally elementary. This same idea had to be applied to the profusion of new particles discovered in the late 1950s and early 1960s. It was intuitively clear that these objects could not all be elementary. QFT was then forced to make a sharp distinction between elementary and composite particles. Not until the 1960s did research on this topic appear in the literature, but no satisfactory criterion for such a distinction was given.[6] Thus within the scope of strong interactions it seemed pointless to explore QFT with the few particles that happened to have been discovered first, with these treated as elementary particles.

In addition, at the beginning of the 1960s, QFT was also branded as empirically inadequate, since it offered no explanation for the existence of a family of particles associated with a Regge trajectory. Related to this were other difficulties: QFT seemed unable to produce analytic scattering amplitudes and also seemed to introduce arbitrary subtraction parameters into dispersion relations.

Against this context, a new trend manifested in the developments of dispersion relations, Regge poles and the bootstrap hypothesis was orchestrated by Geoffrey Chew and his collaborators into a highly abstract scheme, the analytic S-metrix theory (SMT), which, in the 1950s and 1960s, became a major research programme in the domain of strong interactions.

SMT was originally proposed by Heisenberg in (1943a, b; 1944) as an autonomous research programme to replace QFT. His criticism of QFT focused on the divergence difficulties, which suggested to him that a future theory would contain a fundamental length. The purpose of his work on SMT was to extract from the foundations of QFT those concepts which would be universally applicable and hence contained in a future true theory. Lorentz invariance, unitarity, and analyticity[7] were concepts of this nature and were incorporated in SMT. The S-matrix was determined by asymptotic states and hence was a quantity immediately given by experiments. Furthermore, SMT seemed not to encounter any divergence difficulties. Finiteness was, in fact, the main objective of Heisenberg's research programme, and the advent of the renormalization programme removed the necessity of this objective. Together with some other difficulties encountered earlier, renormalization made Heisenberg's SMT recede into the background. However, questions posed by this programme, such as 'what is the connection between causality and analyticity of S-matrix elements?' (Jost, 1947; Toll, 1952, 1956) and 'How does one determine the interaction potential of Schrödinger theory from the scattering data?' led to the dispersion theory programme of Gell-Mann and Goldberger (1954; with W. Thirring, 1954).[8]

The dispersion theory programme provided a dynamical scheme for calculating the S-matrix elements that was lacking in Heisenberg's programme. Dyson's approach to calculating the S-matrix elements (1949a, b), which was also motivated by Heisenberg's programme, was quite different in nature since it explicitly referred to a field theory. The original motivation for the dispersion programme proposed by Gell-Mann, Goldberger, and Thirring was to extract exact results from field theory. The analyticity of the scattering amplitudes and the dispersion relations they obeyed were derived from the assumption of microcausality for observable field operators. The other 'principle' used, namely crossing symmetry, was regarded as a general property satisfied by the Feynman diagrams representing the perturbation expansion of a field theory.

At the Rochester Conference held in 1956, Gell-Mann (1956) claimed that crossing, analyticity, and unitarity would determine the scattering amplitude if suitable boundary conditions were imposed in momentum space at infinite momenta. He further claimed that this would almost be enough to specify a field theory. Thus the dispersion theory programme was considered by Gell-Mann, one of its inventors, as no more than a way of formulating field theory on the mass shell (with imaginary momenta being included). This was the case, although at the Rochester Conference Gell-Mann also suggested that the dispersion programme, if treated non-perturbatively, was reminiscent of Heisenberg's hope of writing down the S-matrix directly instead of calculating it from field theory.

The results obtained in the dispersion programme in fact were quite independent of the details of field theory. Unitarity related the imaginary part of any scattering amplitude to a total cross section that involved squares of scattering amplitudes, and causality (via analyticity) related the real and the imaginary part of the scattering amplitude to each other. In this way a closed, self-consistent set of non-linear equations was generated (which could be solved either perturbatively or in some non-perturbative way) that would enable physicists to obtain the entire S-matrix from fundamental principles, and free them from the conceptual and mathematical difficulties of QFT. It is such an inner logical connection between the dispersion programme and Heisenberg's programme that made Gell-Mann's 'casual' mention of Heisenberg's SMT a strong impetus for the further development of analytic SMT in the 1950s and 1960s.[9]

However, as a dynamical scheme, dispersion theory lacked a crucial ingredient: a clear concept of what in it was the counterpart of the notion of 'force'. The calculational scheme in this theory was carried out on the mass shell, and hence involved only asymptotic states, in which particles were outside each other's regions of interaction. Thus a question that is crucial to enabling the scheme to describe the interaction is how to define an analytic function representing the scattering amplitude. The original dispersion relations formulated by Gell-Mann, Goldberger, and Thirring (1954) explicitly exhibited particle poles, which were mathematically necessary for defining an analytic function, since otherwise by Liouville's theorem the latter would be an uninteresting constant. Initially, however, no emphasis was given to this feature of the theory, let alone a proper understanding of the role played by the singularities of the S-matrix, namely a role in providing a concept of 'force' in the new programme different from that in QFT.

Progress in this direction was made by: (i) combining Goldberger's relativistic dispersion relations (1955a, b) with Chew and Low's theory of pion–nucleon scattering (1956); (ii) combining Gell-Mann and Goldberger's crossing symmetry (1954) with Mandelstam's double dispersion relations (1958); and (iii) extending analytic continuation from linear to angular momentum (Regge, 1958a, b; 1959, 1960). Let us look at each of these advances in turn.

(i) The Chew–Low theory combined Chew's static model for pion–nucleon scattering (1953a, b) with Low's nonlinear integral equations for the scattering amplitudes (1954), and involved only renormalized quantities. It had a solution in the one-meson approximation that satisfied unitarity and crossing; it had a single pole and good asymptotic behavior, but it suffered from the CDD ambiguity.[10] The notion of force was explicit in traditional Lagrangian field theory. However, the Chew–Low formulation, by employing analytic functions, allowed more direct contact with experimental data. The poles of the analytic functions were the key to such contact. The position of the pole was associated with the particle mass, and the residue at the pole was connected with 'force strength' (coupling constant). Making this semi-relativistic model fully relativistic led to a new concept of force: force resided in the singularities of an analytic S-matrix. (Chew, Goldberger, Low, and Nambu, 1957a, b). Furthermore, the dispersion relations themselves were to be understood as Cauchy–Riemann formulae expressing an analytic S-matrix element in terms of its singularities.[11]

(ii) Crossing was discovered by Gell-Mann and Goldberger in (1954) and was at first understood as a relation by which incoming particles became outgoing antiparticles. Mandelstam's double dispersion relations (1958) were an extension of analytic continuation in the energy variable to the scattering angle, or with Mandelstam's variables, in s (the energy variable in the s channel) to t (the invariant energy in the t channel or, by crossing, the momentum transfer variable in the s channel). One of the important consequences of Mandelstam's double dispersion relation is that it converts the concept of crossing into a new dynamical scheme in which the force in a given channel resides in a pole of the crossed channel.

This new scheme in turn led to the bootstrap approach to understanding the hadrons. The double dispersion relations enabled Chew and Mandelstam (1960) to analyse pion–pion scattering in addition to pion–nucleon scattering. They found that a spin 1 bound state of two pions, later named the ρ meson, constituted a force which, by crossing, was the agent for making the same bound state; that is, the ρ meson as a 'force' generated the ρ meson as a particle. The bootstrap concept was introduced as a result of this calculation. It did not take long for Chew and Frautschi (1961a, b) to suggest that, as in the ρ meson case, all hadrons were bound states of other hadrons sustained by 'forces' represented by hadron poles in the crossed channel. The self-generation of all hadrons by such a bootstrap mechanism could be used, hopefully, to self-consistently and uniquely determine all of their properties.

The Mandelstam representation was proved to be equivalent to the Schrödinger equation in potential scattering theory.[12] Furthermore, it was shown that the Schrödinger equation could be obtained as an approximation to the dispersion relations at low energies.[13] This lent support to the conjecture that an autonomous programme that had a well-established non-relativistic potential scattering theory as its limiting case could replace QFT. In this programme, the dynamics were not specified by a detailed model of interaction in spacetime, but were determined by the singularity structure of the scattering amplitudes, subject to the requirement of maximal analyticity. The latter required that no singularities other than those demanded by unitarity and crossing be present in the amplitude.[14] A further difficulty remaining in these bootstrap calculations was posed by the problem of asymptotic conditions. For potential scattering, the difficulty was settled by Tullio Regge.

(iii) In proving a double dispersion relation in potential theory, Regge (1958a, b; 1959; 1960) was able to show that it was possible to continue the S-matrix simultaneously into the complex energy (s) plane and into the complex angular momentum (j) plane. Consequently, the position of a particular pole in the j plane was an analytic function of s, and a fixed pole ($\alpha(s) = $ constant) was not allowed. If at some energy ($s > 0$) the value of Re $\alpha(s)$ passed through a positive integer or zero, one had at this point a physical resonance or bound state for spin equal to this integer. So, in general, the trajectory of a single pole in the j plane as s varied corresponded to a family of particles of different js and different masses. However, when $s < 0$,[15] the Regge trajectory $\alpha(s)$ was shown by Regge to control the asymptotic behavior of the elastic-scattering amplitude, which was proportional to $t^{\alpha(s)}$, with $\alpha(s)$ negative when s was sufficiently negative. Therefore, only in those partial waves with $l < \alpha(s)$ could there be bound states. This implied that the trajectory for $s < 0$ could be detected in the asymptotic behavior of the t channel, and, most importantly, that the divergence difficulty that plagued QFT could at last be avoided.

These results were noted by Chew and Frautschi (1960, 1961c) and were assumed to be true also in the relativistic case. The resulting extension of the maximal-analyticity principle from linear momentum ('first kind') to angular momentum ('second kind') implied that the force in a channel resides in the moving singularity of the crossed channel, and made it necessary to restate the bootstrap hypothesis as follows: all hadrons, as the poles of the S-matrix, lie on Regge trajectories.

This version of the bootstrap hypothesis provided the basis for a dynamic model of hadrons. In this model, also known as nuclear democracy, all hadrons could be regarded as either composite particles or as constituents, or binding forces, depending on the specific process in which they were involved. The basic concern was not with the particles, but with their interaction processes, and the question of the structure of hadrons was reformulated in terms of the structure of hadron-reaction amplitudes.[16]

According to Chew and Frautschi, the criterion for distinguishing composite from elementary particles was whether a particle lay on the Regge trajectory or not: composite particles were associated with a Regge pole moving in the j plane as a function of s, whereas elementary-particle poles were associated with unique angular momenta, and did not admit continuation in j. From this criterion they claimed that

If $j = 1/2$ and $j = 0$ elementary particle poles actually occur in nature, it may be argued that working directly with the S-matrix is simply a technique for evaluating conventional field theory. On the other hand, if all baryon and meson poles admit continuation in the j plane, the conventional field theory for strong interactions is not only unnecessary but grossly misleading and perhaps even wrong.

(Chew and Frautschi, 1961a)

This was a serious challenge to QFT, and the response by Gell-Mann and other field theorists to it led to an attempt at reconciling QFT with SMT, that is, the Reggeization programme.[17]

As a fundamental framework opposed to QFT, SMT was the result of contributions made by a subset of particle theorists in the 1950s and 1960s. The central ideas and principles, however, were mainly provided by Landau and Chew. Landau's contributions in laying the foundations of SMT consisted in: (i) arguing for the failure of QFT to deal with interactions; (ii) helping to recognize the generality of the correspondence between hadrons and poles of the analytic S-matrix;[18] and (iii) formulating the graphical representation and rules for S-matrix singularities.[19] The Landau graph technique was distinct from Feynman's, and was by no means equivalent to perturbation theory: contributions to a physical process from all relevant particles were included in the singularities,[20] and lines in the graph corresponded to physical hadrons representing asymptotic states. No renormalization had to be considered. Landau graphs representing physical processes thus became a new subject of study and an integral part of the foundations of SMT.[21] In fact, Landau's paper (1959) on graph techniques became the point of departure for an axiomatic analytic SMT developed in the 1960s.[22]

However, the unifying force and the real leader of the movement away from QFT towards SMT was Chew. In addition to his original contributions in establishing the new dynamical scheme throughout all of its three stages, Chew also provided some philosophical arguments to support this dynamical model of hadrons. He passionately argued against the atomistic paradigm adopted in QFT, and rejected the idea of arbitrarily designating elementary particles. Influenced by the ideas of Gell-Mann and Landau, and stimulated by his own researches on Regge poles and bootstrap calculations, Chew, during the course of writing his book on the *S-Matrix Theory of Strong Interactions* (1961), gave a talk at a conference at La Jolla in 1961, in which he indicated the connection between his ideas and those of Heisenberg's old SMT. In this famous lecture Chew severed his ties to QFT and adopted SMT.[23] Chew took an extremely radical position, insisting that QFT had to be abandoned and SMT used in strong-interaction physics. In January 1962 at an American Physical Society meeting in New York, Chew (1962a) declared that he was completely committed to the analytic SMT and that he rejected the view held by his 'oldest and closest friends' that 'field theory is an equally suitable language'. For him, 'the basic strong interaction concepts, simple and beautiful in a pure S-matrix approach, are weird, if not impossible, for field theory'.

Chew's anti-field-theoretical position was based mainly on two of his hypotheses: (i) nuclear democracy and the bootstrap hypothesis; and (ii) maximal analyticity. According to (i), all of the hadrons are composite, all of the physically meaningful quantities are self-consistently and uniquely determined by the unitarity equations and the dispersion relations, and no arbitrarily assigned fundamental quantities (such as those related to the elementary particles or basic fields that occurred in QFT) are allowed. The hypothesis of maximal analyticity entails that the S-matrix possesses no singularities other than the minimum set required by the Landau rules, or that all of the hadrons lie on Regge trajectories. At that time the latter property was generally understood to be characteristic of composite particles. According to Chew, the Regge hypothesis provided the only way of realizing the bootstrap idea. He argued that QFT was incompatible with the maximal analyticity principle because it assumed elementary particles that could not produce Regge poles. Since low-mass hadrons, particularly the nucleons, were found to lie on Regge trajectories, Chew could break with the traditional belief that the proton was elementary, and claim that all of strong-interaction physics would flow from maximal analyticity, unitarity, and other SMT principles.

For Chew, QFT was unacceptable chiefly because it assumed the existence of elementary particles that inevitably led to a conflict with nuclear democracy and the bootstrap hypothesis, and with analyticity. The non-existence of elementary particles was so crucial to Chew's anti-field-theoretical position that he even chose this criterion to characterize his new theory.[24] In addition to Chew's arguments for the bootstrap and maximal analyticity, his rejection of the concept of elementary particles was also supported by previous work of Kazuhiko Nishijima (1957, 1958), Wolfhart Zimmermann (1958), and Rudolf Haag (1958). These authors claimed that there existed no difference between a composite particle and an elementary particle as far as scattering theory was concerned, since all particles gave rise to the same poles and cuts, irrespective of their origin. Chew could even appeal to Feynman's principle that the correct theory should not allow a decision as to which particles are elementary.[25] Thus from the La Jolla conference of June 1961 to the Geneva conference of July 1962, Chew exerted great pressure upon field theorists.

As an originator and the major advocate of the bootstrap hypothesis, Chew made the principles of self-consistency and uniqueness the philosophical foundations of SMT. Chew strongly influenced the hadronic physics community in the first half of the 1960s. His philosophical position, though viewed as dogmatic and religious even by some of his close collaborators (for example Gell-Mann and Low),[26] was however supported by some well-confirmed physical results that he and his collaborators had obtained within the framework of SMT. The most convincing of these was that a large number of baryon and meson resonances were found to lie on nearly linear Regge trajectories. Another impressive demonstration of the bootstrap dynamics was the calculations of the mass and width of the ρ resonance, the results of which were found to be close to the experimental values.[27] Chew's reciprocal bootstrap calculation of the nucleon, pion, and (3,3) resonance provided a more complicated example of this kind (1962a).

SMT's success in physics and its philosophical appeal made it extremely popular in the early 1960s. Physicists working on hadronic dynamics could not avoid responding to Chew's challenge to QFT. However, except for some of Chew's most faithful followers, it is difficult to find, even among his collaborators, anyone who was wholeheartedly committed to Chew's radical position. Nevertheless, there were some physicists, such as Blankenbecler, who, in trying to adopt the ideas crucial to Chew's position, actually became supporters of it. Others, such as Gell-Mann, while incorporating some of the ideas advocated by Chew, rejected his position of opposing SMT to QFT, and tried to reconcile the two approaches, even at the price of conceptual inconsistency.[28]

In any case, the conceptual pressure exerted by Chew on the foundations of particle physics was strongly felt by almost every particle physicist. For instance, in the first half of the 1960s, ambitious research on bootstrapping internal symmetries was carried out.[29] Another illuminating example involved Steven Weinberg, who later became a prominent field theorist. In 1964–65, Weinberg (1964a, c; 1965b) tried to extend SMT from hadronic to electromagnetic[30] and gravitational interactions, and succeeded in deriving gauge

invariance and Maxwell's equations, as well as the equivalence principle and
Einstein's equations within the SMT framework. Weinberg's objective was to
'question the need for field theory in understanding electromagnetism and
gravitation', which until then were thought to be the most secure examples of a
field theory. At that time it was 'not yet clear' to Weinberg

whether field theory will continue to play a role in particle physics, or whether it will
ultimately be supplanted by a pure S-matrix theory.

(Weinberg, 1964a)

When in his analysis Weinberg used the language of Feynman diagrams, he
asked his readers to 'recognize in this the effects of our childhood training
rather than any essential dependence on field theory' (*ibid.*).

Weinberg's attitude reflected the prevalent atmosphere of the time, which
was characterized by Dyson a year later as follows:

Many people are now profoundly sceptical about the relevance of field theory to
strong-interaction physics. Field theory is on the defensive against the now fashionable
S-matrix . . . .. It is easy to imagine that in a few years the concepts of field theory will
drop totally out of the vocabulary of day-to-day work in high energy physics.

(Dyson, 1965)

The difference between QFT and SMT in foundational assumptions can be
summarized as follows:

(i)   Processes rather than entities are the basic ontology in SMT; the building blocks
       of QFT are elementary fields or their quanta, elementary particles.
(ii) Force in SMT is represented by the singularity structure, and in QFT by the
       propagation of virtual quanta.
(iii) While in QFT the number of fermions is conserved and that of bosons is not, the
       difference between fermions and bosons in SMT is not so explicit.
(iv) There are only composite particles in SMT, whose parameters can be uniquely
       determined by dynamical equations; in QFT there are elementary particles with
       arbitrary parameters.
(v)   SMT produces analytic amplitudes without any fixed singularities; QFT produces
       amplitudes with fixed singularities, which come from the elementary particles
       with fixed spins. This distinction leads to the difference in their high-energy
       behavior, which is much softer in SMT than in QFT.
(vi) Methodologically, SMT starts its calculation of observable events with observa-
       ble quantities alone, without introducing any unobservable entities. The dy-
       namics of the system is determined by general principles, without introducing
       any unobservable mechanism. As a result, no renormalization is required. How-
       ever, QFT starts its calculations with fundamental fields (sometimes unobserva-
       ble as in the case of quarks and gluons) and gauge groups for fixing the dynamics
       of the fundamental system. It gives a detailed description of interactions in

microscopic spacetime. Here the guiding principle is empirical 'trial and error'. Since no consistent solution has been discovered, the perturbation method and renormalization procedure have to be invoked to obtain results comparable to those of experiments.

(vii) Unification realized in SMT with the bootstrap mechanism is horizontal; in QFT it is vertical with downward reduction and upward reconstruction.

The popularity of SMT in hadronic physics began to decline in the mid-1960s, partly because of the mathematical difficulties encountered in dealing with cuts and in trying to extend the theory beyond the two-body channels, partly because the quark idea was gaining popularity at that time. Yet, conceptually, in terms of its foundational assumptions as compared with QFT, SMT was still very powerful for a long time. The bootstrap idea, which demanded that the set of exchanged hadrons in the *t* channels and *u* channels be the same as the set of resonances and bound hadrons formed in the *s* channel, led to the dual-resonance model. It, in turn, suggested a string picture of hadrons and provided the background for the introduction of supersymmetry and superstring theories.

Historically, ever since SMT came into being, QFT as a fundamental framework had always been under conceptual pressure from SMT, just as the latter had been under pressure from the former. Some physicists tried to reduce QFT to a limiting case of SMT,[31] others viewed the SMT principles as properties of QFT.[32] Weinberg regarded QFT as a convenient way of implementing the axioms of SMT (1985; 1986a, b). Few people, however, have studied the interplay between QFT and SMT as two independent research programmes, which may turn out to be helpful in understanding the conceptual development of particle physics.[33]

8.6 The PCAC hypothesis[34] and current algebra

The second category of response to the crisis in QFT was based on the consideration of symmetry. The symmetry approach was first applied to the domain of the weak and electromagnetic interactions of the hadrons. Later, it was extended to the domain of low energy strong interactions. Phenomenologically, these interactions seemed to be well described by effective Hamiltonians in terms of hadron currents. The hadron currents were supposed to be constructed out of hadron fields, yet in a way whose details had to await a decision about fundamental matters concerning the nature of hadrons, the dynamic laws governing the behavior of hadrons (the field equations), and the theory explaining the laws. In the 1950s and 1960s physicists had little knowledge of precise laws other than electromagnetism, and were unable to solve any of the realistic models proposed to explain the dynamics of hadrons. Thus the effort to develop a theory of strong interactions along the lines of QED was generally abandoned among mainstream physicists, although a proposal was made by Yang and Mills, and some responses to it occurred subsequently. As with SMT, the second trend in hadron physics during that period was to put off analyzing the hadron currents from 'first principles'. Instead, it concentrated on symmetries whose implications were supposed to be extractable independently of dynamic details.

In the late 1950s and early 1960s, a new research programme, known as the PCAC hypothesis and current algebra, was gradually taking shape. In this program, the symmetry properties of a local field theory were taken to transcend the dynamic details of the theory and have validity beyond the first few orders of perturbation. This symmetry programme, like the analyticity programme (SMT), was also a quasi-autonomous program. It was *quasi*-autonomous because it had its origin in QFT. It was extracted from QFT and could be studied with field-theoretical models. The current operators, the primary object for investigation in this programme, were supposed to be constructed, in principle, out of local field operators and could be manipulated in the same way as the field operators were manipulated in QFT: a current operator can be inserted into two physical states to express its action upon the initial state, which causes a transition to the final state. It was quasi-*autonomous* because an independent theoretical push was made to extract physical information from some general property of a physical system (symmetry in this programme compared with analyticity, unitarity, etc. in SMT) without recourse to dynamics (field equations). The point of departure was symmetry. Currents were regarded as, primarily, representations of symmetries instead of being derived from dynamical systems, which were unsolvable in that period.

If our knowledge of the physical world could be derived from a priori mathematical symmetry groups which were exhibited in various model Lagrangians and physical currents, then the Platonists and believers in mathematical mysticism might have found some support from this programme. But this was a misinterpretation or, at least, an illusion. The real nature of this programme, on the contrary, was logico-empirical in character. The invariance group structure, such as U(1), SU(2), SU(3), etc., was by no means a priori, but, instead, was suggested by exact or approximate regularities that appeared in the experimental data. The matrix elements of currents were supposed to describe physically observable processes that took place among real particles.

Such a phenomenological appproach possessed all the strength and weakness that were normally associated with the logico-empiricist methodology. On its positive side, the adherence to the directly observable (S-matrix elements, form factors, etc.) helped it circumvent all the difficulties caused by the introduction of unobservable theoretical structures in a local field theory with the renormalization procedure; and the formal manipulations of algebraic relations (current commutators, etc.) not only simplified real situations and made them manipulable, but also appeared to be universally valid. On its negative side, the neglect of unobservable entities and processes in general, and of microscopic dynamics in particular, hampered progress toward a deep understanding of elementary particles and their behavior. Moreover, without a proper understanding of dynamics, there was no guarantee that purely formal manipulations, which often inclined to oversimplify the real situation, would be universally valid.

It is interesting to note, as we shall examine in some detail later in this section, that the evolution of the symmetry programme soon reached its limit around 1967. Not only had its theoretical potential been almost exhausted, but also a few of its predictions based on formal manipulations were found to be in direct conflict with experiments. The latter convincingly indicated that one of the basic assumptions of the symmetry programme, namely that current commutators were independent of dynamical details, was simply wrong. Great effort had been made to clarify the situation, which soon led to a field-theoretical investigation of current algebra. The result was the profound discovery of anomalies in local field theories (see section 8.7). This discovery rightly hit upon the heart of the notion of renormalization. Not only does the occurrence of gauge anomaly destroy the symmetry, and hence the renormalizability of the theory, but more significantly the concept of anomaly provided a basis for properly understanding the renormalization group equations (see section 8.8), and a critical ingredient for the new concept of renormalization (see section 11.4). The irony for the symmetry programme is striking. It started with a rejection of a local field theory in the sense of expelling dynamic considerations. Its development, however, had prepared the way for a return to the field-theoretical framework in order to critically examine the programme itself. Of course, this was not a simple return to the previous situation, but, as a negation of a negation, a higher level, with a deeper understanding of a richer structure to the local field theory and renormalization.

Currents and their matrix elements; form factors

The idea of exploiting symmetry properties of a dynamic system without any recourse to dynamics, which was simply ineffective in the 1950s, was first suggested by the universal occurrence of physical currents in fundamental physics from the late 1950s on. Mathematically, it was well known, through Noether's work, that a conserved current followed from the symmetry of a dynamical system and constituted a representation of the symmetry group. However, the currents investigated in the symmetry programme were not purely mathematical objects. Rather, they first came from descriptions of weak and electromagnetic interactions. Only afterwards, in the process of investigating these currents, were assumptions concerning the conservation or approximate conservation of the currents made, the physical currents identified with symmetry currents, and the symmetry properties of dynamic systems fully exploited.

The prototype of physicists' reasoning was provided by electromagnetism. The electromagnetic current, either leptonic or hadronic, is conserved owing to the conservation of the electric charge, and thus can be identified with the U(1) symmetry current, to which a physical vector boson field is coupled. In this case, the conserved current is both an expression of the symmetry of the system and a medium for exchanging energy in the process of interaction.

The physicists' conception of weak currents was developed after the pattern of electromagnetic currents, with some necessary modifications. In (1955), S. S. Gershtein and Y. B. Zel'dovich first mentioned and discarded an idea that a conserved vector current of isospin (and the corresponding vector bosons) would be somewhat relevant to beta decay. In (1957), Schwinger speculated more seriously, starting from symmetry considerations, the existence of an isotopic triplet of vector bosons whose universal couplings would give both the weak and electromagnetic interactions: the massive charged Z particles would mediate the weak interactions in the same way as the massless photon mediates the electromagnetic interactions. This was the begining of the intermediate-vector-boson theory of the weak interaction, which is of far-reaching influence and significance.

However, more substantive progress in understanding the nature and structure of the weak currents was directly inspired by a mighty advance in theoretical analysis and experimental confirmation of parity non-conservation in weak interactions. In Fermi's original theory, beta decay was written in the vector (V) current form. The observation of the parity non-conservation in 1957 required an axial vector (A) coupling. Assuming maximal violation of parity, E. C. G. Sudarshan and Marshak (1958), and also Feynman and Gell-Mann (1958), proposed that the weak currents took a V–A form. Feynman and Gell-Mann also suggested a universal Fermi theory in which the effective weak interaction Hamiltonian was written as a product of weak currents. In accordance with the idea discarded by Gershtein and Zel'dovich, and with the more sophisticated idea of Schwinger's, Feynman and Gell-Mann speculated that the vector part of the weak current was identical with the isotopic vector part of the electromagnetic current, which was known to be conserved. Thus a connection between the physical weak currents and SU(2) isotopic symmetry was proposed. The connection was further elaborated by Sidney Bludman in (1958), with an extension from isotopic SU(2) symmetry to chiral SU(2) symmetry[35] and a derivation of weak currents from this symmetry.

Bludman's theory was essentially an attempt at a (Yang–Mills-type) gauge theory of weak interactions, in which massive charged vector fields were coupled to the conserved weak currents. An influential theory of strong interactions with a similar theoretical structure was proposed by Sakurai in (1960). In Sakurai's theory, the strong interactions were described by three conserved vector currents: the isospin current, the hypercharge current, and the baryonic current. These currents were coupled to massive vector bosons. However, before integrating the idea of spontaneous symmetry breaking, which was just taking shape at that time (see section 10.3), into QFT, it was impossible to account for the masses of gauge bosons, and the theory soon degenerated into the vector-meson dominance model. (There will be more discussion about the non-Abelian, or Yang–Mills-type, gauge theories in part III.)

In all these cases (Schwinger; Feynman and Gell-Mann; Bludman; and Sakurai), the hadron currents involved in electromagnetic, weak, and strong interactions were supposed to be well-defined Lorentzian four-vectors, constructed out of fundamental hadron fields but which could not be analyzed in a reliable way. These currents can interact with each other, and their couplings can be realized by exchanging vector quanta. Most importantly, their divergences and canonical commutators express the symmetries of the system. Thus the symmetry properties of the system seemed to be exploitable by manipulating these currents in a formal way. Assuming this to be true was the underlying idea of the symmetry programme.

The matrix elements of hadron currents (or charges) to the lowest order in the coupling constant are direct observables, obeying the dispersion relations and being supposed to be dominated by one pole. Putting aside the kinematic factors, which are easy to determine, the rest of the matrix element can be expressed, by using symmetry principles, in terms of a few unknown scalar functions of four-momentum transfer, the so-called form factors, such as the electromagnetic form factors $F_i(q^2)$ $(i = 1, 2, 3)$ defined by

$$\langle p'|_a j_\mu^{\text{em}}(0)|p\rangle_b = u(p')_a[F_1(Q^2)(p' + p)_\mu + F_2(Q^2)(p' - p)_\mu$$
$$+ F_3(Q^2)\gamma_\mu]u(p)_b, \tag{1}$$

where $q^2 = (p' - p)^2$, and the weak form factors $f_v(q^2)$, $f_m(q^2)$, $f_s(q^2)$, $g_A(q^2)$, $g_P(q^2)$ and $g_E(q^2)$, defined by

$$(2\pi)^3 \langle p'|_{s'p} V_\mu(0)|p\rangle_{sn} = iu(p')_{s'p}[\gamma_\mu f_v(Q^2) + \sigma_{\mu v} Q_v f_m(Q^2)$$
$$+ iQ_\mu f_s(Q^2)]u(p)_{sn} \tag{2}$$

and

$$(2\pi)^3 \langle p'|_{s'p} A_\mu(0)|p\rangle_{sn} = iu(p')_{s'p}\{\gamma_5[\gamma_\mu g_A(Q^2) + iQ_\mu g_P(Q^2)$$
$$+ i(p' + p)_\mu g_E(Q^2)]\}u(p)_{sn}, \tag{3}$$

where V_μ and A_μ are the weak vector current and weak axial vector current respectively, and $q_\mu = (p' - p)_\mu$.

These weak (or electromagnetic) form factors were supposed to contain all the physical information concerning the modifications of the basic weak (or electromagnetic) interactions by virtual strong interactions. There was no way to predict the quantitative details of these form factors, because these were determined by the dynamics of the strong interaction, which was not available in the 1950s and 1960s. Nevertheless, these form factors could still be studied effectively. First, symmetry considerations put strong constraints on them, limiting their number and establishing correlations among them. Second, the pole-dominance hypothesis helped to provide concrete suggestions concerning the contributions to the form factors by virtual particles. For example, the existence of the omega-meson was first conjectured by Nambu (1957a) to explain the electron–proton scattering form factors, and then, in 1962, confirmed by experimental discovery. Third, since the form factors are observables, all the analyses based on symmetry (and/or analyticity) considerations could be tested by experiments. Thus the symmetry programme can be regarded as a fully justified research programme. Now, let us turn to a closer examination of the programme from a historical perspective.

The conserved vector current (CVC) hypothesis

Feynman and Gell-Mann speculated in (1958) that the vector part V_μ of the hadron weak current (J_μ) responsible for nuclear beta decay was conserved. The original motivation was to give an explanation of the observation that the vector current coupling constant G_v $(= Gf_v(0))$ in nuclear beta decay was very close to the vector current coupling constant G in muon beta decay. The observation implied that the renormalization factor $f_v(0)^{36}$ for the vector current coupling constant was very close to 1 and was not renormalized by the strong interaction.[37] A current conserved by the strong interaction must be the symmetry current. That is, it is derivable from a symmetry of the strong interactions. Thus it was quite natural for Feynman and Gell-Mann to identify the current V_μ of beta decay with the isotopic vector current. This means, specifically, that the charged vector current V_μ, which raises electric charge in beta decay, was to be identified with the isotopic raising current $\mathscr{J}^v_{\mu+} = \mathscr{J}^v_{\mu 1} + i\mathscr{J}^v_{\mu 2}$, and V^+_μ with the isotopic lowering current $\mathscr{J}^v_{\mu-} = \mathscr{J}^v_{\mu 1} - i\mathscr{J}^v_{\mu 2}$, whereas the isovector part (j^v_μ) of the electromagnetic current j^{em}_μ $(= j^s_\mu + j^v_\mu)$ was the third component of the same isotopic vector current, $\mathscr{J}^v_{\mu 3}$.

Clearly, this hypothesis makes it possible to relate the known electromagnetic processes to some weak processes through isotopic rotations. For example, since for a pion at rest

$$\langle \pi^+ | \mathscr{J}^v_{03} = j^{em}_0 (I=1) | \pi^+ \rangle = 1, \tag{4}$$

the relevant S-matrix element for pion beta decay must be

$$\langle \pi^0 | V_0 | \pi^+ \rangle = 2^{1/2}. \tag{5}$$

Thus the scale of the weak vector current is fixed. Another example is the relation between the form factors for the vector part of nuclear beta decay and those of electron scattering of the neutron and proton:[38]

$$F_i(Q^2)^+ = F_i(Q^2)^P - F_i(Q^2)^N. \tag{6}$$

From these relations, Gell-Mann (1958) made a definite prediction about 'weak magnetism' $f_m(0)$

$$f_m(0) = F_2(0)^P - F_2(0)^N = (\mu_P - \mu_N)/2M = (1.79 + 1.91)/2M = 3.70/2M, \tag{7}$$

which was subsequently confirmed by Y. K. Lee, L. W. Mo, and C. S. Wu (1963).

The CVC hypothesis occupies a prominent place in the history of particle physics. First, it suggested, though did not explain, the connections among the weak, electromagnetic, and strong interactions, and thus revealed their unitary character: the physical weak currents were derived from the SU(2) isospin symmetry of the strong interaction, and connected with the electromagnetic currents under the operations of the same symmetry. Second, since the conservation of a vector current expresses a conservation law, the CVC hypothesis implied an extension of Yang and Mills's idea from the domain of the strong to the weak interaction. The extension was subsequently embodied in Bludman's gauge theory of the weak interaction with a chiral SU(2) × SU(2) symmetry. As a matter of fact, the success of the CVC hypothesis reinforced physicists' faith in the Yang–Mills theory, and, as a positive feedback,

encouraged Sakurai to propose an extended gauge theory of the strong interaction. Third, the conservation of the weak vector currents $J_{\mu i}^V$ implied that the
associated weak charges Q_i^V are the generators of SU(2) weak isospin symmetry. They generate an SU(2) Lie algebra satisfying the canonical commutation
relations

$$[Q_i^V, Q_j^V] = i\epsilon^{ijk}Q_j^V, \quad i,j,k = 1,2,3. \tag{8}$$

Equation (8) was the starting point of a big project in current algebra. It
deals with the charges of the strangeness-preserving charged weak vector
currents and of the isovector part of the electromagnetic current. The extension
to the entire charged weak vector current, including strangeness-changing
terms, and the entire electromagnetic current was straightforward, based on the
approximate SU(3) symmetry[39] of the strong interaction (Gell-Mann and
Ne'eman, 1964), or on the more sophisticated Cabibbo model (Cabibbo, 1963).
In this case, all the weak and electromagnetic vector currents, including the
isoscalar electromagnetic current, the triplet of isovector electromagnetic and
weak strangeness-conserving currents, and the weak strangeness-changing
currents, belong to the same octet, and the associated charges Q_α^V generate a
larger SU(3) Lie algebra, satisfying

$$[Q_\alpha^V, Q_\beta^V] = if_{\alpha\beta\gamma}Q_\gamma^V, \quad \alpha,\beta,\gamma = 1,2,\ldots,8, \tag{9}$$

where $f_{\alpha\beta\gamma}$ are the SU(3) structure constants. A similar proposal was also made
with respect to the weak axial vector currents $J_{\mu\alpha}^A$ and the associated charges
Q_α^A. In the chiral symmetry limit, a chiral $SU(3)_L \times SU(3)_R$ algebra was
postulated with additional relations

$$[Q_\alpha^A(x_0), Q_\beta^A(x_0)] = if_{\alpha\beta\gamma}Q_\gamma^V(x_0), \tag{10}[40]$$

and

$$[Q_\alpha^V(x_0), Q_\beta^A(x_0)] = if_{\alpha\beta\gamma}Q_\gamma^A(x_0). \tag{11}$$

When Gell-Mann first systematically proposed a current algebra framework
in (1962a), he recommended that these algebraic relations be used to supplement the dispersion relations in calculating the matrix elements of the weak
and electromagnetic currents. Before examining current algebra and its applications, however, we first have to look at another hypothesis, the PCAC
hypothesis, which, although conceptually somewhat independent, was practically often intertwined with current algebra.

The PCAC hypothesis

If the vector part V_μ of the weak current J_μ ($= V_\mu - A_\mu$) was conserved by the strong interaction, then what about the axial part A_μ? The experimental value of the renormalization factor $g_A(0)$[41] for the axial coupling constant was not far away from unity ($g_A/f_V \approx 1.25$). Feynman and Gell-Mann (1958) speculated that if the value were exactly unity, which they thought was not experimentally excluded, then there would be no renormalization for the axial current coupling constant, and the axial current would also be conserved. Thus they suggested trying to construct conserved axial currents and to explore the implications of the symmetry groups which were involved.

Their suggestion was soon taken up by John Polkinghorne (1958) and John C. Taylor (1958), and more systematically by Bludman (1958) and Feza Gürsey (1960a, b). The works by Bludman and Gürsey had a strong symmetry orientation and were pioneered by Schwinger (1957) and Bruno Touschek (1957). Thus the study of the axial current, like that of the vector current, was guided by symmetry considerations, which later led to the formulation of current algebra.

Progress along this line, however, was severely inhibited because the axial currents were prohibited from being exactly conserved by several experimental facts. First, the departure of g_A/f_V from unity indicates that there is a renormalization effect for the axial coupling. Second, as was pointed out by Goldberger and Sam Treiman (1958a), the conservation of the axial current would introduce a large effective pseudoscalar coupling which contradicts experiment. Third, as was pointed out by Polkinghorne, it would imply K-meson parity doublets. Fourth, as was pointed out by Taylor, it would imply a vanishing pion decay constant, which also contradicts experiment.

The real impetus to the study of the axial current actually came from another theoretical tradition, that is, the analyticity tradition. While the speculations on the conserved axial currents seemed futile, Goldberger and Treiman published their dispersion relation treatment of pion decay. In their calculation, the absorptive part of the pion decay constant f_π was regarded as a function of the dispersion variable q^2 (the square of the off-shell pion mass) and involved transitions from the off-shell pion to a complete set of intermediate states (which was dominated by the nucleon–antinucleon pair state, and coupled through the axial current to the leptons). The final result was

$$f_\pi = Mg_A/g_{\pi NN} \qquad (12)$$

where M is the nucleon mass, g_A is the axial vector coupling constant in nucleon beta decay, and $g_{\pi NN}$ is the pion–nucleon strong interaction coupling constant. In addition to its excellent agreement with experiment, the Gold-berger–Treiman relation (12) also exhibits a connection between the quantities of strong and weak interactions. This remarkable success, which had nothing to do with symmetry or the conservation of current, challenged theoreticians to provide an explanation or derivations of it from persuasive hypotheses. And this inspired research along various directions which finally led to the PCAC hypothesis.

Among efforts in searching for underlying hypotheses, Nambu's approach (1960d) occupies a prominent place because it was based on a deeper under-standing of symmetry breaking. Nambu argued that the axial current would be conserved if a theory of the massless nucleon field ψ possessed an invariance under the transformation $\psi \rightarrow \exp{(i\alpha \cdot \tau \gamma_5)}\psi$. If the vacuum state were not invariant under the transformation, however, the nucleon would become massive and the symmetry be apparently broken. But the symmetry would be restored owing to the emergence of a massless bound nucleon–antinucleon pair[42] in a pseudoscalar state, which can be identified as the pion state coupling with the nucleons. This kind of spontaneous symmetry-breaking guarantees the existence of the pion but fails to explain the pion decay: the axial current is still conserved because the existence of the pion restores the symmetry.

Nambu, however, followed a suggestion by Gell-Mann and Maurice Levy (1960) and assumed a small bare nucleon mass of the order of the pion mass. In this way, the γ_5 symmetry is explicitly broken, the pion becomes massive, and the matrix element of the axial current can be expressed in terms of form factors $g_A(q^2)$ and $G_A(q^2)$

$$\langle p'|A_\mu|p\rangle_{sn} = u(p')[ig_A(Q^2)\gamma_\mu\gamma_5 - G_A(Q^2)Q_\mu\gamma_5]u(p), \quad Q = p' - p. \quad (13)$$

Here, $G_A(q^2)$ is dominated by a pole at $q^2 = -m_\pi$, arising from the exchange of a virtual pion between nucleons and the axial current. That is,

$$G_A(Q^2) = 2f_\pi g_{\pi NN}/(Q^2 + m_\pi^2). \quad (14)$$

Nambu further argued that if the axial current were almost conserved (in the sense that when $q^2 \gg m_\pi^2$, its divergence vanished), and the form factors were slowly varying, then the Goldberger–Treiman relation (12) would follow from equations (13) and (14).

Another popular approach was initiated by Gell-Mann and his collaborators (Gell-Mann and Levy, 1960; Bernstein, Fubini, Gell-Mann, and Thirring, 1960). Working on field-theoretical models, such as the gradient coupling model, the σ model, and the non-linear σ model, they constructed, following the example of J. C. Taylor, axial currents whose divergence was proportional to a pion field:

$$\partial A^i_\mu / \partial x_\mu = \mu^2 f_\pi \phi. \tag{15}$$

The crucial hypothesis they made in this approach is that the divergence of the axial current is a gentle operator. This means that its matrix elements satisfy the unsubtracted dispersion relations (with the pion mass variable as its dispersion variable) and vary slowly; at low frequencies they are dominated by the contributions from the one pion intermediate state, and at high frequencies (when pion mass is neglegible) they vanish. Thus the axial current can be said to be almost conserved.

As far as the PCAC hypothesis is concerned, these two approaches are almost equivalent. Yet their theoretical bases are quite different. Nambu's approach is based on an understanding of spontaneous symmetry-breaking (SSB). The occurrence of the massless pion is one of the results of SSB although it also helps to restore the symmetry for the whole system, whereas in Gell-Mann's approach there is no such understanding, although the vanishing pion mass was taken to indicate an approximate symmetry.

It should be clear now that the gentleness hypothesis, which is the essence of the PCAC hypothesis, is conceptually independent of any symmetry consideration. Its importance lies in the fact that it allows continuation of the matrix elements of the axial current from the physical pion mass shell to off-shell ones. Frequently interesting statements can be made when the off-shell pion mass variable goes to zero. The hypothesis gives us hope that these statements may remain more or less true when the pion mass variable moves back to the physical mass shell, and this provides a great predictive power and can be verified by experiments.

Practically, the PCAC hypothesis had its applications often intertwined with the idea of current algebra because of its involvement in low energy pion physics (see below). Yet there were some independent tests. For example, Stephen Adler (1964) showed that the matrix element M of a neutrino-induced weak reaction, $\nu + P \rightarrow L + \beta$ (here P stands for proton, L for a lepton, and β for a hadron system), which, with the help of the CVC hypothesis, can be written as

$$M_{q^2 \rightarrow 0} \sim G \langle \beta | \partial A^{1+i2}_\mu / \partial x_\mu | P \rangle, \quad q = p_\nu - p_L. \tag{16}$$

and the matrix element M_π of the strong pion reaction $\pi^+ + P \rightarrow \beta$ which, by using the reduction formula, can be written as

$$M_\pi = \lim_{q^2 \rightarrow -\mu^2} [(q^2 + \mu^2)/2^{1/2}] \langle \beta | \phi^{1+i2} | P \rangle \tag{17}$$

(where q is the pion momentum and $\phi^{1+i2}/2^{1/2}$ is the π^+ field operator) can be related to each other (by using equation (15) and by continuing M_π off the mass shell in q^2 to $q^2 = 0$, which is supposed to result in an amplitude that is not very different from the physical pion amplitude) in the following way:

$$|M|^2_{q^2 \to 0} \sim G^2 f_\pi^2 |M_\pi|^2. \tag{18}$$

The above example, like the Goldberger–Treiman relation, exhibits a relation between the weak and strong interactions. In fact, the PCAC hypothesis leads to a whole class of relations connecting the weak and strong interactions. These relations allow one to predict the weak interaction matrix element $\langle \beta | \partial_\mu J_\mu^A | \alpha \rangle$ if one knows the strong interaction transition amplitude $T(\pi^+ + \alpha \to \beta)$. By using this kind of connection, Adler showed (1965) that in certain cases where only the Born approximation contributes and $\langle \beta | \partial_\mu J_\mu^A | \alpha \rangle$ can be expressed in terms of weak and strong coupling constants, the weak coupling constants can be eliminated and a consistency condition involving the strong interactions alone can be obtained. A celebrated consistency condition for pion–nucleon scattering obtained by Adler is a non-trivial relation between the symmetric isospin pion–nucleon scattering amplitude $A^{\pi N(+)}$, the pionic form factor of the nucleon $K^{NN\pi}$, and the renormalized pion–nucleon coupling constant g_r:

$$g_r^2/M = A^{\pi N(+)}(\nu = 0, \nu_\beta = 0, k^2 = 0)/K^{NN\pi}(k^2 = 0), \tag{19}$$

(where M is the nucleon mass, $-k^2$ is the (mass)^2 of the initial pion, $\nu = -(p_1 + p_2)k/2M$, $\nu_\beta = qk/2M$, and p_1, p_2, and q are, respectively, the four-momenta of the initial nucleon, the final nucleon, and the final pion) which was shown by Adler to agree with experiment to within 10%.[43]

Among various successful applications and predictions of the PCAC hypothesis, there stood outstandingly an exception, namely, its implication for a class of decays involving neutral pion(s), among which is the $\pi^0 \to 2\gamma$ decay. As pointed out by D. G. Sutherland (1967) and by M. Veltman (1967), the PCAC hypothesis implies that the decay rate would be zero, which contradicts experiment. This provocative failure of the PCAC hypothesis triggered investigations into the theoretical basis of PCAC, and also that of current algebra. The result was a clear realization of anomalies (see section 8.7).

Current algebra

The essential elements of current algebra are a set of equal-time commutation (ETC) relations, first suggested by Gell-Mann in 1961, for the time components of the currents that arise in the electromagnetic and weak interactions of hadrons. Yet for the charge densities to form a closed algebraic system under the ETC relations, the associated physical currents must be the Noether currents associated with some continuous symmetries of a physical system. In the early 1960s, regularities in the data of hadronic physics were observed. Some of them could be accounted for in terms of Regge trajectories, others were expresssed in terms of symmetries. In the second category we can find three types. The first type involves exact geometrical symmetries, which imply the conservation of energy, momentum, etc., and also CPT invariance (an invariance under the combined transformations of temporal reversal, spatial reflection, and charge conjugation). In the second type we find exact internal symmetries, which are associated with the conservation of the electric charge, the baryon number, and so on. The third type is the most interesting case because, first, it referred at that time to the so-called approximate symmetries[44] of the strong interaction (such as those associated with the approximate conservation of isospin and strangeness), and, second, it provided a point of departure for the thinking which led to the idea of current algebra.

Methodologically, current algebra is characterized by its independence of dynamical details. Lagrangians were sometimes used, but the purpose was only to express the symmetries suggested by the experimental data, and to generate currents obeying the canonical commutation relations dictated by these sym- metries. Mathematically, there are two ways to express the symmetries. They can be expressed by the conservation laws. Yet the utility of this approach is quite limited. The trouble is that most symmetries involved in hadron physics are approximate ones, and the problem of determining how approximate they are cannot be solved with this approach. This problem can also, and often, be translated into a physically more precise one, namely, the problem of how to fix the scales of the matrix elements of the currents, which are associated with the approximate symmetries and thus are subject to renormalization by the strong interaction. Conceptually, this problem requires a clarification of the meaning and implications of the approximate symmetries in physics.

In an attempt to solve this problem, both physically and conceptually, Gell- Mann adopted another approach to express a symmetry in terms of Lie algebra, which consists of the generators of the symmetry and is closed under the canonical commutation relations.

Historically, Gell-Mann's approach marked a third phase of the use of com- mutators in modern physics. In the first phase, the canonical, non-relativistic commutator between the position q and the momentum $p \equiv \delta L/\delta q$, $i[p, q] = (h/2\pi i)I$ was introduced by Heisenberg in 1926, to express the condition for quantization. This commutator is independent of the specific form of the Lagrangian L and can be used to derive useful interaction-independent results such as the Thomas–Kuhn sum rule. The second phase was associated with Thirring's approach of expressing the microcausality in terms of the commutator of two field operators, which should vanish for space-like separations of their arguments. In form, Gell-Mann's way of defining symmetries in terms of commutator algebra resembles Thirring's commutator definition for causality, because both of them are adapted to contemporary formulations of QFT. In spirit, however, Gell-Mann was closer to Heisenberg than to Thirring because the set of commutator algebra was designed to exhibit interaction-independent relations in relativistic dynamics, which could be used to extract physical information without solving the equations of the theory.

Mathematically, if there is a finite set of linearly independent operators R_i, and a commutator of any two R_i is a linear combination of the R_i:

$$[R_i(t), R_j(t)] = ic_{ijk}R_k(t), \tag{20}$$

then this system is called a Lie algebra. Physically, this abstract formula has its realization in the isospin charge operators (cf. equation (8) and subsequent discussion). With the CVC hypothesis and isospin symmetry in mind, Gell-Mann suggested an SU(3) algebra in (1962), and an SU(3) × SU(3) chiral algebra in (1964) which consisted of the charges of the vector and axial vector current octets and the symmetry-breaking term u_0 in the energy density (see equations (9), (10), and (11))[45], and thus initiated a big research programme of current algebra.

Conceptually, there were difficulties in understanding Gell-Mann's programme. First, the vector bosons which were supposed to be coupled with the weak currents were absent in that period, and thus the role of the currents was quite mysterious. Second, the currents were supposed to be representations of the SU(3) × SU(3) symmetry. Yet most of its sub-symmetries, except the U(1) gauge symmetry, were badly broken. Thus there was confusion as to what was the real meaning of these approximate symmetries.

Gell-Mann's solutions to these difficulties were both radical and ingenious. First, he considered the approximate higher symmetries in an abstract way and took currents as the primary subjects for investigation. Thus the vector bosons, though helpful in visualizing physical process, were not necessary. Second, the currents were characterized in terms of their ETC relations rather than the symmetries themselves. A subtle yet crucial difference between these two ways of characterizing the currents is this. Formally, the ETC relations depended only on the structure of the currents when regarded as functions of the canonical field variables. If the symmetry-breaking terms in the Lagrangian involved no derivative couplings, then independent of details, the currents would retain the original structure, and the ETC relations would also remain unchanged. In this way, the vector and axial vector currents, even under the breakdown of SU(3) symmetry for the strong interactions, would still have octet transformation properties of SU(3) × SU(3) symmetry, not in the sense that their charges are conserved, but in the sense that their charges satisfy the SU(3) ETC relations. Thus a precise though abstract meaning was specified for the notion of approximate symmetries, in terms of ETC relations, which now can be regarded as an exact property of the strong interactions.

Gell-Mann's novel understanding of the approximate symmetry allows one to derive from a broken symmetry exact relations among measurable quantities. These relations help to fix the scales of the matrix elements of the currents and, especially, to connect those of the axial currents with those of the vector currents. In addition, the hypothesis of universality, concerning the strength of the leptonic and hadronic weak interactions, also acquires a precise meaning. That is, the ETC relations for the total leptonic weak current are the same as those for the total hadronic weak current. With the help of algebraic relations involving u_0, the deviation from equality, namely, the renormalization effect, can also be calculated. Thus a workable research programme was available for the physics community after Gell-Mann's classic papers (1962a, 1964a, b) were published.

In this programme, the central subject is the processes involving pions, because the matrix elements of pions in the chiral limit $(m_\pi \to 0)$ can be calculated by a direct application of the ETC relations, combined with the PCAC notion of pion pole dominance for the divergence of the axial current. For example, the matrix elements of the process $(i \to f + \pi)$ are related, via the PCAC hypothesis and the reduction formula, to

$$\langle f|\partial^\mu A_\mu^a|i\rangle = \mathrm{i}k^\mu\langle f|A_\mu^a|i\rangle. \tag{21}$$

Similarly, processes involving two soft pions can be studied by analyzing the matrix element

$$\langle f|\mathrm{T}\partial^\mu j_\mu^a(x)\partial^\nu j_\nu^b(0)|i\rangle. \tag{22}$$

In pulling out the differential operators, we obtain, in addition to a double divergence of a matrix element of two currents, an equal-time commutator of the two currents, which turns out to be crucial for calculation.

There was no rapid progress until 1965, when S. Fubini and G. Furlan suggested certain techniques for pursuing this programme. Their central ideas were (i) inserting a complete set of intermediate states into the matrix element of the commutator, (ii) separating out the one-nucleon contribution, and (iii) calculating in the infinite momentum frame. By using essentially the same techniques, Adler and William Weisberger obtained, independently, a sum rule expressing the axial form factor g_A in terms of zero-mass π^+ (π^-)–proton scattering total cross sections at center-of-mass energy W:

$$1 - 1/g_A^2 =$$

$$[4M_N^2/g_r^2 K^{NN\pi}(0)^2](1/\pi)\int_{M_N+M_\pi}^{\infty} W[\sigma_0^+(W) - \sigma_0^-(W)]\,dW/(W^2 - M_N^2). \quad (23)$$

This formula yields a prediction $\mid g_A\mid\ = 1.21$, which agrees remarkably with the experimental data $g_A^{exp} \approx 1.23$.

The influence of the Adler–Weisberger sum rule, which can also be interpreted as a low energy theorem on pion–nucleon scattering, was enormous. It provided the most celebrated test of the PCAC–current algebra programme, and thus established a paradigm case for the programme. Its publication inaugurated the golden age, as Sam Treiman once put it, for the programme. Within two years, it was followed by about 500 papers, which had successes in low energy theorems and/or in high energy sum rules.

Among numerous successful applications, however, there were also outstanding failures. The most notable cases were: (i) the Sutherland–Veltman theorem of vanishing $\pi^0 \rightarrow 2\gamma$ decay rate; and (ii) the Callan–Gross sum rule (1969), which predicts that the longitudinal cross section for total electroproduction off protons vanishes in the deep inelastic limit. A thorough examination of these failures, with sophisticated concern about the validity of some of the formal manipulations, led to a profound understanding, in the context of current algebra and the PCAC hypothesis, of anomalous symmetry breaking. This understanding indicates, convincingly, that formal manipulations without any resort to the dynamics that underlies the algebraic relations are unreliable. It also shows that the assumption that the ETC relations are independent of the details of the interactions is simply wrong, thus undermining the whole research programme.

Before turning to this important development, however, some comments on the programme are in order. The first concerns the perspective perceived when the programme first appeared. Since the currents in this programme were supposed to be both the physical weak and electromagnetic currents and the symmetry currents of the strong interaction, it seemed possible to test the symmetry of the strong interaction with the weak and electromagnetic processes. As far as the flavor symmetry, which is an expression of mass degeneracy, is concerned, this expectation was partially fulfilled.

The preliminary successes in this regard raised some higher expectations, concerning the possible relation between current algebra and the Yang–Mills theory. For example, Gell-Mann (1962a; 1964a, b; 1987) claimed that current algebra justified the flavor chiral $SU(3) \times SU(3)$ symmetry as a symmetry of the strong interaction, that it would generate a gauge group for a Yang–Mills theory of the strong interaction, and that at least the ETC relations were a kind of precondition for the Yang–Mills theory. It turns out, however, that the flavor symmetry is irrelevant to the construction of a strong-interaction theory of Yang–Mills type, although it does have some relevance to the low energy dynamics of the strong interaction. It was also hoped (Bernstein, 1968) that the algebraic constraints of ETC-type would help sort out dynamic models of currents. This hope has not been fulfilled either.

However, in leptonic physics, current algebra did play a heuristic role in advancing a theory of Yang–Mills type. By connecting electromagnetic and weak currents with the symmetry of the strong interaction, it helped introduce new quantum numbers, the weak isospin and weak hypercharge, into leptonic physics. Since the weak and electromagnetic currents associated with these quantum numbers were supposed to couple with vector bosons, these conceptual constructions did help to construct an electroweak theory of Yang–Mills type.

The second comment concerns the methodology. The PCAC–current algebra programme deliberately ignored the dynamic details, thus simplifying the complicated situation. It is this simplification that made possible the progress in understanding some low energy aspects of the strong interaction, mainly pion physics, which are related to the spontaneous breakdown of the flavor chiral symmetry in the strong interactions. Yet the deficiencies of this method were soon revealed when investigations went deeper, and the necessity for clarifying the dynamic basis of the programme was gradually realized after 1967.

The clarification took two different directions. The first, initiated by Schwinger (1966, 1967) and Weinberg (1967a, 1968), was to develop a phenomenological approach to the low energy strong interactions, so-called chiral dynamics, so that clues could be found on how to go beyond the soft-pion limit. The idea was quite simple. If one constructed a chirally invariant effective Lagrangian and calculated the lowest order graphs, then the PCAC and ETC relations and their results would be reproduced.

Although both were committed to a phenomenological approach, significant differences existed between Schwinger and Weinberg. In Schwinger's phenomenological source theory, the use of the coupling terms in the lowest order is justified by the very nature of a numerical (i.e. non-operator) effective Lagrangian. Schwinger also argued that as long as the origin of symmetries remained obscure, the phenomenological approach was suitable. For Weinberg, however, the reason for calculating only in lowest order was that to reproduce the results of the PCAC–current algebra approach, the effects of loops were already accounted for by the presence of the form factors f_v and g_A. As to the symmetry aspect, Weinberg, in contrast with Schwinger, felt uneasy at using a symmetry on the phenomenological level when he did not know how to derive it from a fundamental Lagrangian. It is clear that Weinberg was less phenomenologically oriented than Schwinger even when he was also engaged in developing a phenomenological approach.

Historically, the importance of the phenomenological approach lies not so much in its applications to various low energy meson processes, which were often derivable from the formal (non-dynamic) approach, as in its function as a precursor of, and its heuristic value in the development of, the effective field theory, which developed out of a totally different context (see section 11.4). This is especially so if we consider the fact that one of the initiators of the latter programme was again Weinberg.

The second direction for research in clarifying the dynamic basis of the PCAC–current algebra approach was to develop a fundamental understanding of the subject. Activity in this direction lasted for several years, and an accumulation of collective experience in this area eventually led to a consequential awareness of anomalous symmetry breaking, which radically changed the conceptual basis of QFT and our understanding of renormalization. In section 8.7 we shall give an account of activities up until 1969. Later developments, which have revealed the richness of the notion of anomalous symmetry breaking, will be examined in sections 8.8 and 11.4.

8.7 Anomalies

The intensive investigations, in the second half of the 1960s, into the anomalous behavior of local field theories were carried out along three closely related lines: (i) as a response to the formal (independent of dynamical details) manipulations of ETC relations adopted by current algebra, undertaken mainly by Johnson, Low, John Bell, and others; (ii) as a response to the necessity of modifying PCAC relations, made mainly by Veltman, Sutherland, Bell, Roman Jackiw, Adler, and others; and (iii) as part of a study of the renormalization of currents, made mainly by Adler, William Bardeen, Kenneth Wilson, and others. These investigations firmly established the existence of certain anomalous behavior of local field theories, delimited the scope of validity of formal manipulations adopted by the PCAC–current algebra programme, and made it clear that the perturbation theory was the dynamical basis of the programme, thus putting an end to the programme as an autonomous research[46] framework, and paving the way for physicists returning to the framework of local field theories with various exact and broken symmetries, where renormalizability is the most urgent problem to be solved.

Anomalous commutators

Serious doubt about the validity or consistency of current algebra, which was based on the canonical ETC relations, was first cast by Johnson and Low in (1966). The root of the doubt can be traced back to Goto and I. Imamura (1955) and Schwinger (1959), when current algebra did not exist. It was known to Goto, Imamura, and Schwinger that in a relativistic theory with positive metric and for a vector or axial vector current there must be extra terms in the vacuum expectation of the commutators of V_0 with V_j which involve space derivatives of delta functions instead of canonical delta functions, later known as the Schwinger terms.

Schwinger's observation was picked up by Johnson in a historically as well as conceptually important paper on the Thirring model (1961). In that paper, Johnson (i) attacked the problem of how to define products of singular field operators (e.g. currents) at coincident spacetime points; and (ii) demonstrated that the products did not satisfy the canonical ETC relations, owing to their singular nature and the necessity of regularization. In addition to providing a point of departure for later work on the operator product expansions by Wilson (1969, 1970b, c; see below), with the idea of fields having dynamic scale dimensions, Johnson's paper is also historically important because of its extension of Schwinger's observation from the vacuum expectation to the operator (i.e. non-vacuum) structure of the commutators of V_0 with V_j. While the former was concerned only with kinematics, resulting in constant subtractions, the latter deeply touched the dynamics of the local field theories.

With the advent and flourish of current algebra, the non-canonical behavior of the equal-time commutators became a subject for intensive investigations,[47] because of its implications for the consistency of current algebra. Among various investigations to appear in 1966, the most fruitful and most influential seems to have been the Johnson and Low paper.[48] Working in a renormalizable interacting quark model and helped by a technical device, later known as the Bjorken–Johnson–Low (BJL) limit, by which the equal-time commutators could be related to the high energy behavior of the known Green's functions, Johnson and Low calculated, to the lowest non-vanishing order in the meson–baryon coupling, the matrix elements of the commutators between one meson and vacuum (or one meson) for all sixteen Dirac covariants and an arbitrary internal group. With this comprehensive examination, Johnson and Low convincingly demonstrated that in certain cases in which the matrix elements of commutators were related to certain Feynman graphs (such as triangle graphs; see below), the commutators clearly had finite extra terms. The cause for the presence of these extra terms, as pointed out by Johnson and Low,[49] is this. Any local relativistic theory, no matter how convergent, must be singular enough that regularization has to be introduced, which is essentially a limiting procedure. When a calculation involves a worse than logarithmically divergent integral with two or more limits to be taken, which are not legally inter-changeable,[50] there is a limiting ambiguity, and the extra terms result from a careful treatment of this limiting ambiguity.

Stimulated by Johnson and Low's paper, Bell (1967a) illustrated some aspects of the anomalous commutators in the solvable Lee model, with a special emphasis on the cut-off regularization. When the cutoff is finite, Bell suggested, the canonical commutators were correct for zero-energy theorems; and only for sum rules involving infinite energy did the anomalous com-mutators become significant.

Both Low (1967) and Bell suggested that the anomalous behavior of com-mutators would occur in some high energy sum rules, yet only in an abstract way (Johnson and Low), or in an unrealistic model (Bell). The discussion of anomalous commutators remained detached from experiments until 1969, when Curt Callan and David Gross (1969) published their derivation, based on canonical ETC relations proposed by James Bjorken (1966, 1969), of asympto-tic sum rules for high energy longitudinal electro-production cross sections, which was immediately challenged simultaneously by Jackiw and Preparata (1969), and by Adler and Wu (1969), with an anomalous commutator argument of Johnson–Low type.

Modification of PCAC

Another line of investigation of anomalies had closer contact with experiments. As a matter of historical fact, it was directly stimulated by the provocative discrepancies between the PCAC predictions and the observation of meson decays. In late 1966, Sutherland found that the $\eta \to 3\pi$ decay, in the case where one of the pions had vanishing four-momentum ($q = 0$), would be forbidden if it were calculated with the assumptions of PCAC and current algebra.[51] This was in direct contradiction with experiment. Here PCAC was invoked in two ways. First, the axial vector current was brought into the calculation by taking its divergence as the interpolating pion field. Second, the calculated decay amplitude as a function of q^2 was continued from $q^2 = 0$ to $q^2 = m_\pi^2$. One possible explanation for the experimental failure of PCAC suggested by Sutherland was that the continuation might be illegal because of some unknown mechanism, such as the difference caused by the long range nature of the photon interaction as compared with the short range W-meson interaction, both of which were assumed to be responsible for the decay. This suggestion implies modifying PCAC in its second usage. Among investigations along this line we can find the paper by Riazuddin and Sarker (1968).

Sutherland's observation was generalized by Veltman (1967), to the effect that any process containing a π^0, such as $\pi^0 \rightarrow 2\gamma$ and $\omega \rightarrow \pi^0\gamma$, when calculated with PCAC and current algebra, must be forbidden in the limit of zero π^0 four-momentum.[52] From this observation Veltman concluded that the usual PCAC (equation (14)) was incorrect and extra terms involving the gradient of the pion field had to be added, so that the observed decays could be accounted for. Veltman's suggestion implies modifying PCAC in its first usage. The paper by Arnowitt, Friedman, and Nath (1968) was along this line, although the forms of the extra terms suggested there were different from those by Veltman.

Deeply involved in the investigation of current algebra and PCAC, Bell was particularly impressed by Veltman's observation.[53] After much work,[54] Bell, collaborating with Jackiw, published an interesting paper (1969). In this paper they purported 'to demonstrate in a very simple example the unreliability of the formal manipulations common to current algebra calculations'.[55] They also tried 'to provide a scheme in which things are sufficiently well defined for the formal reasoning to be applicable to the explicit calculations', namely, a scheme that 'embodies simultaneously the ideas of PCAC and gauge invariance'. The way Bell and Jackiw compromised these two seemingly contradictory pursuits is intriguing and reveals how deep their study had touched upon the conceptual predisposition of contemporary theoretical physics.

Bell and Jackiw noticed that the $\pi^0 \rightarrow 2\gamma$ decay considered in the σ model posed a puzzle. On the one hand, the PCAC reasoning of the Veltman–Sutherland type implies that the invariant amplitude for $\pi^0 \rightarrow 2\gamma$, $T(k^2)$, vanishes when the off-mass-shell continuation is made with a pion field that is the divergence of the axial vector current, that is, $T(0) = 0$. This conclusion is in conflict both with experiment and with the old perturbative calculation by Steinberger (1949) of triangle graphs contributing to the process, which is in excellent agreement with experiment.[56] On the other hand, the explicit calculation of the same process in the σ model, which has PCAC built in as an operator equation gives just the same result as Steinberger's. This is the puzzle that Bell and Jackiw confronted.

To reveal the origin of the puzzle, Bell and Jackiw went beyond the formal reasoning and examined the details of the calculation. First, they found that the non-vanishing $T(0)$ obtained in the σ model calculation came from a surface term, which was picked up when variables in the linearly divergent integral of the triangle graphs were shifted. This extra term was compatible with PCAC but violated gauge invariance. Here then is a clash between PCAC and gauge invariance. Second, they followed Johnson and Low and argued that the formal manipulations of the integrals of the relevant triangle graphs, though respecting both PCAC and gauge invariance, might fail if the integrands did not converge well. The resulting surface term violated gauge invariance and required regularization. They stressed that the conventional Pauli–Villars regularization, though restoring gauge invariance, spoiled PCAC. The clash between PCAC and gauge invariance appeared again. Bell, in his letter to Adler dated 2 September 1968, remarked:

Our first observation is that the σ model interpreted in a conventional way just does not have PCAC. This is already a resolution of the puzzle.[57]

However, a strong predisposition, which is deeply rooted in the symmetry worship widely shared by many contemporary theoretical physicists and exhibited in the second motivation of the paper we have just mentioned above, made Bell and Jackiw unhappy with this resolution. They tried to solve the puzzle in a different way, dictated by their respect both for PCAC and for gauge invariance.

In line with Bell's previous argument (1967a, b) that the anomalies in commutators are irrelevant and can be avoided by introducing regularization, Bell and Jackiw suggested a new regularization which was dictated by PCAC. The crucial assumption here is that the coupling constant g_1 of the regulator field must vary with the mass of the field m_1 in such a way that $m_1/g_1 = m/g =$ constant even when m_1 goes to infinity and the anomalies caused by the real and auxiliary fields cancel each other. Thus, instead of modifying PCAC, Bell and Jackiw modified the regularization scheme in the σ model, so that the anomalies, along with the clash between PCAC and gauge invariance, and the discrepencies between the formal reasoning and explicit calculation, would be eliminated.

Unfortunately, the consistency of formal reasoning restored by Bell and Jackiw cannot survive scrutiny. As pointed out by Adler (1969), the new regularization would make the strong interactions in the σ model non-renormalizable[57a]. The anomalies, as Adler pointed out, were too deeply rooted in the theoretical structure of local field theories to be removable by any artificial manipulations.

In spite of its uninspired motivation and untenable solution, the Bell–Jackiw paper did hit on something important. Its importance lies both (i) in its relating the anomalies appearing in the calculations of triangle graphs with the observable $\pi^0 \rightarrow 2\gamma$ decay, thus raising enormous interest in, and drawing attention to, the anomalies among physicists; and (ii) in its tying the anomalies in with PCAC, thus pointing in the right direction for understanding the nature of the anomalous behavior of local field theories. The real importance of the paper, however, might not be appreciated as timely and properly had there not been a paper by Adler (1969) that clarified the anomaly situation with a more complete and more elegant formulation. In response to Bell and Jackiw's resolution of the PCAC puzzle, Adler, instead of modifying the σ model so as to restore PCAC, stayed within the conventional σ model and tried to systematize and exploit the PCAC breakdown, thus giving a reasonable account for the $\pi^0 \to 2\gamma$ decay. Yet this response, made in an appendix to his paper, was only an application of his principal ideas reported in the text of his paper (1969). The ideas were developed, independently of the Bell–Jackiw paper, in his study of spinor electrodynamics, which was motivated by renormalization considerations and carried out in a context quite different from the PCAC–current algebra programme.[58]

Renormalization of currents

Adler's dynamic study of the renormalization of the axial vector vertex in QED itself was an important step, taken by a theorist active in the PCAC–current algebra programme, away from the programme, in which physical information embodied in the vertex form factors was extracted through pure symmetry considerations without appeal to dynamics. However, the hangover of the fascination with symmetry can still be easily detected. For Adler, this study was of interest because it was connected with the γ_5 invariance of massless QED. Precisely because the study was carried out with a strong interest in its bearing on symmetry, the clash discovered between the dynamical calculations and the formal algebraic manipulations dictated by symmetry considerations was deliberately exploited to reveal its significance for our understanding of the dynamical symmetries of systems described by local field theories: some symmetries exhibited in the Lagrangians might disappear when radiation corrections represented by loop graphs were taken into account, and extra ('anomalous') terms might appear in the relevant calculations. Thus the scope of validity of symmetry considerations is delimited by the presence or absence of the breakdown of relevant symmetries, which is caused by particular processes of radiative correction. Before exploring the implications of anomalies, however, let us first take a closer look at the steps taken by Adler.

In his perturbative study Adler found that in QED the axial vector vertex generally satisfied the usual Ward identity except in the cases in which the integrals defining the Feynman graphs were linearly divergent or worse, and thus the shift of integration variables was dubious. An inspection shows that the only trouble cases in QED are those involving the triangle graphs.

Employing Rosenberg's explicit expression for the triangle graph (Rosenberg, 1963) and calculating it with the usual Pauli–Villars regularization, Adler found that if the calculation was compatible with gauge invariance, then the axial vector Ward identity failed in the case of the triangle graph, and an extra term appeared.[59] The failure of the axial vector Ward identity in Adler's formulation was the result of a combination of two factors. The first is a special property of the triangle graph, i.e. its linear divergence, which leads to an ambiguous expression for the graph. The second is the constraint imposed by gauge invariance.

In perturbative theory, Ward identities are dictated by, and representations of, invariances and/or partial invariances of the basic Lagrangians. Therefore the breakdown of the axial vector Ward identity had already hinted at a kind of symmetry-breaking. Yet Adler explicated the connection between his discovery of anomalies and symmetry-breaking more explicitly. Observing that the breakdown of the axial Ward identity occurred not only for the basic triangle graph, but also for any graph with the basic triangle graph as its subgraph, Adler showed that the breakdown in the general case can be simply described by replacing the usual axial vector current divergence

$$\partial j_\mu^5(x)/\partial x_\mu = 2im_0 j^5(x) \tag{24}$$

$(j_\mu^5(x)^- = \psi(x)\gamma_\mu\gamma_5\psi(x), \; j^5(x)^- = \psi(x)\gamma_5\psi(x))$ by

$$\partial j_\mu^5(x)/\partial x_\mu = 2im_0 j^5(x) + a_0/4\pi : F^{\xi\sigma}(x)F^{\tau\rho} : \epsilon_{\xi\sigma\tau\rho}, \tag{25}$$

which Adler regarded as his principal result. Applying this result to the case of massless QED, Adler demostrated that the presence of the axial vector triangle graph provided a special mechanism for breaking the γ_5 invariance of the Lagrangian for massless QED. The γ_5 invariance is broken in this case simply because the axial vector current associated with the γ_5 transformation is no longer conserved owing to the extra term $a_0/4\pi : F^{\xi\sigma}(x)F^{\tau\rho} : \epsilon_{\xi\sigma\tau\rho}$.[60]

Applying to the Bell–Jackiw version of the σ model the same idea which leads to equation (25), Adler found that the same extra term also occurred at the right hand side of the PCAC relation. This entails that PCAC must be modified in a well-defined manner in the presence of electromagnetic interactions. Adler's modified PCAC relation gives a prediction for the $\pi^0 \to 2\gamma$ decay ($\tau^{-1} = 9.7$ eV),[61] which is different from those given by Sutherland, by Veltman, and by Bell and Jackiw, and is in good agreement with the experimental value, $\tau_{exp}^{-1} = (7.35 \pm 1.5)$ eV.

An essential though implicit assumption here is that the anomaly is an exact result, valid to all orders and not renormalized by high-order radiative corrections. This assumption follows in Adler's analysis naturally from the fact that radiative corrections to the basic triangle graph always involve axial vector loops with more than three vertices, which in the case of QED are at worst superficially logarithmically divergent, and thus satisfy the normal axial vector Ward identities. In response to the disagreement with this assumption raised by Jackiw and Johnson (1969), Adler and Bardeen (1969) worked out the details of the argument for this assumption. This resulted in a non-renormalization theorem, which is crucial for the anomaly cancellation conditions and for the consistency conditions posed by Julius Wess and Bruno Zumino (1971) and by Gerard 't Hooft (1976a).[62]

As to the renormalization of the axial vector vertex, Adler noticed that the axial vector divergence with the extra term included was not multiplicatively renormalizable, and this significantly complicated the renormalization of the axial vector vertex, although the complications that arise can be circumvented by arranging for anomalies from different fermion species to cancel.[62a]

The conceptual situation related to the anomalous behavior of local field theories was greatly clarified with Adler's criticism of Bell and Jackiw's new regularization, which was supposed to be able to help get rid of the anomalies. This criticism convincingly suggests that the triangle graph anomaly cannot be eleminated without spoiling gauge invariance, unitarity, or renormalizability.[63] With this argument, the triangle anomalies soon became part of the established lore.

The Adler–Bell–Jackiw anomalies, first well established in an Abelian theory, were soon extended, by Gerstein and Jackiw (1969) and by Bardeen (1969), to the non-Abelian cases. This extension has a direct bearing on the construction of the standard model, which is subject to the renormalizability constraint (see section 10.3). While Gerstein and Jackiw argued, in the case of SU(3) × SU(3) coupling, that the only graph other than the VVA triangle graph that had an anomaly was the AAA graph, Bardeen claimed a more general SU(3) anomaly structure. For neutral currents, it agreed with Gerstein and Jackiw's conclusion; for charged currents, anomalies also occurred in the VVVA and VAAA box graphs and in VVVVA, VVAAA, and AAAAA pentagon graphs.

Operator product expansions at short distances (OPE)

The renormalization effects on currents were also attacked, almost simultaneously, by Wilson (1969).[64] Wilson's framework, operator product expansions at short distances (OPE) without involving Lagrangians, and his major concern, scale invariance rather than γ_5 invariance, were quite different from Adler's and Bell–Jackiw's. Yet, his analysis pointed in the same direction for the understanding of the relationship between renormalization and symmetry breaking, with similar or even more significant consequences for later conceptual developments.

Responding to the failure of current algebra in dealing with the short distance behavior of currents, which was known to many since Johnson and Low's work, Wilson tried to develop a new language, OPE, which could give a more detailed picture of the short distance behavior of currents in the strong interactions:

$$A(x)B(x) = \sum_n C_n(x - y)O_n(x). \tag{26}$$

Here, $A(x)$, $B(x)$, and $O_n(x)$ can be any local field operators (elementary fields, currents, energy tensor, etc.); $C_n(x - y)$, which involve powers of $(x - y)$ and logarithms of $(x - y)^2$ and may have singularities on the light cone, contain all the physical information about the short distance behavior of currents.

OPE had its origin in detailed studies of renormalization in perturbation theory.[65] Yet now Wilson developed it on the new basis of broken scale invariance,[66] which was utilized to determine the singularity structure of $C_n(x - y)$. The point of departure was Johnson's work on the Thirring model, in which (i) the canonical commutators were shown to be destroyed by renormalization effects, and (ii) the scale dimensions of fields were shown to vary continuously with the coupling constant. Wilson made the generalization that these two observations also held in the case of his OPE formulation for the strong interactions. The deviation of the scale dimensions of currents from their non-interacting value, and whose presence was one of the major points Wilson made and powerfully argued for in the paper, convincingly suggested a breakdown of scale invariance caused by the renormalization effects. This suggestion was also supported by Johnson and Low's observation, appealed to by Wilson in the paper, that the canonical ETC relations would be destroyed by non-invariant interactions.

One implication of Wilson's analysis seems to be this. The scale invariance of the strong interactions is broken, not by symmetry-breaking terms in the Lagrangian (there is no such Lagrangian in Wilson's scheme), nor by a non-invariant vacuum, but, like the anomalous breakdown of γ_5 invariance, only by some non-invariant interactions introduced in the renormalization procedure. This implication was soon intensely explored by Wilson himself (1970a, b, c; 1971a, b, c; 1972), and also by Callan (1970), Symanzik (1970), Callan, Coleman, and Jackiw (1970), Coleman and Jackiw (1971), and others. The exploration directly led to the revival of the idea of the renormalization group.

As we shall see in the next section, Wilson's attitude towards the anomalous breakdown of scale invariance was quite different from others'. This attitude seems to have its origin in his 1969 paper. While acknowledging the existence of the scale anomaly, which was reflected in the change of the scale dimension

of the currents, Wilson insisted that all the anomalies could be absorbed into
the anomalous dimensions of the currents, so that the scale invariance would
persist in the asymptotic sense that the scaling law still held, although only for
currents with changed scale dimensions.

Wilson's attitude seems to be attributable (i) in part to the influence upon his
work of that by Gell-Mann and Low on the scaling law of bare charge in QED
(1954); (ii) in part to that of the scaling hypothesis of Fisher, Widom, and
Kadanoff in critical phenomena; and (iii) in part to his ambitious attempt to
divorce himself from perturbation theory. The first two points gave him faith in
the existence of scaling laws, or, to use later terminology, in the existence of
fixed points of the renormalization group; the last point made him unable to
analyze the scale anomaly in a thoroughly dynamical way. This and other
aspects of the scale anomaly will be the subject of the next section.

8.8 The renormalization group

The conceptual basis for the revival and reformulation of renormalization
group equations in the early 1970s was the idea of the scale anomaly, or the
anomalous breakdown of scale invariance. Historically, however, the idea of
the scale dependence of physical parameters, within the framework of QFT,
appeared earlier than the idea of scale invariance, and provided the context for
a proper understanding of the scale invariance.

We can trace the early (or even the first) appearance of the idea of the scale
dependence of physical parameters to Dyson's work on the smoothed interaction representation (1951). In this representation, Dyson tried to separate the
low frequency part of the interaction from the high frequency part, which was
thought to be ineffective except in producing renormalization effects. To
achieve this objective, Dyson adopted the guidelines of the adiabatic hypothesis
and defined a smoothly varying charge for the electron and a smoothly varying
interaction, with the help of a smoothly varying parameter g. He then argued
that when g is varied, some modification had to be made in the definition of
the g-dependent interaction, so that the effect caused by the change of the g-
dependent charge could be compensated.

In line with this idea of Dyson's, Landau and his collaborators developed a
similar concept of smeared out interaction in a series of influential papers
(1954a, b, c, d; 1956). In accordance with this concept, the magnitude of the
interaction should be regarded not as a constant, but as a function of the radius
of interaction, which must fall off rapidly when the momentum exceeds a
critical value $P \sim 1/a$, where a is the range of the interaction. As a decreases,
all the physical results tend to finite limits. Correspondingly, the electron's

charge must be regarded as an as yet unknown function of the radius of interaction. With the help of this concept, Landau studied the short distance behavior of QED and obtained some significant results, which have been mentioned in earlier sections. Both Dyson and Landau had the idea that the parameter corresponding to the charge of the electron was scale dependent. In addition, Dyson hinted, though only implicitly, that the physics of QED should be scale independent. Landau, more explicitly, suggested that the interactions of QED might be asymptotically scale invariant.

The term 'renormalization group' made its first appearence in a paper by Stueckelberg and Petermann (1953). In that paper they noticed that while the infinite part of the counter-terms introduced in the renormalization procedure was determined by the requirement of cancelling out the divergences, the finite part was changeable, depending on the arbitrary choice of subtraction point. This arbitrariness, however, is physically irrelevant because a different choice only leads to a different parameterization of the theory. They observed that a transformation group could be defined which related different parameterizations of the theory, and they called it the 'renormalization group'. They also pointed out the possibility of introducing an infinitesimal operator and of constructing a differential equation.

One year later, Gell-Mann and Low, in studying the short distance behavior of QED, exploited the renormalization invariance more fruitfully (1954). First, they emphasized that the measured charge e was a property of the very low momentum behavior of QED, and that e could be replaced by any one of a family of parameters e_λ, which was related to the behavior of QED at an arbitrary momentum scale λ. When $\lambda \to 0$, e_λ became the measured charge e, and when $\lambda \to \infty$, e_λ became the bare charge e_0. Second, they found that by virtue of renormalizability, e_λ^2 obeyed an equation: $\lambda^2 \, \mathrm{d}e_\lambda^2/\mathrm{d}\lambda^2 = \psi(e_\lambda^2, \, m^2/\lambda^2)$. When $\lambda \to \infty$, the renormalization group function ψ became a function of e_λ^2 alone, thus establishing a scaling law for e_λ^2. Third, they argued that as a result of the equation, the bare charge e_0 must have a fixed value independent of the value of the measured charge e; this is the so-called Gell-Mann–Low eigenvalue condition for the bare charge.

In the works of Stueckelberg and Petermann and of Gell-Mann and Low, Dyson's varying parameter g and Landau's range of interaction were further specified as the sliding renormalization scale, or subtraction point. With Gell-Mann and Low, the scale-dependent character of parameters and the connection between parameters at different renormalization scales were elaborated in terms of renormalization group transformations, and the scale-independent character of the physics was embodied in renormalization group equations. However, these elaborations were not appreciated until much later, in the late

1960s and early 1970s, when a deeper understanding of the ideas of scale invariance and renormalization group equations was gained, mainly through the researches of K. G. Wilson, as a result of fruitful interactions between QFT and statistical physics.

The idea of the scale invariance of a theory is more complicated and very different from the idea of the scale independence of the physics of a theory (as vaguely suggested by Dyson), or the idea of the independence of the physics of a theory with respect to the renormalization scale (as expressed by the renormalization group equations). The scale invariance of a theory refers to its invariance under the group of scale transformations. The latter are only defined for dynamical variables (the fields), but not for the dimensional parameters, such as masses. Otherwise, a scale transformation would result in a different physical theory. While the physics of a theory should be independent of the choice of the renormalization scale, a theory may not be scale invariant if there is any dimensional parameter.

In Gell-Mann and Low's treatment of the short-distance behavior of QED, the theory is not scale invariant when the electric charge is renormalized in terms of its value at very large distances. The scale invariance would be expected in this case because the electron mass can be neglected and there seems to be no other dimensional parameter appearing in the theory. The reason for the unexpected failure of scale invariance is entirely due to the necessity for charge renormalization: there is a singularity when the electron mass goes to zero. However, when the electric charge is renormalized at a relevant energy scale, by introducing a sliding renormalization scale to effectively suppress effectively irrelevant low-energy degrees of freedom, there occurs an asymptotic scale invariance. This 'asymptotic scale invariance' was expressed by Gell-Mann and Low in terms of a scaling law for the effective charge. They also took it as the eigenvalue condition for the bare charge, meaning that there is a fixed value for the bare charge, which is independent of the value of the measured charge.

Although there was a suggestion by Johnson in the early 1960s that the Thirring model might be scale invariant, the real advance in understanding the nature of scale invariance was made in the mid-1960s, as a result of developments in statistical physics. Research in this area was also stimulated by the discovery of field-theoretical anomalies in the study of current algebra and in the short-distance expansion of products of quantum field operators.

Here an interaction between QFT and statistical mechanics, which can readily be discerned in the shaping of Wilson's ideas, played an important role in the development. Conceptually, the interaction is very interesting, but also quite complicated. In 1965, B. Widom in (1965a, b) proposed a scaling law for the equation of state near the critical point, which generalized earlier results obtained by J. W. Essam and M. E. Fisher (1963) and Fisher (1964), concerning the relations among the critical exponents. Wilson was puzzled by Widom's work because it lacked a theoretical justification. Wilson was familiar with Gell-Mann and Low's work. Moreover, he had just found a natural basis for the renormalization group analysis, while working to develop a lattice field theory, by solving and eliminating one momentum scale for the problem (1965). Wilson realized that there should be applications of Gell-Mann and Low's idea to critical phenomena. One year later, Leo Kadanoff (1966) derived Widom's scaling law from the idea, which essentially embodied the renormalization group transformation, that the critical point becomes a fixed point for transformations on the scale-dependent parameters. Wilson quickly assimilated Kadanoff's idea and amalgamated it into his thinking about field theories and critical phenomena, exploiting the concept of broken scale invariance.[67]

Wilson had also done some seminal work in 1964 (unpublished) on operator product expansions (OPE), but had failed in the strong coupling domain. After thinking about the implications of the scaling theory of Widom and Kadanoff when applied to QFT, and having investigated the consequences of Johnson's suggestion concerning the scale invariance of the Thirring model (1961), and that of Mack's concerning the scale invariance of the strong interactions at short distances (1968), Wilson reformulated his theory of OPE and based it on the new idea of scale invariance (1969). He found that QFT might be scale invariant at short distances if the scale dimensions of the field operators, which were defined by the requirement that the canonical commutation relations were scale invariant, were treated as new degrees of freedom.[68] These scale dimensions can be changed by the interactions between the fields and can acquire anomalous values,[69] which correspond to the nontrivial exponents in critical phenomena.

The most important implications for the foundational transformations of QFT stemming from the dramatic advances in statistical physics can be summarized with two concepts that have been stressed by Wilson: (i) the statistical continuum limit of a local theory, and (ii) the fixed points of renormalization group transformations.

Concerning the first concept, Wilson noticed that systems described by statistical physics and by QFT embodied various scales. If functions of a continuous variable, such as the electric field defined on spacetime, are themselves independent variables and are assumed to form a continuum, so that functional integrals and derivatives can be defined, then one can define a statistical continuum limit that is characterized by the absence of a characteristic scale. This means that fluctuations in all scales are coupled to each other and make equal contributions to a process. In QED calculations, this typically leads to logarithmic divergences. Thus renormalization is necessary for the study of these systems. That this concept of the 'statistical continuum limit' occupied a central position in Wilson's thinking on QFT in general, and on renormalization in particular, is reflected in his claim that 'the worst feature of the standard renormalization procedure is that it gives no insight into the physics of the statistical continuum limit' (1975).

The second concept is more complicated. As we have noticed, a varying parameter characterizing the physics at various renormalization scales reflects the scale dependence of the renormalization effects. The values of the parameter are related to each other by the renormalization group transformations that are described by the renormalization group equations. Since the late 1960s, it has become recognized that the scale invariance of any quantum field theory is unavoidably broken anomalously because of the necessity of re-normalization. A convincing argument for this statement is based on the concept of 'dimensional transmutation' and runs as follows.

The scale invariance of a theory is equivalent to the conservation of the scale current in the theory. To define the scale current, a renormalization procedure is required, because, as a product of two operators at the same point, the scale current implicitly contains an ultraviolet singularity. However, even in a theory without any dimensional parameter, it is still necessary to introduce a dimen-sional parameter as a subtraction point when renormalizing, in order to avoid the infrared divergences and to define the coupling constant. This necessity of introducing a dimensional parameter, called 'dimensional transmutation' by Sidney Coleman and E. Weinberg in (1973), breaks the scale invariance of the theory. Precisely because of dimensional transmutation, the scale invariance in any renormalized theory is unavoidably broken anomalously, though the effects of this breakdown can be taken care of by the renormalization group equations.

In statistical physics, the renormalization group approach effects connections between physics at different scale levels. By scaling out the irrelevant short-range correlations and by locating stable infrared fixed points, it has made possible the conceptual unification of various descriptions (such as those of elementary excitations (quasi-particles) and collective ones (phonons, plas-mons, spin waves)), the explanation of the universality of various critical behavior, and the calculation of order parameters and critical components. In QFT, the same approach can be used to suppress the irrelevant low-energy degrees of freedom, and to find out the stable ultraviolet fixed point. In both cases, the essence of the approach, as Weinberg has indicated (1983), is to concentrate on the relevant degrees of freedom for a particular problem,[70] and the goal is to find fixed point solutions of the renormalization group equations.

According to Wilson, the fixed point in QFT is just a generalization of Gell-Mann and Low's eigenvalue condition for the bare charge in QED.[71] At the fixed point, a scaling law holds, either in the Gell-Mann–Low–Wilson sense or in Bjorken's sense, and the theory is asymptotically scale invariant. The scale invariance is broken at non-fixed points, and the breakdown can be traced by the renormalization group equations. Thus, with the more sophisticated scale argument, the implication of Gell-Mann and Low's original idea becomes clearer. That is, the renormalization group equations can be used to study properties of a field theory at various energy scales, especially at very high energy scales, by following the variation of the effective parameters of the theory with changes in energy scale, arising from the anomalous breakdown of scale invariance, in a quantitative way, rather than qualitatively as suggested by Dyson or Landau.

It is clear that if the renormalization group equations of a given field theory possess a stable ultraviolet fixed-point solution, then the high-energy behavior of the theory causes no trouble, and the theory can be called, according to Weinberg, an 'asymptotically safe theory' (1978). An asymptotically safe theory may be a renormalizable theory if the fixed point it possesses is the Gaussian fixed point.[72] Weinberg, however, argued, and supported his position with a concrete example of a five-dimensional scalar theory, that the concept of 'asymptotic safety' is more general than the concept of renormalizability, and thus can explain and even replace it (*ibid.*). There may in fact be cases in which theories are asymptotically safe but not renormalizable in the usual sense, if they are associated with a Wilson–Fisher fixed point.

The conceptual developments described in this section can be summarized as follows. In systems with many scales that are coupled to each other and without a characteristic scale, such as those described by QFT, the scale invariance is always anomalously broken owing to the necessity of renormalization. This breakdown manifests itself in the anomalous scale dimensions of fields in the framework of OPE, or in the variation of parameters at different renormalization scales that is charted by the renormalization equations. If these equations have no fixed-point solution, then the theory is not asymptotically scale invariant, and thus is, rigorously speaking, not renormalizable.[73] If, on the other hand, the equations possess a fixed-point solution, then the theory is asymptotically scale invariant and asymptotically safe. If the fixed point is Gaussian, then the theory is renormalizable. But there may be some asymptotically safe theories that are non-renormalizable if the fixed point they possess is a Wilson–Fisher fixed point. With the occurrence of the more fundamental guiding principle of asymptotic safety, which is one of the consequences of the renormalization group approach, the fundamentality of the renormalizability principle began to be seriously challenged (see section 11.4).

8.9 Swings of the pendulum

After a short period of fascination with the spectacular successes of the renormalizable perturbation formulation of QED in the late 1940s and early 1950s, physicists gradually moved away from the theoretical framework of QFT, mainly because of the failures of QFT in addressing various problems outside the domain of electromagnetism. It is interesting to note that the movement was preceded by a first swing from Heisenberg and Pauli's QFT in the late 1920s to Wheeler and Heisenberg's SMT in the late 1930s and early 1940s. But more interesting is the fact that the conversion of physicists to the renormalized perturbative formulation of QFT in the late 1940s was going to be followed by a second swing back to the framework of QFT, although this time the latter took the form of gauge invariant theories.

As we have noted in previous sections, QFT was in a deep crisis in the 1950s. Aside from a handful of physicists working on axiomatic field theory or speculating about renormalizable non-Abelian gauge theory, mainstream physicists were caught up by two trends. In the realm of the strong interactions, the dominant ideas were those about dispersion relations, Regge poles, and the bootstrap hypothesis, which were developed by Chew and his collaborators into an anti-field-theoretical framework, the analytic S-matrix theory. In the realm of weak and hadronic electromagnetic processes, as well as their renormalization by the strong interactions, formal reasoning and algebraic manipulations replaced field-theoretical investigations. In both cases, the field-theoretical framework was discredited, if not abandoned completely.

But the situation began to change from the mid-1960s. In the realm of the strong interactions, with the discovery of more and more complicated singularity structures of the S-matrix in the complex j plane in the 1960s,[74] SMT gradually faded away. Those of Chew's followers who still pursued SMT became an isolated band of specialists, working far from the mainstream of particle physics. By contrast, mainly because of the discovery of the scaling law in deep inelastic scatterings by the experiments that were performed at the Stanford Linear Accelerator Center in 1969, the quark-parton model, and the more sophisticated quantum chromodynamics within the field-theoretical framework were accepted as the fundamental framework for hadronic physics.

In the realm of weak and electromagnetic processes, stimulated by the discovery of anomalies, formal munipulations were replaced by field-theoretical investigations of their dynamic foundations again. Furthermore, in the late 1960s, models for describing the weak and electromagnetic processes in a unified way were proposed, within the framework of a non-Abelian gauge theory. With the proof, by Veltman and 't Hooft, of the renormalizability of non-Abelian gauge theories, a self-consistent unified field theory of the weak and electromagnetic forces was available to particle physicists by 1972. All that remained to be done was to obtain experimental evidence for the existence of the neutral currents predicted by the theory. This appeared in 1973.

The experimental discoveries of the scaling law in the strong interactions and of the neutral currents in the weak interactions made non-Abelian gauge theories, within which these discoveries were predicted or explainable, ex-tremely fashionable in the 1970s. But what had changed the general mood toward the gauge theory (and toward QFT in general) was the proof of its renormalizability. The change in the general mood was so radical that Landau's devastating argument against QFT (though it might not challenge an asympto-tically free Yang–Mills theory, namely, the likely vanishing of the renormalized charges), was simply forgotten.

At a more fundamental level, it turns out in the theoretical development of the last four decades that various theories of quantum fields, with the exception of a few unrealistic models, can only be made renormalizable by introducing the concept of gauge invariance. For this reason, the further development of QFP as a consistent and successful research programme is almost entirely within the framework of a gauge theory.

Notes

1. Note that the field operator I refer to in the text is that in the Heisenberg picture, which has a different physical interpretation from that in the interaction picture.
2. For the early W-meson theory, see Kemmer (1938) and Klein (1938). They encountered a difficulty in explaining the π decay $\pi \rightarrow \mu + \nu$. The modern versions of W-meson theory were suggested by Schwinger (1957) and Glashow (1961) in their earlier attempts to unify electromagnetism with weak interactions. But this only makes sense within the framework of the quark model. (See Weinberg, 1967b; Salam, 1968).
3. This, however, is controversial. See Weinberg (1983), Gell-Mann (1987).
4. Some axiomatic field theorists later became known as constructive field theorists. Wight-man (1978) takes constructive quantum field theory as the offspring of axiomatic field theory, with some difference between them. While the concern of axiomatic field theory is the general theory of quantum fields, constructive field theory starts from specific Lagrangian models and constructs solutions satisfying the requirements of the former. For the early development of axiomatic field theory, see Jost (1965) and Streater and Wightman (1964); for constructive field theory, see Velo and Wightman (1973).
5. This result was not generally accepted. Thus Francis Low (1988) did not, and does not,

believe Landau's proof, because it was based on a leading logarithmic approximation in a perturbative approach, which he felt was not rigorous enough for a proof. John Polkinghorne (1989) held a similar position. Gell-Mann (1987) also held such a position, and argued with Landau on this point when he visited Moscow in 1956. Gell-Mann has also remarked: 'Presumably, in today's language, Landau wanted to say that non-asymptotically free field theory was no good' (*ibid.*).

6. The earliest research on this topic within the framework of QFT occurred in 1957–58 (Nishijima, 1957, 1958; Zimmermann, 1958; Haag, 1958). The result was that there was no difference between a composite particle and an elementary particle as far as the theory of scattering was concerned. Then under increasing pressure from the S-matrix theory side, came the conjecture of '$Z = 0$' as a criterion for the compositeness of particles, which was first suggested and verified for the Lee model (Lee, 1954) by J. C. Houard and B. Jouvet (1960) and M. T. Vaughn, R. Aaron, and R. D. Amado (1961). This topic was fashionable but only discussed for some simplified model field theories. Further references can be found in Weinberg (1965a).

7. Kramers (1944) suggested the importance of analyticity; Kronig (1946) connected analyticity with causality.

8. For a historical review, see Cushing (1986, 1990).

9. Judging from his talk at the Rochester conference, Gell-Mann seemed not to be a clear-cut field theorist at that time. He finished his talk with the prospect that 'maybe this approach [dispersion programme] will throw some light on how to construct a new theory that is *different* [from the old field theory], and which may have some chance of explaining the high energy phenomena which field theory has no chance whatsoever to explain (1956).

   The claim that Gell-Mann's Rochester talk was a strong impetus for the making of SMT is justified by the following facts: (i) Mandelstam's double dispersion relation (1958), which has been generally acknowledged as a decisive step in the making of SMT, started with Gell-Mann's conjecture that dispersion relations might be able to replace field theory; (ii) Chew quoted Gell-Mann's Rochester remarks at the Geneva conference of 1958; (iii) in addition to various contributions to the conceptual development of S-matrix theory, as we mentioned in the text, Gell-Mann was also among the first physicists to recognize the importance of Regge's work and to endorse the bootstrap hypothesis. As Chew remarked in 1962, Gell-Mann 'has for many years exerted a major positive influence both on the subject [SMT] and on me; his enthusiasm and sharp observations of the past few months have markedly accelerated the course of events as well as my personal sense of exitement' (1962a).

   However, Gell-Mann's later position concerning the status of SMT was quite different from Chew's. For Chew, SMT was the only possible framework for understanding strong interactions because QFT was hopeless in this regard. For Gell-Mann, however, the basic framework was QFT, since all of the principles and properties of SMT were actually abstracted from or suggested by QFT, and SMT itself could only be regarded as a way of specifying a field theory, or an on-mass-shell field theory. Gell-Mann's position on the relationship between SMT and QFT was very much in the tradition of Feynman and Dyson, and of Lehmann, Symanzik, and Zimmerman (1955), in which the S-matrix was thought to be derivable from field theory, or at least not incompatible with the field-theoretical framework. But for Chew, the conflict between QFT and SMT was irreconcilable because the notion of elementary particles assumed by QFT led to a conflict with the notions of analyticity and the bootstrap.

10. In the dispersion relation for a partial wave amplitude, L. Castillejo, R. H. Dalitz, and Dyson (CDD) found some additional fixed poles in the amplitude, so it could not be completely determined by the input partial-wave amplitude (1956). The CDD poles normally correspond to input elementary particles, but, as discovered late in the 1960s, they can also emerge from the inelastic cut (see Collins, 1977).

11. For a discussion on the relation between dispersion relations and analytic S-matrix theory, see Chew (1989).

12. Blankenbecler, Cook and Goldberger (1962).

13. Charap and Fubini (1959).
14. Chew and Frautschi (1961a, b).
15. In this case, s represents the momentum transfer squared in the t channel, in which the energy squared is represented by t.
16. For a discussion about the ontological commitment of the S-matrix theory, see Capra (1979).
17. For Gell-Mann's original motivation of Reggeizing elementary particles, see his (1989); for a historical investigation of the Reggeization programme, see Cao (1991).
18. See Chew (1989) and Capra (1985).
19. Landau (1959). In (1960), R. E. Cutkosky gave a further rule for including the unitarity condition.
20. Thus there is a one–many correspondence between Landau graphs and Feynman graphs.
21. This is partly due to the shift of emphasis in SMT from objects (particles) to processes (scatterings or reactions).
22. Stapp (1962a, b; 1965; 1968); Chandler (1968; with Stapp, 1969); Coster and Stapp (1969; 1970a, b); Iagolnitzer and Stapp (1969); and Olive (1964).
23. See Chew (1989) and Capra (1985).
24. The title of Chew's APS talk (January 1962) was 'S-matrix theory of strong interactions without elementary particles' (1962b); the title of his talk at the Geneva conference (July 1962) was 'Strong interaction theory without elementary particles' (1962c).
25. Feynman expressed this principle in a talk given at the Aix-en-Provence conference on elementary particles in 1961. See footnote 2 in Low's paper (1962). This idea of Feynman's was widely quoted by Chew and Frautschi (1961a), Salam (1962a), and many other physicists in the early 1960s.
26. Gell-Mann's remarks in (1987, 1989), Low's in (1988).
27. Chew and Mandelstam (1960), and Zachariasen and Zemach (1962).
28. See Cao (1991).
29. Capps (1963); Cutkosky (1963a, b, c); Abers, Zachariasen and Zemach (1963); and Blankenbecler, Coon, and Roy (1967).
30. Two years earlier, Blankenbecler, Cook and Goldberger (1962) had tried to apply SMT to photons. More discussions on this can be found in Cao (1991).
31. For proving the existence of a causal-field operator from pure SMT, see Lehmann, Symanzik and Zimmerman (1957); for reducing QFT to a limiting case of the dual model, see Neveu and Sherk (1972); for regarding supergravity and super-Yang-Mills theories as limiting cases of superstrings, see Green, Schwarz, and Witten (1987).
32. For deriving S-matrix elements from a field operator, see Lehmann, Symanzik, and Zimmerman (1955); for deriving dispersion representation of quantum Green's functions from the causality requirement, see Nambu (1955).
33. A preliminary study of the Reggeization program, a product of the conceptual interplay between QFT and SMT, can be found in Cao (1991).
34. PCAC is an acronym for 'partially conserved axial current'.
35. This extension presupposes that the axial vector part of the weak currents is also conserved.
36. $f_v(0)$ is defined by $(2\pi)^3 \langle p'| \, V_\mu(0)| \, p \rangle = \mathrm{i}u(0)\gamma_4 f_v(0)u(0)\delta_{\mu 4}$, which is the limiting case of equation (2) when $p = p' \to 0$.
37. A simple proof that the CVC hypothesis implies the non-renormalization of $f_v(0)$ by the strong interactions can be found in Coleman and Mandula (1967).
38. Define

$$T_+ = \int \mathrm{d}r \, (J_{01}^v(r,0) + \mathrm{i}J_{02}^v(r,0)).$$

From the isotopic commutation relations we have $J_{\mu +}^v = [J_{\mu 3}^v, T_+]$. Thus

$$\langle p'|_p J_{\mu +}^v |p\rangle_N = \langle p'|_p [J_{\mu 3}^v, T_+]|p\rangle_N = \langle p'|_p J_{\mu 3}^v |p\rangle_p - \langle p'|_N J_{\mu 3}^v |p\rangle_N$$
$$= \langle p'|_p j_\mu^{em}|p\rangle_p - \langle p'|_N j_\mu^{em}|p\rangle_N.$$

Here the property of T_+ as an isotopic raising operator and $j_\mu^{em} = j_\mu^s + j_\mu^v \,(= J_{\mu 3}^v)$ are used. If we define $F_1(q^2)^+ = f_v(q^2)$, $F_2(q^2)^+ = f_m(q^2)$, then equation (6) follows from equations (1) and (2).

39. Note that unlike the gauge SU(3) symmetry (the color symmetry), the SU(3) symmetry (and its subsymmetry SU(2)), popular in the early 1960s, was a global symmetry, later called the flavor symmetry, and was a manifestation of the mass degeneracies, first of the three Sakatons (the components of the nucleons suggested by Sakata in (1956)) (or proton and neutron in the case of SU(2)), and then of quarks. In the massless limit, the SU(3) or SU(2) flavor symmetry becomes $SU(3)_L \times SU(3)_R$ ($SU(2)_L \times SU(2)_R$) chiral symmetry.

40. We shall not give all the complications involved in the definition of this commutation relation. For the details, see Gell-Mann (1962a, 1964a, b).

41. $g_A(0)$ is defined, in a similar way to $f_v(0)$ (see note 35), in the limiting case of equation (3).

42. The bound state is produced through a mechanism similar to that in the theory of superconductivity, which is responsible for the existence of the collective excitations of the medium. See Brown and Cao (1991).

43. Another important application of the PCAC hypothesis is Adler's rule relating the elastic amplitude of π–N scattering and the radiative amplitude, where an extra massless (soft) pion is emitted at zero energy. Yet this application involves more considerations of symmetry, and is equivalent to the early idea of 'chirality conservation' of Nambu and Lurié (1962), which was derived from the exact γ_5 invariance.

44. We differentiate approximate symmetries from broken symmetries. The former are broken at the classical level, explicitly by a small non-invariant term in the Lagrangian. The latter refer, in addition to explicitly broken symmetries, to symmetries which are spontaneously broken by the non-invariant vacuum state while the Lagrangian is invariant, and also to symmetries which are anomalously broken at the quantum level (see sections 8.7 and 10.3).

45. When the charge algebra (equations (8)–(11)) was extended to the time–time ETC relations, the formal reasoning was thought to be safe. Yet for the space–time commutators, the formal rule had to be modified by some gradients of delta functions, the so-called Schwinger terms (see Goto and Imamura, 1955; Schwinger, 1959). For the algebraic relations involving u_0, see Gell-Mann (1962a).

46. This statement is not true for Wilson (1969). More discussion on his ambitious ideas will be given later in this and the next section.

47. Several works on the subject appeared almost simultaneously with Johnson and Low's paper (1966): Okubo (1966); Bucella, Veneziano, Gatto, and Okubo (1966); Hamprecht (1967); Bell (1967a, b; with Berman, 1967); Polkinghorne (1967); Brandt and Orzalesi (1967).

48. The essential points of that paper were reported by F. Low in his talk 'Consistency of current algebras', given at the then influential Eastern Theoretical Physics Conference held at Brown University, Fall 1966, and published in (Feldman, 1967). In this talk, Low also emphasized that 'in practice [the ETC relation] is a sum rule', and if the integral connected with the sum rule is singular, 'then anomalous behavior can occur'.

49. Also by Hamprecht (1967) and Polkinghorne (1967). It is most clearly analysed by Adler and Wu (1969). See below.

50. An equivalent statement of this condition, in the language of momentum space, is that the integral variables cannot be legally shifted.

51. Current algebra was soon recognized to be irrelevant to the discrepencies discussed here and below.

52. Later study showed that Sutherland's conclusion about $\omega \to \pi^0$ involved some mistakes and could be avoided even within the framework of PCAC and current algebra. See Adler (1969).

53. According to Veltman (1991), there were extensive discussions between Veltman and Bell after Veltman's talk given at the Royal Society Conference on 2 November 1966 (Veltman, 1967), including telephone conversations and the exchange of letters. See also Jackiw (1991).

54. Bell (1967b; with Berman, 1967; with Van Royen, 1968; with Sutherland, 1968).

55. This statement is followed by references to the Johnson–Low paper and other relevant papers.

56. The same calculations on the triangle graphs were also performed by Fukuda and Miyamoto (1949), and by Schwinger (1951), with different theoretical motivations. For a brief account of the motivations, see Jackiw (1972).

57. I am grateful to Professor Stephen Adler for providing me with a copy of this letter and other relevant materials.

57a. However, as Bardeen pointed out (1985): 'As the mass of this regulator fermion becomes large it does not decouple but produces an effective "Wess–Zumino"-like effective action for the chiral field, a mechanism which has recently been rediscovered through the more general study of decoupling.'

58. Although Adler won his early reputation by his prominent contributions to the PCAC–current algebra programme, he decided to do something different after he collaborated with Dashen and finished editing an anthology on current algebra (1968). See Adler (1991).

59. Adler also noticed that because the integral was linearly divergent, its value was ambiguous and depended on convention and the method of evaluation. A different result from what appeared in the text of his paper could be obtained, which would satisfy the axial vector Ward identity but violate gauge invariance. See Adler (1969), footnote 9.

60. Without invoking the triangle graph, Johnson (1963) noticed that in the unrealistic yet exactly solvable Schwinger model (two-dimensional massless QED), the formal γ_5 invariance of the Lagrangian would be lost if the current was defined in a gauge-invariant way. The impossibility of defining a gauge-invariant axial vector current first shown in this simple model was another early example provided by Johnson of the failure of formal reasoning.

61. In extracting the experimental consequence, the smoothness for matrix elements of the naive divergence operator was assumed. Yet the full divergence operator is not a smooth operator since the extra term in equation (25) is manifestly not smooth when its matrix elements vary off the pion mass shell. Cf. Bardeen (1969) and Jackiw (1972). Thus the PCAC in both of its usages is modified in a well-defined way.

62. For the historical details relating to the non-renormalization theorem, see Adler (1970); for its theoretical implications, see Bardeen (1985).

62a. This was historically the first appearance of the idea of anomaly cancellation, although the context was not the gauge theory.

63. Unitarity can be saved if one is willing to take a definition of the measure for the path integrals that spoils Lorentz invariance.

64. Unlike the axial vector currents treated by Adler and by Bell and Jackiw, the currents dealt with by Wilson are the SU(3) × SU(3) currents, the most general ones appearing in the current algebra analysis of physical processes.

65. Following this statement Wilson gives references to Valatin (1954a, b, c, d); Zimmermann (1958, 1967); Nishijima (1958); Haag (1958); and Brandt (1967). This statement is also substantiated by Wilson's drawing heavily on Johnson's work (1961).

66. Regarding this concept, Wilson refers to his predecessors: Wess (1960) for free-field theories; Johnson (1961) for the Thirring model; and Kastrup and Mack (in Mack 1968) for strong interactions. In addition, Wilson was also inspired by the scaling hypothesis then recently developed by Essam and Fisher (1963), Fisher (1964), Widom (1965a, b), and Kadanoff (1966; with collaborators, 1967) for dealing with critical phenomena in statistical mechanics. Cf. Wilson (1969, 1983) and section 8.8.

67. Wilson (1983) in his Nobel lecture vividly described how the progress in statistical physics in the mid-1960s, especially the works of Widom and Kadanoff, had influenced his thinking in theoretical physics.

68. (a) Wilson further corroborated his dynamical view of the scale dimension of field operators with an analysis of the Thirring model (Wilson 1970a) and the $\lambda\varphi^4$ theory (Wilson 1970b).

(b) Even the dimensions of spacetime have become new degrees of freedom since then, at least in an instrumentalist sense. In statistical physics, the ϵ-expansion technique was introduced; in QFT dimensional regularization. Both techniques are based on this new conception of spacetime dimensions. While realistic field-theoretical models ultimately have to be four dimensional, and only toy models can be two dimensional, in statistical physics, however, two-dimensional models are of great relevance in the real world.

69. It is worth noting that there is an important difference between Wilson's concept of the asymptotic scale invariance of QFT at short distances and that of Bjorken (1969). While Bjorken's scaling hypothesis about the form factors in deep inelastic lepton–hadron scattering suggests that the strong interactions seem to turn off at very short distances, Wilson's formulation of OPE re-establishes scale invariance only after absorbing the effects of interactions and renormalization into the anomalous dimensions of fields and currents. But this is just another way of expressing logarithmic corrections to the scale invariance of the theory. Thus Bjorken's ideas were soon fitted into the framework of a non-Abelian gauge theory (QCD) and re-expressed as asymptotic freedom, while Wilson's idea has found its applications in other areas (cf. section 10.2).

70. Weinberg (1983) also noticed that sometimes the relevance problem is more complicated than simply choosing an appropriate energy scale, and involves turning on collective degrees of freedom (e.g. hadrons) and turning off the elementary ones (e.g. quarks and gluons).

71. Wilson further generalized Gell-Mann and Low's result by introducing a generalized mass vertex which breaks the scale invariance of the theory (1969). Callan (1970) and Symanzik (1970) elaborated Wilson's generalization and obtained an inhomogeneous equation, in which the renormalization group function ψ was exact rather than asymptotic as in the case of Gell-Mann and Low. When the inhomogeneous term was neglected, the Gell-Mann–Low equation was recovered.

72. The Gaussian fixed point corresponds to a free massless field theory for which the field distributions are Gaussians.

73. QED is regarded as perturbatively renormalizable only because the breakdown of perturbation theory at ultra-relativistic energy, as pointed out by Gell-Mann and Low, is ignored.

74. Cf. P. D. B. Collins (1977).

Part III

The gauge field programme for fundamental interactions

Modern gauge theory started with Yang and Mills's proposal about isotopic gauge invariance of the strong interactions. The Yang–Mills theory, a non-Abelian gauge theory, emerged totally within the framework of the quantum field programme, in which interactions are transmitted by field quanta and realized through localized coupling between field quanta. Physically, it obtained impetus from the charge independence of the strong nuclear forces, but at the same time was constrained by the short-range character of the same forces. Methodologically, it was driven by the desire of having a universal principle to fix a unique form of couplings among many possibilities. Physicists took some interest in the Yang–Mills theory, in part because they thought it renormalizable, but soon abandoned it because there seemed no way to have a gauge-invariant mechanism to account for the short-range character of the nuclear forces. (Chapter 9)

The difficulty was overcome, first in the early 1960s, by the discovery of the spontaneous breakdown of symmetry (section 10.1), and then in the early 1970s by the discovery of asymptotic freedom (section 10.2). With the proof, by Veltman and 't Hooft, of the renormalizability of non-Abelian gauge theories (section 10.3), a seemingly self-consistent conceptual framework was available to the particle physics community.

Conceptually, the framework is very powerful in describing various fundamental interactions in nature, and in exploring novel, global features of field theories that were supposed to be local, which features have direct bearings on our understanding of the structure of the vacuum and the quantization of charges (section 10.4). Thus a new research programme in fundamental physics was initiated: the gauge field programme (GFP).

According to GFP, all kinds of fundamental interactions can be characterized by gauge potentials. The programme acquired a fresh appearance in the so-called standard model (section 11.1). The success of the standard model encouraged further extension of the programme. Great efforts were made, but no success, either in unifying the electroweak forces with the strong forces, nor in gauging gravity, has been achieved. Thus the universality of the gauge principle remains controversial (section 11.2).

The vicissitudes of its empirical adequacy notwithstanding, the ontological implications of GFP are quite interesting. It is arguable that the notion of gauge fields provides a basis for a geometrical understanding of QFT. This implies that GFP, if its potential can be completely realized, embodies a synthesis of QFP and GP (section 11.3).

Since the late 1970s, however, GFP has failed to make further progress, both in terms of explaining or predicting new properties of elementary particles and in terms of solving the conceptual difficulties it faces. In part because of this stagnation, in part because of the changing theoretical context, the conceptual development of 20th century field theories has pointed in a new direction: effective field theory (section 11.4).

9

The route to gauge fields

This chapter is devoted to examining the physical and speculative roots of the notion of gauge fields, reviewing early attempts at applying this attractive notion to various physical processes, and explaining the reasons why these heroic attempts failed.

9.1 Gauge invariance

The idea of gauge invariance, as we mentioned in section 5.3, originated in 1918, from Weyl's attempt to unify gravity and electromagnetism, based on a geometrical approach in four-dimensional spacetime (1918a, b). Weyl's idea was this. In addition to the requirement of GTR that coordinate systems have only to be defined locally, the standard of length, or scale, should also only be defined locally. So it is necessary to set up a separate unit of length at every spacetime point. Weyl called such a system of unit-standards a gauge system. In Weyl's view, a gauge system is as necessary for describing physical events as a coordinate system. Since physical events are independent of our choice of descriptive framework, Weyl maintained that gauge invariance, just like general covariance, must be satisfied by any physical theory. However, Weyl's original idea of scale invariance was abandoned soon after its proposal, since its physical implications appeared to contradict experiments. For example, as Einstein pointed out, this concept meant that spectral lines with definite frequencies could not exist.

Despite the initial failure, Weyl's idea of a local gauge symmetry survived, and acquired new meaning with the emergence of quantum mechanics (QM). As is well known, when classical electromagnetism is formulated in Hamiltonian form, the momentum P_μ, is replaced by the canonical momentum $(P_\mu - eA_\mu/c)$. This replacement is all that is needed to build the electromagnetic interaction into the classical equations of motion, that is, to fix the form of the electromagnetic interaction. The form of the canonical momentum is motivated by Hamilton's principle. The essential idea here is to obtain both Maxwell's equations and the equations of motion for charged particles from a single physical principle. A key concept in QM is the replacement of the momentum P_μ in the classical Hamiltonian by an operator $(-ih\partial/\partial x_\mu)$. So, as Vladimir Fock observed in (1926), QED could be based on the canonical momentum operator $-ih(\partial/\partial x_\mu - ieA_\mu/hc)$. Soon after that Fritz London (1927) pointed out the similarity of Fock's work to Weyl's earlier work. Weyl's idea of gauge invariance would then be correct if his scaling factor ϕ_μ were replaced by $(-ieA_\mu/hc)$. Thus instead of a scale change $(1 + \phi_\mu \, dx_\mu)$, we can consider a phase change $[1 - (ie/hc)A_\mu \, dx_\mu] \approx \exp[-(ie/hc)A_\mu \, dx_\mu]$, and which can be thought of as an imaginary scale change. In his (1929), Weyl put all of these considerations together and explicitly discussed the transformation of the electromagnetic potential $A_\mu \to A'_\mu = A_\mu + \partial_\mu \Lambda$ (a gauge transformation of the second kind) and the associated phase transformation of the wave function of a charged particle $\psi \to \psi' = \psi \exp(ie\Lambda/hc)$ (a gauge transformation of the first kind).

Here the essential clue was provided by the realization that the phase of a wave function could be a new local variable. The previous objections to Weyl's original idea no longer applied because the phase was not directly involved in the measurement of a spacetime quantity, such as the length of a vector. Thus in the absence of an electromagnetic field, the amount of phase change could be assigned an arbitrary constant value, since this would not affect any observable quantity. When an electromagnetic field is present, a different choice of phase at each spacetime point can then easily be accommodated by interpreting the potential A_μ as a connection that relates phases at different points. A particular choice of phase function will not affect the equation of motion since the change in the phase and the change in the potential cancel each other exactly. In this way the apparent 'arbitrariness' formerly ascribed to the potential is now understood as the freedom to choose any value for the phase of a wave function without affecting the equation. This is exactly what gauge invariance means.

In applying Weyl's results to QFT, Pauli, in his influential review article (1941), pointed out that while the invariance of a theory under the global phase change (a gauge transformation of the first kind with Λ independent of its spacetime position) entails charge conservation, the local gauge invariance (local in the sense that Λ is dependent on its spacetime position) is related to the electromagnetic interaction and determines its form by the replacement $[\partial/\partial x_\mu \to (\partial/\partial x_\mu - ieA_\mu/hc)]$, the so-called minimal electromagnetic coupling.

Three points are clear now. First, the correct interpretation of gauge invariance was not possible without quantum mechanics. Second, there is a striking similarity in theoretical structure between gauge theory and general relativity, i.e. the similarity between the electromagnetic field, local phase functions, electromagnetic potential, and gauge invariance on the one hand, and the gravitational field, local tangent vectors, affine connection, and general covariance on the other.

Third, as Michael Friedman has pointed out (1983), the notion of invariance is different from the notion of covariance. A theory is invariant under a transformation group if some geometrical objects are invariant under the action of the group; a theory is covariant if, under the action of a transformation group, the forms of the equations in the theory are unchanged. Considering this subtle distinction, the concept of gauge invariance in many cases refers actually to gauge covariance, and the symmetry group involved in a gauge theory is actually the covariance group, not the invariance group.[1] Incidentally, this remark applies throughout the rest of this volume. Normally, invariance and symmetry are used in the sense of covariance. For example a theory is said to be symmetrical, or invariant, under the transformations between different representations of relevant observable degrees of freedom, if the forms of the equations or laws in the theory are unchanged under the action of the transformation group. But when we say that only those quantities that are gauge invariant are physically observable, we are using the concept of gauge invariance in its proper 'invariance' sense.

9.2 The gauge principle of fixing forms of interactions

The modern idea of a gauge theory began with the speculation published by Yang and Mills in (1954a, b). The heart of their speculation, or any gauge theory, is a gauge symmetry group that determines the dynamics of the theory. It may be helpful to give some definitions. A symmetry is global if its representation is the same at every spacetime point, and local if it is different from point to point. Thus Lorentz symmetry is a global symmetry, and general coordinate transformation symmetry is a local one. Further, symmetry is external if the relevant observable degree of freedom is spatio-temporal in nature, and internal otherwise. Thus Lorentz symmetry is an external symmetry, but the phase symmetry in quantum electrodynamics is an internal one.

The Yang–Mills theory emerged entirely within the framework of the quantum field programme, and was motivated by two considerations. First, more and more new particles were discovered after the Second World War, and various possible couplings among those elementary particles were being proposed. Thus Yang and Mills felt it necessary to have some principle to choose a unique form out of the many possibilities being considered. The principle suggested by Yang and Mills is based on the concept of gauge invariance, and is thus called the gauge principle. Second, in choosing a proper gauge symmetry, Yang and Mills were driven by curiosity to find the consequences of assuming a law of conservation of isospin, which was thought to be the strong interaction analogue of electric charge.[2]

Mathematically, the conservation of isospin implies a generalization of local phase symmetry to a general localized internal symmetry. Physically, it is a reformulation of the empirical discovery of the charge independence of nuclear force. It assumes, following Heisenberg, that the proton and neutron are but two states in an abstract isospin space of one and the same particle. Now charge conservation is related, in quantum electrodynamics, to phase invariance. So, by analogy, one might assume that the strong interaction would be invariant under isospin rotation. Isospin invariance entails that the orientation of isospin is of no physical significance. The differentiation between a neutron and a proton is then a purely arbitrary process. The arbitrariness, however, is global in character, namely that once a choice is made at one spacetime point the choices at all other points are fixed. But as Yang and Mills realized, this is not consistent with the requirement in QFT that every field quantity should only be defined locally.

In order to obtain a local isospin-invariant theory, Yang and Mills were guided by an analogy to QED. In QED, the local phase invariance is preserved by introducing the electromagnetic potential A_μ, which counteracts the variation of the phase in the wave function of a charged particle. In an entirely similar manner, Yang and Mills introduced a gauge potential B_μ, which under an isospin rotation transforms as $B_\mu \rightarrow B'_\mu = S^{-1} B_\mu S + (i/\epsilon) S^{-1} \partial S / \partial x_\mu$, to counteract the spacetime dependence of the isospin rotation S of a nucleon wave function. Here $S = \exp(-i\theta(x)L)$ is an element of the local isospin group, a non-Abelian Lie group: L has three components L_i, which are the generators of the group. Thus, in the Yang–Mills theory, while the strong interaction analogue of the electric charge is the isospin, the corresponding analogue of the photons is the three vector gauge bosons B^i_μ. The results Yang and Mills obtained are significant.

First, it is easy to show that the potential B_μ must contain a linear combination of L_i: $B_\mu = 2b_\mu \cdot L$. This relation explicitly displays the dual role of the Yang–Mills potential as both a four-vector potential in spacetime and an SU(2) isospin vector operator. That is, Yang–Mills potentials are non-Abelian potentials. Generalization to other non-Abelian potentials when other internal quantum numbers from isospin, such as SU(3) color, are considered is conceptually straightforward.

Second, the field equations satisfied by the twelve independent components b_μ^i of the potential B_μ and their interaction with any field having an isospin charge, are essentially fixed by the requirement of isospin invariance, in the same way that the electromagnetic potentials A_μ and their interaction with charged fields are essentially determined by the requirement of phase invariance. Thus we have a powerful principle for determining the forms of the fundamental interactions, namely the gauge principle. This principle, absent from the older formulations of QFT, is the corner-stone of the gauge field programme.

9.3 Early attempts

Yang and Mills proposed their speculation about an isospin-invariant field theory and tried to use it to account for the strong interactions. The apparent attraction of such a gauge-invariant theory was that it might be renormalizable. But as far as the strong interactions were concerned, the theory was obviously untenable. The finite range of the strong interaction required, as shown by Yukawa, that it be mediated by quanta of non-zero mass. But massive quanta ruined the gauge invariance and the renormalizability as well. On the other hand, in order to preserve the gauge symmetry, as Pauli pointed out, its gauge bosons had to be massless.[3] But then these gauge bosons could not be responsible for the short-range interactions.

In (1956), Ryoyu Utiyama proposed, by postulating the local gauge invariance of systems under a wider symmetry group, a more general rule for introducing, in a definite way, a new field that has a definite type of interaction with the original fields. While Yang and Mills confined their attention to the internal symmetry, Utiyama's treatment extended to the external symmetry as well. According to Utiyama, then, the gravitational field and gravitational interaction could also be incorporated into the general framework of a gauge theory. Utiyama's idea is controversial,[4] but has some bearing on quantum gravity (see section 11.2).

A more consequential effort was made by Schwinger (1957), who proposed a Yang–Mills-type gauge theory, based on an SU(2) group, in which the electromagnetic and weak interactions can be accounted for in a unified way: the neutral vector gauge boson (photon), which was supposed to be coupled to the electromagnetic current, and the charged vector gauge bosons ($W^\pm$), which were supposed to be coupled to the conserved weak currents, formed a triplet of the symmetry group SU(2). The huge mass splitting between the massless photon and massive $W^\pm$ bosons in the triplet was explained away by introducing an auxiliary scalar field.

Schwinger's idea was picked up by Bludman (1958), Salam and Ward (1959), and Sheldon Glashow (1961). Later, it was also elaborated by Salam and Ward (1964), Weinberg (1967b), and Salam (1968).

Bludman's work is interesting in three respects. First, he extended the isotopic symmetry to the chiral SU(2) symmetry so that the parity non-conservation in the weak interaction could be accounted for. As a bonus of this move, he became the first physicist to suggest a 'charge-retention Fermi interaction', later to be rephrased as the neutral current, in addition to the traditional charge-exchange interactions. Second, he accepted Schwinger's idea that leptons, like hadrons in the strong interactions, should carry weak isospin in the weak decays, and, to execute this idea, assigned the known leptons a doublet representation of the chiral SU(2) group. This assignment was inherited by later investigators and became part of the structure in the standard model. Third, he suggested that the charged and neutral vector bosons should be described by massive Yang–Mills fields. His suggestion opened up a new field of investigation, the massive Yang–Mills theory. Bludman also observed that the charged vector bosons had singular non-renormalizable electromagnetic interactions. This observation provoked intensive research on the renormalizability of gauge theories with charged gauge bosons.[5]

Glashow (1961) followed Schwinger's step in pursuing a unification of weak and electromagnetic interactions. He argued, however, that within the SU(2) group structure it was impossible to describe the parity-conserving electromagnetic current and the parity-violating weak currents in a unified way. Thus he extended Bludman's work, in which the SU(2) symmetry was responsible only for the weak interations, to include electromagnetism by introducing an Abelian group U(1) in addition to SU(2). Thus the underlying group structure should be SU(2) × U(1), and the resulting Lagrangian for the interactions of leptons had four vector bosons: two $W^\pm$ bosons coupled to the charged weak currents, one W^0 boson coupled to the neutral weak current, and a photon coupled to the electromagnetic current. Among four vector bosons, Glashow also introduced a photon–W^0 boson mixing angle, later being called the Weinberg angle. Esthetically, Glashow's model was appealing. Phenomenologically, judging from later developments, it turned out to be the best model for unifying the electromagnetic and weak interactions, and was strongly supported by experiments. Nevertheless, one of the basic theoretical difficulties of Yang–Mills-type theories, namely, having a massive gauge boson without destroying gauge invariance, remained unaddressed.

The discovery of spontaneous symmetry breaking and the Schwinger–Anderson–Higgs mechanism (see section 10.1), which was incorporated into the models proposed by Weinberg (1967b) and Salam (1968), solved this difficulty. Yet for all these models, a fundamental difficulty of non-Abelian gauge theories, that is, their renormalizability, remained unaddressed. Weinberg speculated that his model would be renormalizable because of gauge invariance. Yet for reasons that will be clear in section 10.3, this speculation was technically untenable and conceptually empty.

An influential theory of the strong interactions, with a theoretical structure of Yang-Mills type, was proposed by Sakurai in 1960. Sakurai was the first theorist to explore the consequences of the gauge principle for the strong interactions in a detailed physical model rather than by purely speculating. He proposed that the strong interactions are mediated by five massive vector gauge bosons, which are associated with the gauge group $SU(2) \times U(1) \times U(1)$, and thus are coupled with three conserved vector currents: the isospin current, the hypercharge currents, and the baryonic current.

Well aware of the mass difficulty, Sakurai looked for a dynamical mechanism to produce the mass. One possibility was that the mass arose from the self-interaction of the gauge bosons. He wrote: 'It might not be entirely ridiculous to entertain the hope that an effective mass term which seems to violate our gauge principle may arise out of the fundamental Lagrangians which strictly satisfy our gauge principle' (1960). This sentence anticipates precisely the idea of spontaneous breakdown of symmetry and the so-called Higgs mechanism, which later allowed gauge theory to give a consistent and successful account of the electroweak interactions.[6]

Sakurai's anticipation of the spontaneous breakdown of symmetry and the Higgs mechanism was a bold step. He recorded in a footnote to his paper that 'this point [about generating the mass from the Lagrangian] has been criticized rather severely by Dr. R. E. Behrends, Professor R. Oppenheimer, and Professor A. Pais' (*ibid.*). Another note is more interesting:

Several critics of our theory have suggested that, since we are not likely to succeed in solving the mass problem, we might as well take [the equations] with massive *B* fields as the starting point of the theory, forgetting about the possible connection with the gauge principle. This attitude is satisfactory for all practical purposes. However, the author believes that in any theory efforts should be made to justify the fundamental couplings on a priori theoretical grounds.

(Ibid.)

Among those critics was Gell-Mann. Gell-Mann himself, at first, was attracted by the Yang–Mills theory, and tried to find a 'soft-mass' mechanism that would allow the supposed 'renormalizability' of the massless theory to persist. He failed. As he later recalled (1987), he lost interest in the Yang–Mills theory in the early 1960s, and even persuaded other committed gauge theorists, including Sakurai, to abandon the gauge principle in strong interactions, and to adopt instead the 'vector meson dominance' model, a more 'practical' alternative. Gell-Mann's early fascination with the Yang–Mills theory, however, was not without consequences. Theoretically, his current algebra is conceivable only within the framework of a Yang–Mills theory, in which vector (and/or axial vector) bosons were coupled with vector (and/or axial vector) currents of fermions. The commutation rules are the consequence of the symmetry of the gauge-invariant theory. By taking away the bosons because of failing in solving the gauge boson mass problem, what was left were currents satisfying current commutation rules.

The logical connection between current algebra and the Yang–Mills theory was keenly perceived and fruitfully exploited by some physicists in the mid-1960s, when current algebra achieved its magnificent successes, among which the most spectacular was the success of the Adler–Weisberger relation. For example, Martinus Veltman (1966) succeeded in deriving the Adler–Weisberger relation and other consequences of the current algebra and PCAC programme by a set of equations, which he called divergence equations. In the divergence equations, the equations of the current algebra and PCAC programme were extended to include higher-order weak and electromagnetic effects, by means of the replacement of ∂_μ by $(\partial_\mu - W_\mu \times)$, where W_μ denoted a vector boson. Responding to Veltman's work, Bell (1967b) presented a formal derivation of the divergence equations from the consideration of gauge transformations, thus making it clear, at least to some field-theory-oriented physicists like Veltman, that the successes of the current algebra and PCAC programme must be a consequence of gauge invariance. Perceiving the logical connection between current algebra and the Yang–Mills theory this way, Veltman (1968b) took the success of the Adler–Weisberger relation as experimental evidence for the renormalizability of non-Abelian gauge theories, with a hidden assumption that only renormalizable theories could successfully describe what happened in nature. As we shall see in section 10.3, Veltman's confidence in the renormalizability of the Yang–Mills theory was consequential.

Notes

1. Friedman also defined the symmetry group of a theory as the covariance group of the standard formulation, which is a system of differential equations for the dynamic objects (of the theory) alone. Since in gauge theory all objects are dynamic, the symmetry group and the covariance group are degenerate.
2. Cf. C. N. Yang (1977, 1983).
3. The reason for the masslessness of gauge bosons is as follows. To obtain the correct equation of motion for the vector gauge field from the Euler–Lagrange equations, the mass of the potential must be introduced into the Lagrangian through a term of the form $m^2 A_\mu A^\mu$. The

term is clearly not gauge invariant because the extra terms that arise from the transformation of the potential are not canceled by the transformation of the wave function of the fermion. Thus the gauge fields are required to be massless in a Yang–Mills theory, or more generally in any gauge theory. This is an inescapable consequence of local gauge invariance.

4. See, for example, the debate between T. W. B. Kibble (1961) and Y. M. Cho (1975, 1976). For more about the external gauge symmetry, see also Hehl *et al.* (1976), Carmeli (1976, 1977) and Nissani (1984).

5. See, for example, Glashow (1959), Salam and Ward (1959), Salam (1960), Kamefuchi (1960), Komar and Salam (1960), Umezawa and Kamefuchi (1961), and many other papers.

6. Sakurai's paper (1960) was submitted for publication before the papers on spontaneous breakdown of symmetry by Nambu and Jona-Lasinio (1961a, b) and by Goldstone (1961). There will be more discussion on the spontaneous breakdown of symmetry in section 10.1.

10

The formation of the conceptual foundations
of gauge field theories

For a gauge invariant system of quantum fields to be a self-consistent framework for describing various interactions, some mechanisms for short-range interactions must be found (sections 10.1 and 10.2) and its renormalizability proved (section 10.3). In addition, non-Abelian gauge theories have exhibited some novel features, which have suggested certain interpretations concerning the structure of the vacuum state and the conditions for the quantization of physical parameters such as charges. Thus a new question, which never appeared in the traditional foundational investigations of (Abelian-gauge-invariant) QED or other non-gauge-invariant local field theories, has posed itself with a certain urgency, attracted intense attention, and become a favorite research topic among a sizable portion of mathematics-oriented physicists in recent years. This is the question of the global features of non-Abelian gauge field theories (section 10.4). This chapter will review the formation of these conceptual foundations of gauge theories, both as a theoretical framework and as a research programme, and will register some open questions that remain to be addressed by future investigators.

10.1 Mechanisms for short-range interactions (I):
spontaneous symmetry breaking

The original Yang–Mills theory failed to be an improvement on the already existing theories of strong nuclear interactions, because it could not reproduce the observed short-range behavior of the nuclear force without explicitly violating gauge symmetry. A major obstacle to be overcome in the further development of gauge theories was then the need to have a consistent scheme with massive gauge quanta while retaining (in some sense) gauge invariance. One way of solving the problem is so-called spontaneous symmetry breaking (SSB).[1]

What is SSB? Suppose a Lagrangian is invariant under a symmetry group G, and there is a degenerate set of minimum energy states (ground states) which constitute a G multiplet. If one of these is selected as 'the' ground state of the system, the symmetry of the system is said to be spontaneously broken. In this case, the symmetry is still there as far as the underlying laws are concerned. But the symmetry is not manifested in the actual states of the system. That is, the physical manifestations of a symmetric physical theory are quite asymmetric. A simple example of SSB can be found in the phenomenon of the ferromagnet: the Hamiltonian of a ferromagnetic system has rotational invariance, but the ground state of the system has the spins of the electrons aligned in some arbitrary direction, and thus shows a breakdown of rotational symmetry, and any higher state built from the ground state shares its asymmetry.

The idea that certain systems of quantum fields may show SSB was clearly explored from the late 1950s, first by Nambu on his own (1959), and then with his collaborator G. Jona-Lasinio (1961a, b), and then by Jeffrey Goldstone (1961) and some others. For this reason, many physicists tend to believe that SSB made its first appearance around 1960 in the writings of the above authors. Yet, in fact, the idea of SSB is much older than particle physics, and its rediscovery and integration into the scheme of gauge theories around 1960 had a long history and, in particular, resulted from an active interplay between the fascinating developments in condensed matter physics and particle physics.

Early ideas of SSB

The modern realization of the importance of symmetry in physics began with Marius Sophus Lie (with Engel, 1893) and Pierre Curie (1894). While Lie was concerned primarily with the symmetry of the laws of nature (i.e., the equations), Curie was interested in the symmetry of the physical state (i.e., the solutions of the equations).[2] Curie did not have the modern idea of SSB, but he knew that for certain phenomena to occur, 'it is necessary that some symmetry elements be missing', and even asserted that 'asymmetry is what creates phenomena' (*ibid*.). Classical examples of SSB are many; some of them were known even before Curie's paper of 1894.[3] A common feature of these examples of SSB is that they occur in non-linear systems that have bifurcation points for some critical value of a parameter, such as the Reynolds number in hydrodynamics, or the square of the angular momentum in the problem of a rotating self-gravitating fluid.

Poincaré (1902), in discussing, within the context of statistical thermodynamics, how microscopic phenomena obeying Newton's time-reversible laws can lead to apparently irreversible macroscopic phenomena, pointed out, at a deeper level, another aspect of SSB: the system must consist of many elements. Thus, although the system 'cycles', after a long time, arbitrarily close to its initial state, as required by the ergodic theorem, in actual large physical systems the 'Poincaré cycle' time is unrealizably long.

Heisenberg's paper on ferromagnetism (1928) offered a now classic illustration of SSB in quantum theory. He assumed an obviously symmetric form for the interaction energy of a pair of electron spins, namely,

$$H_{ij} = J_{ij} S_i \cdot S_j,$$

(where S_i and S_j are spin operators of atoms on neighboring lattice sites of a ferromagnet). However, the invariance under the SU(2) spin rotation group is destroyed by a magnetic perturbation, leading to a ground state with all spins aligned in the same direction. This state is describable by a macroscopic parameter $\langle S \rangle$, the magnetization. Although there are an infinite number of directions along which this alignment can occur, and all possess identical energy, the system under consideration is so large that neither quantum nor thermal fluctuations can cause a transition to a different ground state in any conceivable finite time. Thus the SSB phenomenon associated with a ferromagnet can be understood in terms of a 'large-number argument', and be characterized by the non-zero value of a macroscopic parameter, namely the magnetization $\langle S \rangle$.[4]

In 1937, Landau generalized the idea of a non-vanishing macroscopic symmetry-breaking parameter, in connection with his theory of phase transitions (1937), and especially with the theory of superconductivity he worked out with V. L. Ginzburg (1950). Landau's work put SSB in a new light. First, in his discussion of continuous phase transitions, Landau showed that SSB occurs whenever the different phases have different symmetry, thus establishing its universality; he also indicated its physical cause: the system seeks an energetically more favored state. Second, he introduced the concept of the characteristic order parameter, which vanishes in the symmetrical phase and is non-zero in the asymmetric phase.

In the case of superconductivity, Landau introduced the 'effective wave function' ψ of the superconducting electrons as the characteristic function, and wrote a phenomenological equation for its description, with this important consequence: if ψ does not vanish, then SSB must take place and the superconducting state characterized by ψ is asymmetric.

On 21 November 1958, Landau, in support of Heisenberg's non-linear unified field theory, wrote:

The solutions of the equations will possess a lower symmetry than the equations themselves. From the asymmetry, the electromagnetic and perhaps also the weak interactions follow . . .. This is more a programme than a theory. The programme is magnificent but it must still be carried out. I believe this will be the main task of theoretical physics.[5]

Heisenberg's new concept of the vacuum

In the late 1950s, Heisenberg made a crucial contribution to the rediscovery of SSB.[6] The crucial question for Heisenberg's non-linear unified field theory programme was how to derive phenomena in different domains of interaction, possessing different dynamical symmetries, from the equations of the underlying field, which possess a higher symmetry than the phenomena themselves. To address this issue, Heisenberg turned his attention to the concept of the vacuum, whose properties underlie the conceptual structure of any field theory.

At the 1958 Rochester Conference on high-energy nuclear physics held in Geneva, Heisenberg invoked the idea of a degenerate vacuum to account for internal quantum numbers, such as isospin and strangeness, that provide selection rules for elementary particle interactions (1958).[7]

In an influential paper submitted in 1959,[8] Heisenberg and his collaborators used his concept of a degenerate vacuum in QFT to explain the breaking of isospin symmetry by electromagnetism and weak interactions. They remarked that 'it is by no means certain a priori that [the theory] must give a vacuum state possessing all the symmetrical properties of the starting equation' (*ibid.*). Furthermore, 'when it appears impossible to construct a fully symmetrical "vacuum" state', it should 'be considered not really a "vacuum", but rather a "world state", forming the substrate for the existence of elementary particles. This state must then be degenerate', and would have 'in the present theory a practically infinitely large isospin'. Finally, they took this 'world state' to be 'the basis for the symmetry breaking, which is, e.g., experimentally expected in the breaking of the isospin group by quantum electrodynamics, [because the relevant excited states can "borrow" some isospin from the "world states"]' (*ibid.*).

Heisenberg's degenerate vacuum was at the time widely discussed at international conferences.[9] It was frequently quoted, greatly influenced field theorists, and helped to clear the way for the extension of SSB from hydro-dynamics and condensed matter theory to QFT. However, Heisenberg never reached a satisfactory understanding of the origin, mechanism, and physical consequences of SSB, nor did he give a convincing mathematical formulation of it.[10]

The inspiration from superconductivity

A more effective approach to an understanding of SSB was inspired by new developments in the theory of superconductivity in the late 1950s. To explain the fact that the measured specific heat and thermal conductivity of super-conductors become very small at low temperatures, Bardeen proposed a model with a gap in the energy spectrum just above the lowest energy value (1955). This model soon developed by himself, in collaboration with Leon N. Cooper and J. Robert Schrieffer, into a microscopic theory of superconductivity, the BCS theory (1957). In this theory, the gap comes about via a definite microscopic mechanism, namely Cooper pairing: the correlated Cooper pairs of electrons, having opposite momenta and spins and with energy near the Fermi surface, are caused by an attractive phonon-mediated interaction between electrons; and the 'gap' is simply the finite amount of energy required to break this correlation.

Gauge invariance and the BCS theory

The BCS theory was able to give a quantitative account of the thermodynamic and electromagnetic properties of superconductors, including the Meissner effect (the complete expulsion of magnetic flux from the interior of a super-conductor), but only in a specific gauge, the London gauge (with div $A = 0$). There would have been no objection to doing the calculation in a particular gauge, were the BCS theory gauge invariant; but it is not. Thus the derivation of the Meissner effect from BCS theory, and BCS theory itself, became the subjects of lively debate in the late 1950s.[11]

Bardeen, P. W. Anderson, David Pines, Schrieffer, and others tried to restore the gauge invariance of the BCS theory, while retaining the energy gap and the Meissner effect.[12] An important observation made by Bardeen during the debate was that the difference between the gap model and the gauge-invariant theory lay only in the inclusion in the latter of electrostatic potential. The inclusion would give rise to longitudinal collective modes, but had nothing to do with the magnetic interactions associated with the transverse vector potential.

Anderson and others refined Bardeen's contention. Anderson showed that collective excitation of longitudinal type restores the gauge invariance. When the long-range Coulomb forces are included, the longitudinal excitations resemble high-frequency plasmons and do not fill the gap near zero energy. Anderson's assertion can be rephrased as follows: (i) gauge invariance requires the presence of massless collective modes; and (ii) long-range forces re-combine these massless modes into massive ones. Taken over into QFT, point

(i) corresponds to the Nambu–Goldstone mode and point (ii) to the Higgs mechanism. However, there had first to be a bridge between superconductivity and QFT and that bridge was constructed largely by Nambu.

Nambu's approach to the gauge invariance of the BCS theory

In the late 1950s, Nambu adopted Bogoliubov's reformulation of the BCS theory and his treatment of it as a generalized Hartree–Fock approximation,[13] thus clarifying the logical relations among gauge invariance, the energy gap, and the collective excitations. More importantly, he had provided a suitable starting point for exploring the fruitful analogy between superconductivity and QFT.

Bogoliubov's treatment of superconductivity describes elementary excitations in a superconductor as a coherent superposition of electrons and holes (quasi-particles) obeying the Bogoliubov–Valatin equations:

$$E\psi_{p+} = \epsilon_p\psi_{p+} + \phi\psi^{+}_{(-p)-}, \quad E\psi^{+}_{p-} = -\epsilon_p\psi^{+}_{(-p)-} + \phi\psi_{p+}, \tag{1}$$

with $E = (\epsilon_p^2 + \phi^2)^{1/2}$. Here ψ_{p+} and ψ_{p-} are the wave functions of an electron with momentum p and spin $+$ or $-$, so that $\psi^{+}_{(-p)-}$ effectively represents a hole of momentum p and spin $+$; ϵ_p is the kinetic energy measured from the Fermi surface; ϕ is the energy gap arising from the phonon-mediated attractive force between the electrons. The theory cannot be gauge invariant because the quasi-particles are not eigenstates of charge. Nambu's work, however, shows that the failure of gauge invariance is not a defect of the BCS–Bogoliubov theory, but is deeply rooted in the physical reality of the superconductor, since the energy gap is gauge dependent.[14]

Nambu systematically applied QED techniques to the BCS–Bogoliubov theory. He wrote down the equation for the self-energy part, which corresponded to the Hartree–Fock approximation, and he obtained a non-perturbative solution in which the gauge-dependent ϕ field turned out to be the energy gap. He wrote down the equation for the vertex part, and found that when he included the 'radiative corrections', it was related to the self-energy part, thus establishing the 'Ward identity' and the gauge invariance of the theory. He made the important observation that one of the exact solutions of the vertex-part equation, namely the collective excitation of the quasi-particle pairs, which led to the Ward identity, was the bound state of a pair with zero energy and momentum.

In Nambu's consistent physical picture of superconductivity, a quasi-particle is accompanied by polarization of the surrounding medium (radiative correction), and overall charge conservation and gauge invariance result when both are considered together. The quantum of the polarization field manifests itself as a collective excitation of the medium, formed by moving quasi-particle pairs. Thus the existence of these collective excitations (the bound states) appears to be the logical consequence of gauge invariance, coupled with the existence of a gauge-dependent energy gap.

From superconductivity to QFT

Nambu's work on superconductivity led him to consider the possible application to particle physics of the idea of non-invariant solutions (especially in the vacuum state). Nambu first expressed the crucial point at the Kiev Conference of 1959:

An analogy [exists] between the problem of γ_5-invariance in field theory and that of gauge invariance in superconductivity as formulated by Bogoliubov. In this analogy the mass of an observed particle corresponds to the energy gap in a superconductive state. The Bardeen–Bogoliubov description of superconductivity is not gauge invariant because of the energy gap. But several people have succeeded in interpreting the Bardeen–Bogoliubov theory in a gauge invariant way. In this way one could also handle the γ_5-invariance. Then, for example, one easily concludes the existence of a bound nucleon–antinucleon pair in a pseudoscalar state which might be identified as the pion state.

(1960)

Nambu summarized the analogy a year later at the Midwest Physics Conference of 1960, with the following table of correspondences:

Superconductivity	Elementary particles
Free electron	Bare fermion (zero or small mass)
Phonon interaction	Some unknown interaction
Energy gap	Observed nucleon mass
Collective excitation	Meson (bound nucleon pair)
Charge	Chirality
Gauge invariance	γ_5 invariance (rigorous or approximate)

At the mathematical level, the analogy is complete. There is a striking similarity between the Bogoliubov–Valatin equations (equations (1)) and the equations in γ_5-invariant Dirac theory:

$$E\psi_1 = \sigma \cdot p\psi_1 + m\psi_2, \quad E\psi_2 = -\sigma \cdot p\psi_2 + m\psi_1, \qquad (2)$$

with $E = \pm(p^2 + m^2)^{1/2}$ and where ψ_1 and ψ_2 are the two eigenstates of the chirality operator γ_5. By utilizing the generalized Hartree–Fock approxima-tions, the nucleon mass m can be obtained as the self-mass of an initially massless nucleon, which destroys the γ_5 invariance.[15] To restore this invariance requires massless collective states, which can be interpreted as pseudoscalar bound states of nucleon pairs, an effect that comes from the self-interaction of the fundamental field.[16]

It is of interest to note the impact of Dirac and Heisenberg on Nambu's pursuing this analogy. First, Nambu took Dirac's idea of holes very seriously and viewed the vacuum not as a void but as a plenum packed with many virtual degrees of freedom.[17] This plenum view of the vacuum made it possible for Nambu to accept Heisenberg's concept of the degeneracy of the vacuum, which lay at the very heart of SSB.[18] Second, Nambu was trying to construct a composite particle model and chose Heisenberg's non-linear model, 'because the mathematical aspect of symmetry breaking could be mostly demonstrated there', although he never liked the theory or took it seriously.[19] By noting the mathematical similarity of Bogoliubov's formulation of superconductivity and Heisenberg's non-linear field theory, Nambu introduced a particular model of the pion–nucleon system, taking over into QFT almost all of the results about SSB obtained in the superconductivity theory. These included asymmetric (superconducting-type) solutions, degeneracy of the vacuum state, and sym-metry-restoring collective excitations.

An exception to these parallel concepts was the plasmon, which in super-conductivity theory arises from the combined effect of collective excitations and long-range forces. In Nambu's field-theoretical model, the plasmon had no analogue, both because of the constraint imposed by pion decay[20] and because no analogous long-range force existed. Nevertheless, Nambu anticipated that his theory might help to solve the dilemma of Yang–Mills theory, for he wrote:

Another intriguing question in this type of theory is whether one can produce a finite effective mass for the boson in a gauge invariant theory. Since we have done this for the fermion, it may not be impossible if some extra degree of freedom is given to the field. Should the answer come out to be yes, the Yang–Mills–Lee–Sakurai theory of vector bosons would become very interesting indeed.

(1960c)

The final integration of SSB into the framework of gauge theory

Nambu's work on SSB relied heavily on the analogy of Heisenberg's non-linear field theory with Bogoliubov's version of the superconductivity theory. But the analogy is not perfect, since the energy gap in superconductivity is relative to the Fermi sphere, and thus is a non-relativistic concept. To complete the analogy requires a momentum cutoff in Heisenberg's relativistically invariant non-linear theory. A cutoff is necessary in Heisenberg's theory also because the theory is not renormalizable. The next step, then, would be to find a renormalizable model in which SSB could be discussed without arbitrary cutoffs, that is, a step beyond the analogy.

Beyond the analogy

Soon after reading Nambu's work in preprint form, Goldstone took this step (1961). Goldstone's work generalized Nambu's results. Abandoning the non-linear spinor theory, Goldstone considered a simple renormalizable QFT with a single complex boson field ϕ, with a Lagrangian density

$$L = (\partial \phi^* / \partial x_\mu)(\partial \phi / \partial x_\mu) - V[\phi] \tag{3}$$

with $V[\phi] = \mu^2 \phi^* \phi - (\lambda/6)(\phi^* \phi)^2$. For $\mu^2 < 0$ and $\lambda > 0$, the potential term $V[\phi]$ has a line of minima in the complex ϕ plane defined by $|\phi|^2 = -3\mu^2/\lambda$. These minima are the possible non-zero expectation values of the boson field in the vacuum state, showing that the vacuum state is infinitely degenerate. Goldstone called these vacuum solutions abnormal, or superconductive, solutions. Evidently, the quartic term containing λ is essential for obtaining the non-zero vacuum expectation value of the boson field, in accordance with our understanding that it is the non-linear interaction that provides the dynamical mechanism for the existence of stable asymmetric solutions.

There is obvious invariance under the transformation $\phi \rightarrow \exp(i\alpha)\phi$, and the solution can be specified by giving the value of the phase α. Goldstone placed a superselection rule on the phase by forbidding superpositions of states built upon different vacua. Hence one must choose a particular phase, and that destroys the symmetry. If we definine a new field ϕ' by $\phi = \phi' + \chi$, χ being real and $\chi^2 = |\phi|^2 = -3\mu^2/\lambda$, and with $\phi' = 2^{-1/2}(\phi_1' + i\phi_2')$, the Lagrangian becomes

$$L = 1/2[(\partial \phi_1'/\partial x_\mu)(\partial \phi_1'/\partial x_\mu) + 2\mu^2 \phi_1'^2] + 1/2(\partial \phi_2'/\partial x_\mu)(\partial \phi_2'/\partial x_\mu)$$
$$+ (\lambda/6)\chi \phi_1'(\phi_1'^2 + \phi_2'^2) - \lambda/24(\phi_1'^2 + \phi_2'^2)^2. \tag{4}$$

The massive field ϕ_1' in (4) corresponds to oscillations in the direction χ, and the massless field ϕ_2' corresponds to oscillations of the direction of ϕ. The mass of ϕ_2' must be zero: because of the phase symmetry, there can be no dependence of the energy upon α. Goldstone conjectured that when a continuous symmetry is spontaneously broken, there must exist spinless particles of zero mass. Later these were called Goldstone, or Nambu–Goldstone, bosons.[21]

Several proofs of this conjecture were devised by Goldstone, Salam, and Weinberg (1962). These proofs, while valid for classical fields, cannot be regarded as rigorous in the quantum case for two reasons. First, the factors in the current that is supposed to be conserved are not commutable. Second, the integral of the time component of the current, whose commutation relations with the boson field are crucial for the proofs, may not be convergent. Nevertheless, after this paper appeared, the Nambu–Goldstone massless boson resulting from the spontaneous breaking of a continuous symmetry became accepted as an integral part of some field theory models.

Initially, Goldstone had ambiguous feelings about his result. He wrote: 'A method of losing symmetry is of course highly desirable in elementary particle theory'. However, the non-existence of the massless bosons that accompany this SSB seemed to make the method useless unless 'something more complicated' than Goldstone's simple models could somehow eliminate this difficulty (1961).[22]

Anderson's solution

The final integration of SSB into the framework of a gauge theory took two further steps. The first was taken by Marshall Baker and Sheldon Glashow in (1962), extending the idea of SSB to the case of non-Abelian gauge symmetries. In a model with SU(3) symmetry, they found that without introducing any symmetry-breaking terms into the Lagrangian, solutions existed that had only the lower symmetries of isospin and hypercharge. The importance of incorporating SSB into a gauge theory with massive gauge bosons was soon realized by Anderson, Peter Higgs, T. W. B. Kibble, and others.

Anderson took the second step in (1963). He started from a seminal paper by Schwinger (1962a), in which it was argued that a vector gauge boson need not necessarily remain massless if its self-coupling was sufficiently strong. Anderson appreciated the potential importance of Schwinger's suggestion. Using his familiarity with the superconductivity example of SSB, Anderson found that the plasmon mode in superconductivity was a

physical example demonstrating Schwinger's contention that under some circumstances the Yang–Mills type of vector boson need not have zero mass.

(1963)

As we noticed before, in a proper interpretation of the BCS theory of superconductivity, there are no zero-mass modes. In it, the fermion mass is finite because of the energy gap, whereas the massless Goldstone boson becomes a plasmon of finite mass by interaction with the appropriate gauge field, i.e., the electromagnetic field. By analogy, Anderson suggested that the way is now open for a degenerate-vacuum theory of the Nambu type without any difficulties involving either zero-mass Yang–Mills gauge boson or zero-mass Goldstone boson. These two types of bosons seem capable of 'cancelling each other out' and leaving finite mass bosons only.

(Ibid.)

Anderson stressed that 'the only mechanism . . . for giving the gauge field mass is the degenerate vacuum type of theory'. And the only mechanism for solving the Goldstone zero-mass difficulty is the Yang–Mills theory with massless gauge bosons (*ibid.*). The Schwinger–Anderson mechanism (now known as the Higgs mechanism), as realized and elaborated in QFT by P. W. Higgs (1964a, b), F. Englert and R. Brout (1964), G. S. Guralnik, C. R. Hagen, and T. W. B. Kibble (1964) in the case of an Abelian symmetry, and by Kibble in the case of non-Abelian symmetries, soon became an integral part of the conceptual framework of gauge theories.[23]

The rediscovery of SSB in the late 1950s, and its incorporation into the gauge-theoretical framework, came about because of theoretical developments in various branches of physics: the BCS theory of superconductivity in the Bogoliubov form,[24] the debate on its gauge invariance, and the recognition of its asymmetric ground state; Heisenberg's composite model of elementary particles based on his non-linear spinor theory, and his idea of a degenerate vacuum; and the dispersion relations, which facilitated the recognition of the Nambu–Goldstone boson. The final integration of SSB into the framework of gauge theories occurred as one solution to the problem of having a mechanism for the short-range nuclear forces, which troubled the original Yang–Mills theory.

Adding SSB has changed the conceptual basis of QFT profoundly. As Silvan S. Schweber once pointed out: 'One of the legacies of the 1947–1952 period emphasizing covariant methods was to impose too rigid an interpretation of relativistic invariance'. This rigid attitude 'led to a tacit assumption that the Heisenberg state vector describing the vacuum state of a relativistic quantum field theory was always a nondegenerate vector and that this vector always possessed the full invariance properties of the Lagrangian' (1985). Nambu testifies to the accuracy of this judgment, recalling about his work on SSB: 'Perhaps the part that required the most courage on my part was to challenge the dogma of axiomatic theorists concerning the properties of the vacuum' (1989), that is, to take a plenum rather than a void view of the vacuum.[25]

10.2 Mechanisms for short-range interactions (II): asymptotic freedom

Another possible mechanism for describing the short-range behavior of the nuclear forces while retaining the gauge invariance of a field theory, which has not been well established even now, was suggested, within the general framework of renormalization group analysis, by a combination of ideas of 'asymptotic freedom' and 'color confinement'.

The general context is the quark model of hadrons (the strongly interacting particles: nucleons and mesons). In the quark model, hadrons are described as bound states of a small number of spin 1/2 elementary quarks. This entails that quarks, rather than hadrons, are the true source of the strong nuclear forces, and the overall force between hadrons, the observed strong nuclear force, is merely a residue of the more powerful interquark force.

The suggestion is that if the interquark force in a gauge-invariant theory, described by the effective coupling between quarks and massless gauge bosons (gluons), tends to diminish at short distances (asymptotic freedom) and diverge at large distances (caused by the masslessness of the gluons) so that the quarks and gluons of a hadron, as charged (or colored) states, are confined within the hadron and cannot be 'seen' by other hadrons and their constituents (color confinement), then there will be no long-range force between hadrons, although the exact mechanism for the short-range interactions between hadrons is still wanting.

Conceptually, asymptotic freedom is a consequence of the renormalization group behavior of fields (the dependence of physical parameters, such as the coupling constant, on the (momentum) scale being probed), combined with the non-linearity (the self-coupling) of nonabelian gauge fields (or any nonlinear fields). As we mentioned in section 8.8, the renormalization group analysis was revived by Wilson, and further elaborated by Callan and Symanzik. The resulting Callan–Symanzik equations simplify the renormalization group analysis and can be applied to various physical situations, such as those described by Wilson's operator product expansion.

Both Callan and Symanzik played a crucial role in the development of the idea of asymptotic freedom. Soon after his work on the renormalization equations, Symanzik (1971), attempting to construct an asymptotically free theory to explain Bjorken scaling (see below), employed the equations to investigate the running behavior of the coupling constant in the ϕ^4 theory. He found that the magnitude of the coupling constant increased with the increase of the external momenta, if the sign of the coupling constant was taken such that in the corresponding classical theory the energy was bounded from below. Symanzik argued that if the sign of the coupling constant were reversed, then the opposite would happen, and at very high external momenta, there would be asymptotic freedom. The trouble with Symanzik's argument is that since the 'wrong' sign of the coupling constant is possible only when the energy is not bounded from below, such an asymptotically free system would have no ground state and would thus be unstable, as was pointed out by Coleman and Gross (1973).

Stimulated by Symanzik, Gerard 't Hooft became very interested in asymptotic freedom. 'T Hooft was the first physicist to establish asymptotic freedom in a gauge theory context (1972c). He reported his result at the Marseille conference of June 1972, in a remark following Symanzik's lecture (1972) on the subject, that a non-Abelian gauge theory, such as one with an SU(2) group and less than eleven fermion species, would be asymptoticaly free. The announcement was quite well known and was publicized in (1973).

Another line of development, in this subject, beginning with Callan's work, had closer connections with hadronic phenomena. In (1968), Callan and Gross proposed a sum rule for the structure functions that could be measured in deep inelastic electron–proton scattering. Then Bjorken noticed that this sum rule would imply the scaling of the deep inelastic scattering cross section (1969). Expressing the cross section in terms of the structure functions W_i, Bjorken's scaling relations

$$mW_1(v, q^2) \rightarrow F_1(x), \quad vW_2(v, q^2) \rightarrow F_2(x), \quad vW_3(v, q^2) \rightarrow F_3(x) \qquad (5)$$

(where m is the nucleon mass, $v = E - E'$ with E (E') the initial (final) lepton energy, $q^2 = 4EE' \sin^2(\theta/2)$, θ being the lepton scattering angle, and $0 \leqslant x$ $(= q^2/2mv) \leqslant 1$) indicated that the structure functions did not depend individually on v and q^2, but only on their ratio x, which would be held fixed when $q^2 \rightarrow \infty$.

Bjorken's prediction was soon confirmed by the new experiments at SLAC.[26] These developments set in motion a variety of intense theoretical activities (most of which were pursued within the frameworks of light-cone current algebra and the renormalization group), which contributed significantly to elucidating the structure and dynamics of hadrons.

The physical meaning of Bjorken scaling was soon uncovered by Feynman in his parton model of the hadron, an intuitive picture of deep inelastic scattering (1969). According to the parton model, a hadron is seen by a high-energy lepton beam as a box filled with point-like partons, and the inelastic lepton–nucleon scattering cross section in the scaling region (a region of high energy or short distances) equals the incoherent sum of elastic lepton–parton cross sections.[27] Since the incoherence of lepton–parton scattering means that the lepton sees the nucleon as a group of freely moving constituents, Bjorken scaling requires that, in a field theory context, the short-distance behavior of parton fields be described by a non-interacting theory.

It was in their effort to have a consistent understanding of Bjorken scaling that Callan (1970) and Symanzik (1970) rediscovered the renormalization group equations and presented them as a consequence of a scale invariance anomaly.[28] Callan and Symanzik then applied their equations to Wilson's operator product expansion,[29] which was extended to the light cone, a relevant region for deep inelastic scattering, for analyzing the deep inelastic scattering.[30] From these analyses it became clear that once interactions were introduced into a theory, no scaling would be possible.

A question arose in this context. If there were no interactions among partons, hadrons would be expected to break up easily into their parton constituents, then why had no free partons ever been observed?

In fact, this question was not well posed, since scaling involves only the short-distance behavior of the constituents of hadrons while the breaking up of hadrons would be a long-distance phenomenon.

Yet, even within the scope of short-distance behavior, serious doubts were also cast upon scaling, on the basis of a renormalization group argument. As we mentioned in section 8.8, according to Wilson, scaling holds only at the fixed points of the renormalization group equations. This entails that at short distances, the effective couplings will approach fixed-point values. Since these values are generally those of a strongly coupled theory, the result will be large anomalous scaling behavior that is quite different from the behavior of a free field theory. From a Wilsonian perspective (see note 67 of chapter 8), then, it was expected that the non-canonical scaling (with the effects of interactions and renormalization being absorbed into the anomalous dimensions of fields and currents), which is indicative of a non-trivial fixed point of the renormalization group equation, would appear at higher energies.

In fact, the opinion that the scaling observed at SLAC was not a truly asymptotic phenomenon was rather widespread among theoreticians. The opinion was reinforced by the fact that the observed scaling set in at rather low momentum transfers (so-called 'precocious scaling'). As late as October 1973, Gell-Mann and his collaborators still conjectured that 'there might be a modification at high energies that produces true scaling [Wilsonian scaling]'.[31]

In such a climate, David Gross, a disciple of Geoffrey Chew but later converted to Gell-Mann's current algebra programme, decided, at the end of 1972, 'to prove that local field theory could not explain the experimental fact of scaling and thus was not an appropriate framework for the description of the strong interactions' (Gross, 1992). Gross pursued a two-step strategy. First, he planned to prove that the vanishing of the effective coupling at short distances, or asymptotic freedom, was necessary to explain scaling; and second, to show that no asymptotically free field theories could exist.

The first step was, in fact, soon taken by Giorgio Parisi (1973), who showed that scaling could only be explained by an asymptotically free theory, and then furthered by Callan and Gross (1973), who extended Parisi's idea to all renormalizable field theories, with the exception of non-Abelian gauge theories. The Callan–Gross argument was based on the idea of the scale dimension first suggested by Wilson (see section 8.8). First, they noticed that the composite operators that dominated the amplitudes for deep inelastic scattering had canonical dimensions. Then, they proved that the vanishing anomalous scale dimensions of the composite operators, at an assumed fixed point of the renormalization group, implied the vanishing anomalous dimensions of the elementary fields. This in turn implied that the elementary fields must be free of interactions at this fixed point since, usually, only those fields that are free have canonical scale dimensions. They concluded that Bjorken scaling could be explained only if the assumed fixed point of the renormalization group was at the origin of coupling space, that is, that the theory must be asymptotically free.

The second step of Gross's argument failed. The failure, however, led to a rediscovery, independently of 't Hooft's, of the asymptotic freedom of a non-Abelian gauge theory, which could be used to explain the observed scaling, thus, ironically, establishing the appropriateness of local field theories, at least in the case of a non-Abelian gauge theory, as a framework for the description of the strong interactions.

At first, Gross's judgment that no theory could be asymptotically free sounded reasonable, since in the case of QED, the prototype of QFT, the effective charge grew larger at short distances, and no counter-example, which existed in 't Hooft's work, was known to him.

The reason for the effective charge in QED to grow larger, as a result of charge renormalization, is easy to understand. The underlying mechanism for charge renormalization is vacuum polarization. The vacuum of a relativistic quantum system, according to Dirac, is a medium of virtual particles, in the case of QED, of virtual electron–positron pairs. Putting a charge, e_0, in the vacuum will polarize it, and the polarized vacuum with virtual electric dipoles will screen the charge, so that the observable charge e will differ from e_0 as e_0/ϵ, where ϵ is the dielectric constant. Since the value of ϵ depends on the distance r, the observable value of the charge is describable by a running effective coupling $e(r)$, which governs the force at a distance r. As r increases, there is more medium that screens, and thus $e(r)$ decreases with increasing r, and correspondingly increases with decreasing r. Thus the β function, which is simply minus the derivative of $\log(e(r))$ with respect to $\log(r)$, is positive.

In the spring of 1973, since non-Abelian gauge theories attracted much attention from physicists, mainly because of the work by Veltman and 't Hooft (see section 10.3), Gross, in making his case, had to examine the high-energy behavior of non-Abelian theories. At the same time, Coleman and his student H. David Politzer, for different reasons, carried out a parallel investigation. To do this it was sufficient to study the behavior of the β functions in the vicinity of the origin of coupling constant space. The calculation of the β function in a non-Abelian gauge theory[32] indicated that, contrary to Gross's expectation, the theory was asymptotically free.

The reason for the different behavior of the non-Abelian (color) gauge theory (see section 11.1) from that of the abelian gauge theory like QED lies in the non-linearity or the self-coupling of the non-Abelian gauge fields. That is, it lies in the fact that, in addition to the color-charged quarks, the gauge bosons (gluons) are also color charged. The same screening as in the Abelian case happens in the non-Abelian theory as far as the virtual quark–antiquark pairs are concerned. Yet the color-charged gluons make an additional contribution to the vacuum polarization, tending to reinforce rather than neutralize the gauge charge.[33] The detailed calculation indicated that if the number of quark triplets is less than seventeen, the anti-screening from the virtual gluons wins out over the screening due to the virtual quark–antiquark pairs, and the system is asymptotically free.

Coleman and Gross (1973) also established that no renormalizable field theory that consisted of theories with arbitrary Yukawa, scalar, or Abelian gauge interactions could be asymptotically free, and that any renormalizable field theory that is asymptotically free must contain non-Abelian gauge fields. This argument significantly restricted the selection of a dynamical system for describing the strong interactions.

From their understanding of the peculiar behavior of the effective coupling in non-Abelian gauge theories (at short distances it tends to diminish and at large distances it becomes extremely strong), and informed by the observations of scaling, Gross and Wilczek in (1973b) discussed the possibility of having a non-Abelian gauge theory with an unbroken symmetry as a framework for describing the strong interactions. They argued that the original difficulty of the Yang–Mills theory, that is, having strongly coupled massless vector mesons mediating long-range interactions, disappeared in asymptotically free theories.

Color confinement

The reason for this, they argued, is as follows. The large-distance behavior of the color-charged ingredients of a hadron (quarks and gluons) in this case would be determined by the strong coupling limit of the theory, and be confined within a hadron. Thus at large distance, except for color-neutral particles (color singlets or hadrons), all color-charged states would be suppressed and could not be 'seen' from outside of a hadron. And this is consistent with the absence of long-range interactions between hadrons.

It should be noted that although, heuristically, the idea of confinement can be understood in these terms, and many efforts were attempted,[34] there has never appeared a rigorous proof of confinement.

10.3 Renormalizability

No model built within the framework of a non-Abelian gauge theory (with or without massive gauge bosons, asymptotically free or not) would have any predictive power unless it were proved to be renormalizable. Thus for the gauge theory to act as a successful research programme its renormalizability had to be proven. When Yang and Mills proposed their non-Abelian gauge-invariant theory for the strong interactions, however, its renormalizaability was assumed. The assumption was solely based on a naive power-counting argument: after quantization and having derived the rules for the Feynman diagrams, they found that although the elementary vertices had three types (because of the self-coupling of the charged gauge bosons) instead of only one type as in QED, 'the "primitive divergences" were still finite in number' (1954b). Thus, according to Dyson's renormalization programme, there was no reason to worry about its renormalizability. This conviction had met no challenge until 1962 when Feynman began to investigate the situation (see below). In fact, even after the publication of Feynman's work (and other work stimulated by it), this conviction was still shared by many physicists, such as Weinberg (1967b) and Salam (1968).

With the discovery of the non-conservation of parity in the weak interactions in 1957, and the establishment of the vector and axial vector couplings for the weak interactions in 1958, there appeared many speculations about intermediate vector mesons W and Z transmitting the weak interactions (see sections 8.6 and 9.3). It was tempting to identify the vector mesons with the gauge bosons of a non-Abelian gauge theory. Since the short-range character of the weak interactions required the mediating charged mesons be massive, the result of the identification was the so-called massive Yang–Mills theory. The appearance of the massive Yang–Mills theory opened up two lines of investigation. In addition to the search for a 'soft mass' mechanism, through which the gauge bosons would acquire mass while the theory would remain gauge invariant, great efforts were made to prove the renormalizability of the massive charged vector theory.

The long process of investigating the renormalizability of non-Abelian gauge theories started with a remark by Bludman: a difficulty in interpreting the non-Abelian gauge bosons as real particles was that the charged vector bosons had singular non-renormalizable electromagnetic interactions. Bludman's remark was based on a result that was well established in the early years of meson electrodynamics.[35]

Yet it was well known that the non-renormalizable character of the charged vector meson persisted even in dealing with interactions other than the electromagnetic ones. In discussing the vector interactions of a vector meson with a nucleon, for example, Matthews (1949) noticed a significant difference between the cases of neutral and charged vector mesons: while the former could be shown to be renormalizable, this was not the case for the latter. Mathews's observation was subsequently refined by Glauser (1953), Umezawa (1952), and Umezawa and Kamefuchi (1951) with the emphasis on the conservation of the source current, which was supposed, implicitly by Mathews and explicitly by Glauser and by Umezawa and Kamefuchi, to hold only in the case of the neutral vector meson, but not if the vector meson were charged. At this stage of the discussion, it should be noticed that the non-conservation of the charged current was taken to be the main source of the additional terms which destroyed the renormalizability of charged vector meson theories.

Bludman's remark posed a serious problem to Schwinger's gauge programme, according to which unstable massive and charged vector mesons with their universal gauge couplings were responsible for all phenomena of the weak interactions. An immediate response to Bludman's remark came from Glashow (1959), one of Schwinger's disciples, who was followed by Salam, Umezawa, Kamefuchi, Ward, Komar, and many others.[36] Conflicting claims concerning the renormalizability of charged vector meson theories were made, but there was no progress in understanding the real situation until a paper by Lee and Yang (1962).

To study the issue systematically, Lee and Yang started by deriving the rules for Feynman diagrams from a massive charged vector meson theory. They found that there were certain vertices that were divergent and relativistically non-covariant. By introducing a limiting process (adding a term $[-\xi(\partial_\mu^* \phi_\mu^*)(\partial_\nu \phi_\nu)]$ to the Lagrangian, which depended on a positive parameter $\xi \to 0$), the Feynman rules became covariant, yet the theory continued to be divergent in a non-renormalizable way. To remedy this, they introduced a negative metric that made the parameter ξ take on the role of a regulator. The resulting theory for $\xi > 0$ was both covariant and renormalizable. The price for this achievement was that unitarity of the theory was destroyed by the introduction of a negative metric.

The lack of unitarity, in Lee and Yang's formalism, was physically manifested in the appearance of non-physical scalar mesons. Since unitarity, whose
physical meaning is the conservation of probability, is one of a few basic
requirements that any meaningful physical theory must satisfy, the difficulty
first faced by Lee and Yang, that is, how to remove these non-physical degrees
of freedom and achieve unitarity, became the central issue in renormalizing
non-Abelian gauge theories, which was pursued in the following years.

The next landmark in the renormalization of non-Abelian gauge theories
was Feynman's work (1963).[37] Formally, Feynman's subject was gravitation.
But for the sake of comparison and illustration, he also discussed another non-
linear theory, in which interactions were also transmitted by massless particles,
that is, the Yang–Mills theory. The central theme of Feynman's discussion was
to derive a set of rules for diagrams (in particular the rules for loops) that
should be both unitary and covariant, so that further investigations on the renormalizability of the rules could proceed systematically.

Feynman's discussion of unitarity was carried out in terms of trees and loops
which were connected to each other through his tree theorem. According to
the theorem, a closed loop diagram was the sum of the corresponding tree
diagrams in a physically attainable region and on the mass shell, which were
obtained by opening a boson line in the loop. In addition, Feynman specified
that in calculating a loop, in each tree diagram for which a massless boson line
was opened, only real transverse degrees of freedom of the boson should be
taken into consideration so that gauge invariance could be guaranteed. The
physical motivation behind the tree theorem was Feynman's deep desire to
express physical quantities, which in QFT are usually defined in terms of
virtual fields (in loop diagrams), in terms of actual measurable quantities
involved in real physical processes.

According to Feynman, the unitarity relation is one connection between a
closed loop diagram and the corresponding physical set of trees, but there are
many other connections. In fact, he found that the rules for loops derived from
the usual path integral formalism were not unitary. The roots of the failure of
the rules for loops to be unitary were clearly exposed by Feynman in terms of
breaking a loop into a set of trees. In a covariant formalism (which is necessary
for renormalization), a virtual line in a loop carries longitudinal and scalar
degrees of freedom, in addition to transverse polarization. When it is broken,
becomes an external line, and represents a free particle, it carries transverse
(physical) degrees of freedom only. Thus, to maintain Lorentz covariance (and
match the loop with the trees), some extra non-physical degrees of freedom
have to be taken into consideration. Feynman then was forced to face an extra
particle, a 'fake' or 'fictitious' particle that belonged to the propagator in a

loop and had to be integrated over, but now instead became a free particle. Thus unitarity was destroyed, and Feynman realized that 'something [was] fundamentally wrong' (1963). What was wrong was a head-on crash between two fundamental physical principles: Lorentz covariance and unitarity.

In order to recover unitarity, Feynman proposed to 'subtract something' from the internal boson line in a loop. This insightful suggestion was the first clear recognition in the conceptual development of the subject that some ghost loop had to be introduced so that a consistent formalism of non-Abelian gauge theories could be obtained. This insight was immediately caught upon by a perspicacious physicist who attended Feynman's lecture and requested Feynman to 'show in a little more detail the structure and nature of the fictitious particle needed if you want to renormalize everything directly with the loops' (*ibid.*). This alert physicist was Bryce DeWitt, who himself was going to make substantial contributions to the development of the subject (see below).

In addition to his combinatorial verification of the unitarity of the rules for diagrams,[38] Feynman also discussed the massive case and tried to obtain the massless Yang–Mills theory by taking the limit of zero mass. The motivation was twofold. First, he wanted to avoid the infrared problem of the massless theory, which was worse than in the case of QED because of the self-coupling among gauge bosons. Second, it is impossible to write a manifestly covariant propagator for a massless vector particle that has only two states of polarization, because in extending away from the mass-shell in a covariant way, the massless vector particle obtains a third state of polarization. This is the same clash we just mentioned between covariance and unitarity in the case of a massless vector theory.

In this context, there are two points worthy of note. First, it turned out to be a mistake to assume that the massless theory can be obtained in the limit of zero mass (see below), because the longitudinal polarization of the massive vector boson does not decouple in the limit of zero mass. Second, Feynman tried to find 'a way to re-express the ring diagrams, for the case with $\mu^2 \neq 0$, in a new form with a propagator different from [equation (10), i.e., $(g_\nu^\mu - q^\mu q_\nu / \mu^2)/(q^2 - \mu^2)]$[39] . . . in such a form that you can take the limits as μ^2 goes to zero'. To this end he introduced a series of manipulations that were taken over by later investigators. In any case, this was the first explicit suggestion of finding a gauge different from the usual one so that the Feynman rule for the propagator of the massive vector particle could be changed. The result of this suggestion is the so-called Feynman gauge, in which the vector propagator takes the form $[g_\nu^\mu/(q^2 - \mu^2)]$. The Feynman gauge is a form of renormalizable gauge in which a massive Yang–Mills theory turns out to be renormalizable, because the $q^\mu q_\nu / \mu^2$ term that causes all the problems is absent.

Inspired by Feynman's work, DeWitt (1964, 1967a, b, c) considered the question of a choice of gauge and the associated ghost particle (the 'fictitious particle' suggested by Feynman). Although the issue of unitarity was not clearly discussed, the correct Feynman rules, in the case of gravitation, were established in the Feynman gauge. In addition, the ghost contribution in DeWitt's rules was expressed in the form of a local Lagrangian containing a complex scalar field, which obeyed Fermi statistics.[40]

In the case that the gauge symmetry was spontaneously broken, F. Englert, R. Brout, and M. Thiry (1966) also derived many Feynman rules in the Landau gauge, in which the vector propagator takes the form $[(g_\nu^\mu - q^\mu q_\nu/q^2)/(q^2 - \mu^2)]$, and found the corresponding ghost. On the basis of the form of the vector propagator, they suggested that the theory be renormalizable.

L. D. Faddeev and V. N. Popov (1967), starting from Feynman's work and using the path integral formalism, derived Feynman's rules in the Landau gauge. In their work, the ghost came out directly. But their technique was not sufficiently flexible to discuss a general gauge and to understand the previously established rules in the Feynman gauge. In particular, the ghost introduced by them, unlike DeWitt's ghost, had no orientation, and a simple vertex structure could be obtained only in the Landau gauge. Since neither the choice nor the ghost rules in their work were expressed in terms of local fields in the Lagrangian, the connection between the Lagrangian and the Feynman rules was not very clear. Another defect of the work was that the question of unitarity was left unaddressed. Since the roots of the ghost, as indicated above, lie in the requirement for unitarity, and unitarity in Faddeev and Popov's procedure of introducing ghost as a compensation to the breaking of gauge invariance is not a trivial issue, this defect left a big gap in their reasoning.

The results of DeWitt and of Faddeev and Popov were rederived by Mandelstam in (1968a, b) with an entirely gauge-invariant formalism of field theory. Thus in the summer of 1968, the Feynman rules for massless non-Abelian gauge theories were established and were, at least for some specified choices of gauge, those of a renormalizable theory in the sense of power counting, so that these choices of gauge were called renormalizable gauges. However, whether the infinities could be removed, order by order in a perturbative way, in a gauge-invariant way was not clear. In addition, unitarity was not a settled issue, and the infrared problem still plagued physicists.

At this juncture, a heroic attack on the renormalizability of non-Abelian gauge theories was launched from Utrecht, The Netherlands, by Veltman (1968a, b). Veltman knew very well that the current opinion took the charged massive vector boson theories, which were revived as a consequence of the V–A theory of the weak interaction, as hopelessly divergent. Conceiving the success of current algebra as a consequence of the gauge invariance of a field theory, however, Veltman considered that success, in particular the success of the Adler–Weisberger relation, as experimental evidence of the renormalizability of non-Abelian gauge theories with massive vector bosons, and began detailed investigations to verify his conviction.

Phenomenologically, taking vector mesons to be massive was natural because the vector mesons responsible for the short-range weak interactions must be massive. Theoretically, a massive theory can bypass the infrared problem because 'the zero-mass theory contains horrible infrared divergences', which prohibit the study of an S-matrix with incoming and outgoing soft bosons. Following Feynman, Veltman also believed, wrongly, that the massless theory could be obtained in the limit of zero mass from the massive theory.

The massive theory was obviously non-renormalizable mainly because of the term containing $k_\mu k_\nu/M^2$ in the vector field propagator $(\delta_{\mu\nu} + k_\mu k_\nu/M^2)/(k^2 + M^2 - i\epsilon)$. Informed by previous work by Feynman, Dewitt, and others, Veltman noticed that 'as is well known this term is modified if one performs a gauge transformation', because with a change of gauge the Feynman rules would also be changed. His objective was to find a technique for changing gauge, so that the Feynman rules would be changed in such a way that the vector propagator would take the form of $\delta_{\mu\nu}/(k^2 + M^2 - i\epsilon)$,[41] and the theory would be renormalizable at least in the sense of power counting.

Veltman called the technique he had invented the free-field technique and the crucial transformation the Bell–Treiman transformation, although neither Bell nor Treiman was responsible for it. The essential idea was to introduce a free scalar field without interaction with the vector bosons. By replacing the vector field (W^μ) with some combination of the vector and scalar fields ($W^\mu + \partial^\mu\phi/M$), and adding new vertices in such a way that the scalar field remained a free field, Veltman found that while the Feynman rules for the new theory were different (the propagator for the vector field was replaced by that for the combination), the S-matrix, and thus the physics, remained the same. He also found that the combination could be chosen in such a way that its propagator would take the form

$$(\delta_{\mu\nu} + k_\mu k_\nu/M^2)/(k^2 + M^2 - i\epsilon) - k_\mu k_\nu/M^2/(k^2 + M^2 - i\epsilon)$$
$$= \delta_{\mu\nu}/(k^2 + M^2 - i\epsilon), \quad (6)$$

which would lead to less divergent Feynman rules. The price for this achievement was the introduction of new vertices involving ghosts.[42]

Veltman's free-field technique was developed in the language of diagrams, stimulated by Chew's S-matrix philosophy. Later, it was reformulated in the path integral formalism by D. Boulware in (1970), and also by E. S. Fradkin and I. V. Tyutin (1970), who introduced the possibility of a general choice of gauge through the introduction of a term in the Lagrangian and established a procedure by which the Feynman rules for a general gauge could be obtained.

With a powerful means for changing gauge and changing Feynman rules, Veltman obtained an astonishing result, namely that the Yang–Mills theory with an explicit mass term turned out to be one-loop renormalizable with respect to power counting (that is, up to quadratic terms for the self energies, linear terms for the three-point vertex, and logarithmic terms for the four-point function). Before this breakthrough, claims and statements concerning the renormalizability of gauge theories rarely went beyond speculations. The breakthrough destroyed the entrenched opinion that the charged massive vector boson theory was hopelessly non-renormalizable, and stimulated further investigations on the subject.

Conceptually and psychologically, Veltman's (1968b) was extremely important because it had convinced physicists that by changing gauge, a set of renormalizable Feynman rules would be obtainable. Physically, however, many gaps remained to be filled. To properly renormalize a theory with a gauge symmetry, one has to regularize it in accordance with that symmetry. In Veltman's case (1968b), the theory had no symmetry because the mass term definitely destroyed the assumed symmetry. In addition, he had no gauge-invariant regularization scheme. For these reasons, the theory was not renormalizable in a fuller sense even in the one-loop case.

The situation for two or more loops was even worse. It was soon discovered that the Feynman rules (obtained by modifying the massless rules with the addition of a mass in the denominator), that were valid at the one-loop level, definitely violated unitarity and led to non-renormalizable divergences for two-loop diagrams.[43] Boulware in his path integral treatment also indicated that the non-renormalizable vertices occurred at the level of two or more loops. Since in the massless theory such vertices were absent, it was assumed that they would also cancel out one way or the other. However, A. Slavnov and Faddeev (1970) established that the limit of zero mass of the massive theory was already different from the massless theory at the one-loop level, or as shown by H. van Dam and Veltman (1970) for gravitation even at the tree level.

Thus further progress depended on a proper understanding of what the precise Feynman rules for two-or-more-loop diagrams were and how to verify the unitarity of the theory described by such rules. In general, the problem of unitarity in the case of non-Abelian gauge theories is more complicated than in other cases. The Feynman rules in a renormalizable gauge are not manifestly unitary. There are unphysical degrees of freedom or ghosts, which are required to compensate for gauge dependence introduced by the gauge-fixing term in the Faddeev–Popov formalism. In addition, there are unphysical polarization states of the gauge fields in their internal propagation in the covariant formalism. A theory is unitary only when the contributions from these unphysical degrees of freedom are absent. It turns out that there would be no such contributions if certain relations or Ward identities between Green's functions involving the gauge and ghost fields were valid. That is, Ward identities will ensure the exact cancellation between the contributions from the two types of unphysical degrees of freedom. Thus for the condition of unitarity to be satisfied, the establishment of Ward identities is crucial.

Originally, the identities were derived by Ward (1950) and extended by Fradkin (1956) and Takahashi (1957) as one of the consequences of gauge invariance in QED:

$$k_\mu \Gamma_\mu(q,p) = S_F^{-1}(q) - S_F^{-1}(p), \quad k = q - p, \tag{7}$$

where Γ_μ and S_F are the irreducible vertex and self-energy parts respectively. The identities say that if the polarization vector $e_\mu(k)$ of one or more external lines is replaced by k_μ while all other lines are kept on mass shell and provided with physical polarization vectors, the result is zero.

In the case of non-Abelian gauge theories the situation is much more complicated. In addition to those similar to ones appearing in QED, there is a whole set of identities involving only the non-Abelian gauge fields and the associated ghosts. Furthermore, since these identities do not directly relate irreducible vertices and self-energy parts, they have less direct physical meaning than those in QED. By using Schwinger's source technique and the Bell–Treiman transformation, however, Veltman (1970) succeeded in deriving generalized Ward identities for the massive non-Abelian gauge theories.[44] These identities, together with the cutting equations obtained earlier also by Veltman (1963a, b), can be used to show the unitarity of a non-Abelian gauge theory containing ghost loops. They also play a crucial role in carrying through the actual renormalization process because the renormalized theory must be shown to be unitary.

By the summer of 1970, therefore, Veltman knew how to derive Feynman rules from diagrams, how to have different (unitary or renormalizable) rules by changing gauges through the Bell–Treiman transformation, and how to verify unitarity by using generalized Ward identities and sophisticated combinatorial technique that he had developed over the years. All these are crucial ingredients for renormalizing non-Abelian gauge theories. What Veltman actually achieved, however, was merely an illustration of the one-loop renormalizability of massive vector theories, which was solely based on brute force manipulations and naive power-counting arguments without a rigorous proof. For him, the theory was non-renormalizable once one went beyond one loop. A more serious gap in Veltman's project was the lack of a proper regularization procedure. A gauge-invariant regulator method is necessary for demonstrating that all subtractions can be defined in a consistent way without violating the gauge symmetry. Thus when Gerard 't Hooft, one of his students studying high-energy physics under his supervision, showed interest in the Yang–Mills theory, he suggested to 't Hooft the problem of finding a good regulator method to be used with the Yang–Mills theory. And this suggestion soon led to a final breakthrough.

In (1971a), 't Hooft presented a new cutoff scheme for one-loop diagrams, in which a fifth dimension was introduced and the gauge symmetry of the theory was respected, and applied this method to the massless Yang–Mills theory. He extended, elegantly, the method of Faddeev and Popov (1967) for general gauge fixing and ghost generation, and that of Fradkin and Tyutin (1970) for deriving Feynman rules in a general gauge, and presented Feynman rules containing oriented ghosts in a very large class of gauges in terms of a local unrenor-malized Lagrangian,[45] which opened up the way for a complete renormaliza-tion programme. By using Veltman's combinatorial methods, 't Hooft was able to establish the general Ward identities and, with the help of them, unitarity. However, 't Hooft's proof of the renormalizability of the massless Yang–Mills theory was valid for one loop only because the new cutoff method worked up to one loop only.

In 1971, a massless Yang–Mills theory was conceived as an unrealistic model, and only the massive theory, through charged vector bosons, was considered physically relevant. Since the massive theory in Veltman's previous investigations (1969, 1970) appeared to be non-renormalizable, he really wished to extend 't Hooft's success and to have a renormalizable theory of massive charged vector bosons. Equipped with the techniques learned from Veltman and inspired by Benjamin Lee's work on the renormalization of the σ model, 't Hooft confidently promised a solution.

'T Hooft's basic idea was to introduce spontaneous symmetry breaking into Yang–Mills theories. This idea was motivated by the following advantages. First, the vector bosons can acquire mass through a Higgs-type mechanism. Second, the infrared difficulty of the massless theory can be thereby circum-vented. Third, and most importantly, the renormalizability of the massless theory can be preserved because, as was demonstrated by the work of Lee and others on the σ model[46], the renormalizability of a theory was not spoiled by spontaneous symmetry breaking. Once this step was taken, the techniques used for the massless theory and those developed by Veltman for the one-loop renormalizability of the massive theory could be taken over to verify the renormalizability of the massive Yang–Mills theory ('t Hooft, 1971b).

This paper was immediately cheered as being one 'which would change our way of thinking on gauge field theory in a most profound way'[47] and, together with 't Hooft's (1971a) paper, was universally recognized as a turning point, not only for the fate of gauge theories, but also for that of QFT in general and for many models built thereon, thus radically changing the course of fundamental physics, including particle physics and cosmology.

While this assessment is basically true, to complete the renormalization programme, however, two remaining gaps had to be filled. First, a gauge-invariant regularization procedure, which was to be valid not only up to the one-loop level or to any finite order in perturbation theory, was of vital importance because subsequently the subtraction procedure had to be carried through. This gap was filled by 't Hooft and Veltman in (1972a), in which a systematic procedure, called dimensional regularization, was proposed, based on an extension of 't Hooft's gauge-invariant cutoff (in which a fifth dimension of spacetime was introduced), by allowing the dimensionality of spacetime to have non-integer values and defining this continuation before loop integrations were performed.[48]

Second, the Ward identities, unitarity, and renormalization had to be established, order by order, and rigorously. In (1972b), 't Hooft and Veltman presented a formal combinatorial derivation of the Ward identities and, using them, a proof of unitarity and renormalizability for non-Abelian gauge theories, thus laying down a firm foundation for the further development of the gauge theory programme.

The development of the Veltman–'t Hooft programme has made it crystal clear that maintaining gauge invariance is vital to the proof of renormalizability of any vector boson theory, for two reasons. First, the establishment of unitarity of a renormalizable set of Feynman rules required Ward identities, a consequence of gauge invariance. Second, the rules in the renormalizable gauge are equivalent to those in the unitary gauge only when the theory is gauge invariant. This conclusion has severely constrained the further development of QFT whenever boson fields are involved.[49]

10.4 Global features

Traditionally, field theories study the local behavior of the physical world. In the case of gauge theories, the local character of gauge potentials is dictated by local interactions, which determine the laws of conservation. Yet, with the rise of the gauge-theoretical framework, certain otherwise mysterious physical phenomena have become understandable within this framework and, more importantly, have revealed certain global features of gauge potentials, global in

the sense that their topological behavior is shared by all topologically equivalent (deformable) states (or an equivalent class) of gauge potentials, and
different classes have different behaviors.

One such phenomenon is the equality of the absolute charges of the proton
and the electron. Experimentally, they are equal to a high degree of accuracy.
Theoretically, Dirac (1931) tried to understand the equality by proposing the
existence of magnetic monopoles, which entailed the quantization of charge.
Although monopoles have never been detected,[50] the proposal does hit on the
heart of the theoretical structure of gauge-invariant field theories. Further
explorations of this aspect of gauge theories have revealed the global features
of gauge potentials and shed some light on the quantization of physical
parameters, with or without implying the existence of monopoles. Another
attempt was made by Yang (1970), who, following Schwinger, tried to relate
the quantization of the electric charge to the compactness of the gauge group, a
global feature of the group.[51]

Another such phenomenon is the famous Aharonov–Bohm effect. In the
experiment suggested by Yakir Aharonov and David Bohm (1959) and performed by R. G. Chambers (1960), the outcome (the fringe shift caused by
additional phase shifts in the wave functions of the electron beams) in a doubly
connected region outside a cylinder, where the electromagnetic field strength
$F_{\mu\nu}$ vanishes everywhere, turned out to be dependent on the loop integral of
A_μ, $\alpha = e/hc \oint A_\mu \, dx^\mu$ (or more exactly on the phase factor $e^{i\alpha}$ rather than the
phase α), around an unshrinkable loop.

The implications of the Aharonov–Bohm effect are many. Most important
among them are two. First, it established the physical reality of the gauge
potential A_μ. Second, it revealed the non-local, or global, character of electromagnetism.

The traditional view took $F_{\mu\nu}$ as physically real, responsible for observable
effects, and giving an intrinsic and complete description of electromagnetism.
In contrast, the potential A_μ was taken only as auxiliary and fictitious, without
physical reality, because it was thought to be arbitrary and unable to produce
any observable effects. The Aharonov–Bohm effect has clearly shown that in
quantum theory $F_{\mu\nu}$ by itself does not completely describe all electromagnetic
effects on the wave function of the electrons. Since some of the effects can only
be described solely in terms of A_μ, while $F_{\mu\nu}$ itself can always be expressed in
terms of A_μ, it is reasonable to assume that the potential is not only physically
real in the sense that it can produce observable effects, but also has to be taken
as more fundamental than the electromagnetic field strength. Given the substantiality of the electromagnetic field, the potential must be substantial so that
the field can acquire its substantiality from the potential. This is a significant

implication of the Aharonov–Bohm effect for the ontological commitment of the gauge field programme.

Second, the Aharonov–Bohm effect observed in the region where the field strength is absent has also shown that it is the relative change in the phase of the wave function of the electrons produced by the potential that is physically observable. Yet, as I shall discuss presently, the change is not produced by any local interaction of any specific potential with the electrons. Rather, it is dictated by a certain global property of the potential, which is specified by a pure gauge function and is unique to a specific gauge group. That is, in terms of its ontological foundations, the gauge potential in its 'vacuum' or 'pure gauge' configuration is unique. Thus the original objection to the reality of the gauge potential, namely its arbitrariness, carries no weight in this regard.

Before elaborating the second point in a larger theoretical context, which is the major topic of this section, let me mention one more physical phenomenon, which would be mysterious without taking into consideration the global character of gauge theories. That is the Adler–Bell–Jackiw anomalies discovered in 1969 (see section 8.7). Associated with these anomalies in an effective Lagrangian of a non-Abelian gauge theory is an additional term, which is proportional to $P = -(1/16\pi^2)\, \mathrm{tr}\, F^{*\mu\nu} F_{\mu\nu} = \partial_\mu C^\mu$ $[C^\mu = -(1/16\pi^2)\epsilon^{\mu\alpha\beta\gamma} \mathrm{tr}\,(F_{\alpha\beta}A_\gamma - 2/3 A_\alpha A_\beta A_\gamma]$ (cf. equation (25) of section 8.7). This gauge-invariant term has no effect on the equations of motion because it is a total divergence of a gauge-variant quantity C. This means that, first, the breaking of gauge symmetry by anomalies has a quantum origin because classical dynamics is entirely characterized by the equations of motion. Second, it suggests that the anomalies have connections with the large-distance behavior of gauge potentials. More precisely, it suggests that what is broken by the anomalies is the symmetry under certain finite, rather than infinitesimal, gauge transformations.

Common to all these phenomena is the existence of finite gauge transformations that (unlike the infinitesimal ones), as parallel displacements, are represented by path-dependent or non-integrable phase factors.[52] In the case of a Dirac magnetic monopole, that the gauge transformations are non-integrable is easy to understand. The very existence of the monopole poses a singularity for the gauge potential, which makes the loop integral of the potential around the singularity undefinable. In the case of the Aharonov–Bohm effect, the existence of the cylinder plays the same role as Dirac's monopole: the loop integral in the phase factor has to be performed around an unshrinkable loop. In the case of anomalies, the additional term mentioned above is invariant under infinitesimal transformations, but, with certain boundary conditions, will pick up an additional term under finite transformations for a non-Abelian theory $(A \rightarrow S_U A S_U^+ = U^{-1}(r)AU(r) - U^{-1}(r)\nabla U(r))$, which is determined by the large-distance properties of the gauge potential $A_\mu(r \rightarrow \infty) \rightarrow U^{-1}\partial_\mu U$:

$$\omega(U) = 1/24\pi^2 \int dr \epsilon^{ijk} \operatorname{tr}(U^{-1}(r)\partial_i U(r)U^{-1}(r)\partial_j U(r)U^{-1}(r)\partial_k U(r)), \quad (7)$$

where $U(r)$ is the pure gauge function with large-distance asymptotes I (or $-I$). The difference, in this case, between the finite and infinitesimal gauge transformations lies in a term involving the gauge function $U(r)$. Thus a closer examination of the gauge function seems worthwhile.

In fact, a deep understanding of the global features of gauge theories did come from such an examination of the gauge function $U(r)$. Since the three-dimensional spatial manifold with points at infinity is identified as topologically equivalent to S^3, the surface of a four-dimensional Euclidean sphere labeled by three angles, mathematically, the functions U provide a mapping of S^3 into the manifold of the gauge group. From appendix A2 we know that such mappings fall into disjoint homotopy classes, labeled by the integers; gauge functions belonging to different classes cannot be deformed continuously into each other; and only those homotopically trivial (with winding number equal to zero) can be deformed to the identity. That is, Π_3 (a non-Abelian gauge group) $= \Pi_3(S^3) = Z$. Furthermore, the additional term $\omega(U)$ picked up by the Wess–Zumino term (representing the anomaly effects) under the finite transformations for a non-Abelian theory $(A \rightarrow S_U A S_U^+ = U^{-1}(r)AU(r) - U^{-1}(r)\nabla U(r))$ is, as can be seen from appendix A2, precisely the analytic expression for the winding number of the finite gauge transformations.

The concept of finite gauge transformations, whose characteristic labeled by the winding number is uniquely determined by the global gauge transformation U_∞ ($= \lim_{r \rightarrow \infty} U(r)$), is rich in implications. First, in any realistic non-Abelian gauge theory with chiral source currents, where the gauge transformations are not trivial and the chiral anomaly is inevitable, the gauge invariance is broken at the quantum level and renormalizability is destroyed, unless some arrangement for the source currents is made so that the additional contributions picked up by the anomalous term cancel each other and the gauge invariance is restored. This requirement of the cancellation of anomalies has put a severe restriction upon model building (see section 11.1).

Second, when the three-dimensional gauge transformations are not trivial (such as, but not restricted to, the case of non-Abelian gauge theories), if the action I is invariant against small but not large gauge transformations, then the requirement of gauge invariance in certain situations entails the quantization of physical parameters. To see this we only have to remember that, in the path integral formulation, the theory is determined by $\exp(iI/h)$, the gauge invariance of a theory against finite gauge transformations can be achieved by requiring that the change of its action ΔI caused by a finite gauge transformation is equal to $2n\pi h$, n being an integer. In the case of Dirac's point magnetic monopole, where the $U(1)$ gauge transformations are non-trivial because of the existence of the singularity caused by the monopole and $\Delta I = 4\pi eg$, the requirement of gauge invariance amounts to the requirement $4\pi eg/c = 2n\pi h$, or $2ge/hc = n$, which is the Dirac quantization condition for the electric charge. More examples about the quantization of physical parameters as the result of gauge invariance against non-trivial gauge transformations can be found in Jackiw (1985).

From the above discussion we know that a gauge-invariant theory with a simple non-Abelian gauge group, whose transformations are non-trivial, always contains quantization conditions. Then what about these quantization conditions if the symmetry group is spontaneously broken by a scalar field to a U(1) group, whose transformations, identical to the electromagnetic gauge transformations, are trivial? 'T Hooft (1974) and A. Polyakov (1974) found that in the case of an SU(2) gauge group the quantization conditions would remain because, instead of a singular Dirac monopole, there would be a smooth SU(2) monopole, a configuration which would be regular at the origin, but far away from the origin would appear as a monopole. Later, static solutions with finite classical energy of the smooth monopole were found and were believed to be stable against perturbations. Physicists tried to identify the smooth monopole with a soliton of non-linear field theories in three spatial dimensions, and to describe its quantum states semi-classically.[53]

The 't Hooft–Polyakov monopole and its generalizations must exist in every grand unified theory because all of them are based on a simple symmetry group, which must be broken spontaneously with a U(1) factor surviving. Thus the lack of experimental evidence for the existence of the monopoles and their problematic theoretical implications (such as proton decay) have put an important constraint on model building.

Another interesting implication of the response of a non-Abelian gauge theory to the finite gauge transformations involves a new conception of the structure of the vacuum.

In a pure non-Abelian gauge theory, after one fixes the gauge (such as using the gauge $A_0 = 0$), there remains some residual gauge freedom, such as those finite gauge transformations $U(r)$, depending on the spatial variable only. Thus the vacuum defined by $F_{\mu\nu} = 0$ restricts the gauge potentials to 'pure gauge' and zero-energy configurations $A_i(r) = -U^{-1}(r)\,\partial_i U(r)$, with lim $U(r \rightarrow \infty) = U_\infty$, where U_∞ is a global (position-independent) gauge transformation. Since Π_3 (a compact non-Abelian gauge group) $= Z$ (see appendix A2), there must be an infinity of topologically distinct yet energetically degenerate vacua for such a non-Abelian gauge theory, which can be represented by $A^{(n)} = U_n^{-1} \nabla U_n$, n extending from $-\infty$ to $+\infty$. Thus a quantum state of a physical vacuum should be a superposition of wave functionals $\Psi_n(A)$: $\Psi(A) = \sum_n e^{in\theta} \Psi_n(A)$, with each $\Psi_n(A)$ situated near the classical zero-energy configuration $A^{(n)}$. They must be gauge invariant against homotopically trivial transformations. But the non-trivial ones shift n: $S_n \Psi_n' = \Psi_{n+n'}$, where S_n is the unitary operator that implements a finite gauge transformation U_n belonging to the nth homotopy class. Since $S_n \Psi(A) = e^{-in\theta} \Psi(A)$, the physical vacuum is characterized by θ, the famous vacuum angle, and the vacuum is called the θ vacuum.[54]

The pure gauge potentials $A^{(n)} = U_n^{-1} \nabla U_n$ are analogous to infinitely degenerate zero-energy configurations in a quantum mechanical problem with a periodic potential, with the emergence of a phase in the response of a topological vacuum state to a finite gauge transformation analogous to a shift in the periodic potential problem, whereby a wave function acquires a phase. As in the quantum mechanical problem, the degeneracy between different topological vacua is lifted if there exists tunneling between them. In pure non-Abelian gauge theories, such tunneling is due to instantons.[55] When the massless fermions are included in the theory, the tunneling is suppressed, the vacua remain degenerate, and the θ angle has no physical meaning.[56] In a non-Abelian gauge theory, such as quantum chromodynamics, where fermions are not massless (see section 11.1), the θ angle should be observable theoretically, although it has never been observed in experiment.[57]

In sum, the concept of gauge invariance against large or finite transformations that are homotopically non-trivial opens up a new perspective for gauge theories, helps to reveal or speculate on certain novel features of non-Abelian gauge theories which are global in nature, and thus provides a powerful means for combining a local conception of the physical world described by field theories with a global one.

10.5 Open questions

Although solid foundations have been laid down for gauge theories in the last four decades, many conceptual questions remain open, betraying the truth that the foundations are neither monolithic nor flawless. The most important among them are three.

I. The ontological status of the Goldstone–Higgs scalar system

As we have seen, as far as the basic ideas are concerned, there is no difference between Goldstone's work on spontaneous symmetry breaking (SSB) and Nambu's. Yet, Goldstone takes the scalar bosons as elementary particles and explores the conditions and results of SSB in this boson system, while in Nambu's framework, the scalar bosons are derivative because these are composite modes that appear only as a result of symmetry breaking in a fermion system. An advantage of Goldstone's model is its renormalizability. This makes it much easier to find conditions for the existence of asymmetric solutions to a non-linear system. More interesting than this, however, are some new features brought about by the introduction of an elementary scalar system in the study of SSB.

First, an indication of symmetry breaking in Goldstone's boson system is the occurrence of an incomplete multiplet of massive scalar particles, so-called Higgs bosons. In Nambu's framework, no massive spinless boson is possible without explicit symmetry breaking. Thus Goldstone's approach, compared with Nambu's, has brought out a surplus theoretical structure, the massive scalar boson. It is surplus because it is not required by the fermion system for symmetry restoring.

Second, the introduction of an elementary scalar system in the study of symmetry breaking imposes a double structure on symmetry breaking. That is, the SSB of the fermion system is not specified by its own non-linear dynamical structure. Rather, it is induced, through Yukawa coupling and gauge coupling, by the symmetry breaking in a primary system of scalar bosons. This double structure of symmetry breaking has brought a peculiar feature to the standard model (see section 11.1). That is, apart from a theoretically fixed dynamical sector, which explains and predicts observations, there is an arbitrarily tunable sector, which makes a phenomenological choice of actual physical states.

The introduction of the Goldstone–Higgs scalar system has opened up new possibilities for our understanding of the physical world, from the spectrum and interactions of elementary particles to the structure and evolution of the whole universe. In particular, the Higgs field is intimately involved in two of the most revolutionary aspects of the gauge theory programme. First, it suggests for the first time how mass, not just the mass of gauge quanta, but also that of fermions, is generated through an interaction: a gauge interaction for gauge quanta, the so-called Yukawa interaction between Higgs fields and fermion fields for fermions. The Yukawa interaction is gauge invariant but cannot be derived from the gauge principle. That is, though not in conflict with the gauge principle, it is itself outside the framework of gauge theory. In fact, understanding the nature of the Yukawa interaction requires a new direction in field-theoretical study. Second, the surplus structure of the Higgs field plays the role of a new type of vacuum, analogous to the old-fashioned ether, which acts like a continuous background medium and pervades all of spacetime. In certain models, it is suggested to be responsible for the cosmological constant and cosmic phase transitions. Then the question is where this ether comes from or what the origin of the Higgs field is.

Thus to successfully carry out the gauge theory programme, a thorough and detailed understanding of the nature and dynamics of the Goldstone–Higgs scalar system seems to be even more important than the theory of gauge fields itself, just as the understanding of gauge fields is more important for carrying out the quantum field programme than the theory of quantum fields itself. Then, a serious question is raised concerning the ontological status of this scalar system.

Notice that there is a big difference between the scalar system and a superconducting system. In the superconducting system, the asymmetric phase, the symmetric phase, and the phase transition between the two phases are all real. In the scalar system, however, the Goldstone boson is non-physical, the Higgs boson escapes our observation, and the symmetric solution attracts little or no attention from physicists. The non-physical Goldstone scalars, together with other non-physical scalars, covariant ghosts, and Faddeev–Popov ghosts, are deeply entangled in the theoretical structure of non-Abelian gauge theories, that is, in the description of gauge bosons. An inevitable physical question about these non-physical degrees of freedom concerns their relation to nature: are they representative of physical reality, or just auxiliary constructions for coding some information without direct physical reference?

The instrumentalists take them as *ad hoc* devices for building models, such as the standard model, so that the required observations, such as W particles and neutral currents, can be obtained. They do not take all the implications of these *ad hoc* devices, including the Higgs boson, seriously. But then the instrumentalists have to face further questions. What is the status of the information coded in these constructions? Is it possible for us to have direct access to the information without resort to the fictitious devices? That is, can we reconstruct a self-consistent theory, with the synthesizing power and the powers of explanation and prediction equal to those of the standard model, without all the non-physical degrees of freedom, which are deeply entrenched in the standard model?

If we take a realist position, then the search for the Higgs particle, for the symmetric solution of the scalar system, and for the agent that drives the system from its symmetric phase to its asymmetric phase will be serious physical problems. Moreover, there is another question concerning the nature of the scalar particles: are they elementary or composite?

Some physicists feel that only in a phenomenological model can the scalar particles be taken as elementary, and in a fundamental theory they should be derived from fermions. For them, Goldstone's approach was a retreat from Nambu's more ambitious programme, and the idea of dynamical symmetry breaking, including the idea of technicolor, seems to be more attractive than Goldstone's approach. Yet the primary role that the scalar system has played in the standard model seems to support an alternative view, which was extensively explored in the late 1950s by K. Nishijima (1957, 1958), W. Zimmermann (1958), and R. Haag (1958), and in the early and mid-1960s by Weinberg (1965a) and others, that, as far as the theory of scattering is concerned, there is no difference between elementary and composite particles.

II. No new physics?

Traditionally, physicists have believed that QFT tends to break down at short distances and new physics will sooner or later appear with more and more powerful experimental instruments exploring higher and higher energy regimes (see section 8.3).[58] With the discovery of asymptotic freedom, however, some physicists argue that since in an asymptotically free theory the bare coupling is finite and vanishes, there are no infinities at all, and this has reasserted the consistency of four-dimensional QFT. In particular, the decrease of the effective coupling for high energies means that no new physics will arise at short distances (Gross, 1992).

All these statements are very attractive. The problem, however, is that there is no rigorous proof of the existence of asymptotically free gauge theories in four dimensions. Such a proof has to be based on the renormalization group argument and requires a proof of the existence of the ultraviolet stable fixed point. Yet no proof of the existence of the fixed point has ever been achieved mathematically. The difficulty also shows up in its flip side, the infrared dynamics or low-energy physics, in an even harsher way: there is no rigorous solution to color confinement in a four-dimensional spacetime, and the explanation of low-energy pion–nucleon interactions by an asymptotically free theory seems almost unattainable.[59]

III. Is renormalizability proved?

All proofs of the renormalizability of gauge theories with massive gauge bosons involve at least one extra particle, the Higgs particle. Thus an indispensable part of the proof of their renormalizability ought to have a statement on the Higgs mass. Yet up to now, there has been no such statement at all, and we have no clue as to its magnitude from experiments. Even worse, physicists are divided in their opinions concerning the very existence of the Higgs particle or the possibility of finding it. Many of them feel that the world is more complicated than we conceived with the concept of the Higgs particle, although it is good enough for parameterizing the present experimental situation (Veltman, 1992). Thus, in a fuller sense, the proof of the renormalizability of massive gauge theories is still in a dubious state.

At a deeper level, a rigorous proof of the renormalizability of a quantum field theory can only be obtained through the renormalization group approach. If a theory has a fixed point for the renormalization group, then the theory is renormalizable, but not otherwise. Yet, sharing the bad situation with the proof of asymptotic freedom, the proof of the existence of the fixed point has never been achieved mathematically. Thus the claim that the renormalizability of non-Abelian gauge theories is proved is somewhat exaggerated. Furthermore, recent developments have suggested the legitimacy of non-renormalizable interactions in QFT (see section 11.4), thus posing serious questions for the understanding of the consistency of QFT and the role of the proof of renormalizability in constructing a consistent theory of quantum fields.

Notes

1. The phrase was coined by M. Baker and S. Glashow (1962) because, they remarked, the mechanism for symmetry breaking did not require any explicit mass term in the Lagrangian to manifestly destroy the gauge invariance. A more detailed historical account of the subject can be found in Brown and Cao (1991).
2. E. P. Wigner emphasized the importance of the distinction between physical states and physical laws, and asserted that symmetry principles 'apply only to the second category of our knowledge of nature, to the so-called laws of nature'. (Houtappel, Van Dam, and Wigner, 1965; see also Wigner's earlier articles, 1949, 1964a, b). Wigner's remark exaggerates the distinction, since physical states (for example, states of elementary particles) may constitute representations of symmetry groups. In fact, Wigner's book (1931) concerns the application of group theory to atomic states. Adopting Wigner's restrictive statement of the applicability of symmetry principles would have blocked the line of reasoning that led to SSB. As we shall see below, the concept of SSB rests entirely on a comparison of the symmetry of equations describing a physical system with the more restricted symmetry of the system's physical states.
3. They can be found in L. A. Radicati (1987).
4. In the case of the ferromagnet, this parameter is also a constant of motion, a simplifying feature not typical of SSB. The antiferromagnetic case, first treated by P. W. Anderson in (1952) is more typical: the macroscopic parameter, although not a constant of motion, is

still related to the generators of the spin group, and indicates the degree of broken symmetry.

5. Quoted by Brown and Rechenberg (1988).

6. That he should have done so was no accident, for as early as 1922, as a first-year student at the University of Munich, he had worked on the problem of turbulence. Later, at the suggestion of his teacher Arnold Sommerfeld, Heisenberg took as his thesis problem the onset of turbulence, an example of SSB. See Cassidy and Rechenberg (1985).

7. The idea is that the vacuum possesses a very large isospin and other internal quantum numbers, or to use Gregor Wentzel's term (1956), a large number of all kinds of 'spurions', so that internal quantum numbers can be extracted from the vacuum as needed. This idea seemed unnatural even to Wentzel, and received Pauli's severest criticism. At the Geneva conference, Gell-Mann addressed Heisenberg, rephrasing Pauli's objections as follows: 'Depending on how many particles you want to describe, N, you need 2^N vacua . . . . I suppose you can do it, but it seems to be complicated and very much like adding another field'. See Gell-Mann's discussion remark attached to Heisenberg's (1958).

8. Dürr, Heisenberg, Mitter, Schlieder, and Yamazaki (1959).

9. At the Rochester Conference of 1960, Heisenberg advanced these views and again advocated the replacement of 'vacuum' by 'world state' (1960).

10. In his paper on non-linear field theory (1961), Heisenberg retreated from his earlier more radical position on the degeneracy of the vacuum, and began to appeal to the idea of the large-distance fluctuations of a non-degenerate vacuum. For this reason, and also because he failed to connect his work with that of Nambu and Goldstone, or to relate it to the Yang–Mills theory, Heisenberg's influence on the development of SSB declined. After his ambitious speculative non-linear field theory produced no significant results, Heisenberg dropped out of the mainstream of particle physics. At the present time, his major contributions to the development of SSB appear to be almost completely forgotten.

11. The debate actually went back to 1956, a year before the appearance of the BCS theory. Then M. J. Buckingham (1957) challenged Bardeen, who, in (1955), had claimed that the energy-gap model would yield the Meissner effect, assuming the matrix elements for optical transitions were the same as they would have been with no gap. Buckingham proved that they were the same only in the London gauge, and not in general. Max R. Schafroth in (1958) argued against the BCS derivation of the Meissner effect. He pointed out that as early as 1951 he had already proved that 'any error in the theory which involves a violation of gauge invariance will generally produce a spurious Meissner effect which vanishes when gauge invariance is restored' (Schafroth, 1951).

12. Bardeen (1957), Anderson (1958a, b), Pines and Schrieffer (1958), Wentzel (1958, 1959), Rickaysen (1958), Blatt, Matsubara, and May (1959).

13. Bogoliubov (1958), Nambu (1960).

14. It should be noted that the superconducting phase is only a part of the complete system, whose overall description must be gauge invariant.

15. γ_5 is the product of the four Dirac matrices γ_μ and anticommutes with each of them. It therefore anti-commutes with the Dirac operator D in the Dirac equation $D\psi = m\psi$, since D is linear in the γ_μ, but it commutes with the mass m. Thus only the massless Dirac equation is γ_5-invariant.

16. Nambu and Jona-Lasinio (1961a, b).

17. In his paper at the Midwest Conference, he said that 'the Fermi sea of electrons in a metal is analogous to the Dirac sea of electrons in the vacuum, and we speak about electrons and holes in both cases' (1960c).

18. Nambu stated very clearly: 'Our γ_5-invariance theory can be approximated by a phenomenological description in terms of pseudoscalar mesons . . . . The reason for this situation is the degeneracy of the vacuum and the world built upon it' (Nambu and Jona-Lasinio, 1961a).

19. Cf. Nambu (1989).

20. In superconductivity, the Coulomb interaction promotes the collective excitations into massive plasmons. Nambu could have applied similar reasoning to obtain a pion mass. But such a spontaneous breakdown of the exact symmetry, although it permits the pion a non-

zero mass, implies conservation of the axial vector current, which in turn implies a vanishing pion decay rate, in contradiction to observation. Constrained by this physical consideration, Nambu gave up the analogy and accepted the idea of PCAC, proposed by Gell-Mann and Maurice Levy and others (see section 8.6). Thus Nambu adopted an explicit breakdown of γ_5 invariance in the form of a small bare-nucleon mass (Nambu and Jona-Lasinio, 1961a).

21. Gell-Mann claimed that he and his collaborators in their work on PCAC, whose divergence is dominated by the low-mass pion pole, found the Nambu–Goldstone boson independently. He claimed in (1987) that 'according to PCAC, as the divergence of the [axial] current goes to zero, m_π tends to zero rather than m_n. In the limit, there is a realization of the "Nambu-Goldstone" mechanism, which those authors were developing independently around the same time'. In (1989) he claimed once again that 'in the limit of exact conservation, the pion would become massless, and this is a realization of the Nambu–Goldstone mechanism.'

    The conflation of the limiting case of PCAC with the Nambu–Goldstone mechanism is somewhat misleading. It is misleading because the physical ideas underlying the two are different. The Nambu–Goldstone mechanism is based on the idea of degenerate vacuum states that are stable asymmetric solutions to a non-linear dynamical system. So the symmetry is broken at the level of solutions rather than dynamical law. In the case of PCAC, neither non-linearity nor the degeneracy of the vacuum is its characteristic feature, and the symmetry is broken at the level of the dynamical equation. An illuminating fact is that in the framework of PCAC, when the symmetry is broken, there is no massless spinless boson. Once the massless boson is obtained (by taking the conservation limit of PCAC), there is no symmetry breaking at all. In sharp contrast with this situation, the massless scalar bosons in the Nambu–Goldstone mechanism occur as the result of symmetry breaking. These massless bosons are coexistent with the asymmetric solutions, so that the symmetry of the whole system can be restored. In sum, while the Nambu–Goldstone boson is the result of SSB in a theory possessing a continuous symmetry group, the massless pion in the work of Gell-Mann and his collaborators has nothing to do with SSB. Examining the papers by Gell-Mann and Levy (1960) and by J. Bernstein, S. Fubini, Gell-Mann, and W. Thirring (1960), I find no suggestions, in these papers, about the degeneracy of the vacuum, the existence of an asymmetric solution, or the restoration of symmetry by the massless field. Their distinction of the massive, or possibly massless, pion took place in an entirely different context, unrelated to the Nambu–Goldstone boson.

22. A significant difference exists between the approaches of Nambu and Goldstone to find 'superconductive' solutions of QFT. Nambu's approach is 'microscopic', and comparable to the BCS–Bogoliubov theory. Goldstone's is 'macroscopic', in the sense of 'phenomenological', and comparable to the Landau–Ginzburg theory. In the Goldstone model, bosons come directly as collective oscillations, just as the gap parameter does in the Landau–Ginzburg theory; they are not derived from the interactions of primary fields at the microscopic level. This macroscopic approach to SSB has become standard in the standard model, and helps to explain its peculiar features: one part of the model gives the dynamics, while the other, SSB, makes a phenomenological choice of the actual physical states. The first part, theoretically fixed, predicts observable consequences; the second is arbitrarily tunable. In some respects, Goldstone's work retreats from Nambu's more ambitious programme. The retreat may be strategically valuable, however, in view of our ignorance of some fundamental ingredients of nature and their interactions.

23. Gell-Mann claims that the Higgs mechanism is a solution to the soft mass problem (1987, 1989). The idea of soft mass was a response to the major preoccupations of the time with renormalizability and symmetry breaking, including obtaining approximate global symmetries. Yet the original formulation of the soft mass mechanism was simply to add a gauge boson mass term to a gauge-invariant Lagrangian, a term which destroyed both the gauge invariance and renormalizability. Only in later developments, which were based on the idea of SSB and Schwinger's idea of strong coupling, was a mechanism found by which gauge

bosons acquired masses without breaking the symmetry of the Lagrangian. However, the masses that gauge bosons acquired through this mechanism were by no means soft. On the contrary, they were extremely hard.

24. As I mentioned above, the Landau–Ginzburg theory of superconductivity (Ginzburg and Landau, 1950) already explicitly used SSB and contained SSB features that would be developed and integrated into the framework of gauge theories. For example, the phase mode of the complex scalar field ψ (the order parameter whose meaning was clarified as the 'energy gap' in the BCS theory) introduced by Landau and Ginzburg to describe the cooperative state of electrons in a superconductor turned out to be the prototype of the massless Nambu–Goldstone mode, and the amplitude mode turned out to be the prototype of the Higgs mode. However, as the Landau–Ginsburg theory was a macroscopic theory, its relevance escaped the notice of particle physicists. The analogy between the theoretical structures describing superconductivity and QFT became apparent only after the publication of the microscopic theory of superconductivity, the BCS theory, in 1957.

25. This is one of those recurrent dialectical shifts in the conceptual foundations of physics. We can follow it, on the classical level, from Descartes to Newton and from Maxwell to Einstein, and, on the quantum level, from Heisenberg and Schrödinger to Dirac. Dirac filled the vacuum with a sea of negative energy electrons and other fermions, from which the members of the Copenhagen school and their followers struggled to free it (see Weisskopf, 1983). From the late 1940s, it had been fashionable to regard the vacuum as a void, until a plenum view of the degenerate vacuum, associated with SSB, came once again to dominate.

26. Bloom *et al.* (1969).

27. For an easy-to-understand explanation of the technical problem of how to derive the Bjorken scaling from the parton model, see Pais (1986).

28. The title of Callan's (1970) was 'Bjorken scale invariance in scalar field theory'; that of Symanzik's (1970) was 'Small distance behavior in field theory and power counting'.

29. Symanzik (1971) and Callan (1972).

30. See Jackiw, Van Royen, and West (1970), Leutwyler and Stern (1970), Frishman (1971), and Gross and Treiman (1971).

31. Fritzsch, Gell-Mann, and Leutwyler (1973).

32. It was carried out by Politzer (1973) and by Gross and his student Frank Wilczek (1973a, b).

33. A contribution of the same type is made by $W^\pm$ and Z bosons, which are generated through the Schwinger–Anderson–Higgs mechanism and carry weak charges. But the contribution, in comparison with those made by the massless gluons, is greatly suppressed by the large masses of these massive bosons. On the other hand, Gross and Wilczek (1973a) at first worried about the possibility that the spin-0 Higgs boson would cause additional screening and destroy the asymptotic freedom in a spontaneously broken non-Abelian gauge theory. But if the Higgs boson were very massive, then the same suppression mechanism would make its contribution negligible.

34. See, for example, Wilson (1974) and Creuz (1981).

35. See Umezawa and Kawabe (1949a, b), Feldman (1949), Neuman and Furry (1949), Case (1949), Kinoshita (1950), Umezawa and Kamefuchi (1951), and Sakata, Umezawa, and Kamefuchi (1952).

36. See note 4 of chapter 9.

37. The publication (1963) was based on his tape-recorded lecture at the Conference on Relativity Theories of Gravitation, which was held in Jablonna, Poland, July 1962.

38. Since Feynman took unitarity as a connection between loops and trees, he was unable to discuss the issue beyond one loop.

39. In this form of the vector propagator, which is obtained in the so-called unitary gauge, the term containing $q^\mu q_\nu / \mu^2$, as had long been recognized, makes non-renormalizable contributions to the S-matrix.

40. Thus the ghost propagator has an orientation or an arrow, which was not mentioned by Feynman. Veltman later commented on DeWitt's contribution to the concept of the ghost in

gauge theories: 'Somewhat illogically this ghost is now called the Faddeev–Popov ghost' (1973).
41. As we noticed before, this is a change from the unitary gauge to the Landau gauge.
42. There is a certain apparent similarity between the free field technique and the Stueckelberg formalism in massive QED (1938) in the sense that both introduce an additional scalar field. The differences between the two are nevertheless quite substantial. First, the longitudinal component in the Stueckelberg formalism of massive QED decouples, but the new Feynman rules resulting from the Bell–Treiman transformation include new vertices and a ghost field. Second, Stueckelberg's scalar field transforms in a certain way under gauge transformations, while Veltman's scalar field is the gauge variation itself. Third, since the gauge invariance is broken by the introduction of a mass term for the electromagnetic field, and partially restored by the scalar field, the Stueckelberg approach may result in the discovery of the Higgs mechanism, but not the Faddeev–Popov ghost, which compensates the breaking of gauge invariance. I am grateful to Professor Veltman for clarifying this complicated issue.
43. Veltman (1969) and Reiff and Veltman (1969).
44. Later, Slavnov (1972) and Taylor (1971) derived the generalized Ward identities for massless Yang–Mills theories by using the path integral formalism, which are known as Slavnov–Taylor identities. These identities are similar to Veltman's for the massive theories, and can be put down also for theories whose gauge symmetry is spontaneously broken. They contain all the combinatorial content of the theory because from them the symmetry of the Lagrangian can be deduced.
45. These explicit ghost Feynman rules were in contrast with Faddeev and Popov as well as with Fradkin and Tyutin, who wrote the ghost part of the Lagrangian in terms of a determinant.
46. Lee (1969), Gervais and Lee (1969), and Symanzik (1969, 1970).
47. Lee (1972).
48. Dimensional regularization respects most symmetries, but not scale and chiral symmetries involving the γ_5 matrix because both symmetries are dimension specific (the γ_5 matrix can only be defined in four-dimensional spacetime).
49. In fact, gauge invariance is only one of the two crucial factors for the renormalizability of field theories. The other is the dimensionality of the coupling constant. Making a Higgs-type gauge theory of weak interaction renormalizable is possible only because the dimension (M^{-2}) of the coupling constant G_w in direct Fermi coupling can be transferred to the propagator of the intermediate massive meson ($\sim q^\mu q_\nu / M^2 q^2$). This transference, however, is not applicable to the quantum theory of gravitation. In a gauge theory of gravitation, the gauge quanta mediating the gravitational interaction (gravitons) are massless, because of the long-range character of gravitation. Despite work by Veltman and 't Hooft, therefore, the renormalizability of the quantum theory of gravitation remains a serious challenge to theoretical physicists.
50. Except for one single report of an observation of a monopole by B. Cabrera (1982).
51. Yang (1970) argued: '[in] a space-time-independent gauge transformation on charged fields ψ_j of charge e_j:

$$\psi_j \rightarrow \psi'_j = \psi_j \exp{(ie\alpha)}. \tag{1}$$

... If the different e_j's ($= e_1, e_2, \ldots$) of different fields are not commensurate with each other, the transformation (1) is different for all real values of α, and the gauge group must be defined so as to include all real values of α. Hence, the group is not compact.
'If, on the other hand, all different e_j's are integral multiples of e, a universal unit of charge, then for two values of α different by an integral multiple of $2\pi/e$, the transformations (1) for any fields ψ_j are the same. In other words, two transformations (1) are indistinguishable if their α's are the same modulo $2\pi/e$. Hence the gauge group as defined by (1) is compact.'
52. See Wu and Yang (1975).
53. See Jackiw and Rebbi (1976) and Colman (1977).
54. See Callan, Dashen, and Gross (1976), and Jackiw and Rebbi (1976).
55. Belavin, Polyakov, Schwartz, and Tyupkin (1975).

56. See 't Hooft (1976a, b), Jackiw and Rebbi (1976).
57. The analysis of the stringent experimental limits on the neutron's dipole moment shows that $\theta \leqslant 10^{-9}$. See Crewther, Divecchia, Veneziano and Witten (1979).
58. Such as Schwinger (1948b, 1970) and Gell-Mann and Low (1954).
59. Although some progress towards understanding these difficulties as consequences of QCD, has been made within the framework of lattice gauge theory by using various effective action methods. See Wilson (1974), and Kogut and Susskind (1975).

11

The gauge field programme (GFP)

The Utrecht proof of the renormalizability of gauge-invariant massive vector-meson theories which appeared in 1971, as observed by influential contemporary physicists, 'would change our way of thinking on gauge field theory in a most profound way (Lee, 1972) and 'caused a great stir, made unification into a central research theme' (Pais, 1986). More precisely, with the exhilaration of such a great change in perspective, confidence had quickly built up within the collective consciousness of the particle physics community that a system of quantum fields whose dynamics is fixed by the gauge principle was a self-consistent and powerful conceptual framework for describing fundamental interactions in a unified way. An immediate outcome of the new perspective was the rise of the so-called standard model, consisting of the electroweak theory and quantum chromodynamics (QCD) (section 11.1). With the encouragement of the empirical successes of the model and the power of its underlying concepts, efforts were made to extend the model to a grand unification and to gravity, assuming the universality of the gauge principle (section 11.2).

The emergence of the gauge field programme (GFP) suggests a dialectical comprehension of the developments of 20th century field theories: taking GFP as a synthesis of the geometrical programme and the quantum field programme. Some justifications for such an outlook are given in section 11.3. But the dialectical developments of field theories have not ended with a Hegelian closure (a final theory or a closed theoretical framework). Some discussions on the stagnation of GFP and on a new direction in the field theoretical research appear in section 11.4.

11.1 The rise of the standard model

The term 'standard model' was first coined by Pais and Treiman in (1975), with reference to the electroweak theory with four quarks. In its later usage, it referred to a six-quark picture of hadrons together with leptons, with its dynamics described by the electroweak theory and QCD. From a historical perspective, the rise of the standard model was the result of three lines of development: (i) the establishment of a conceptual framework, (ii) a proof of its consistency, and (iii) model building. *The general conceptual framework* within which the standard model was built was the quark model for hadrons combined with the ideas of gauge coupling and symmetry breaking. The quark model, originally suggested by Gell-Mann (1964a) and George Zweig (1964a, b) with three quark species (later called flavors) forming a fundamental basis of an SU(3) group as a way of incorporating the broken SU(3) symmetry among hadrons, allows quarks (parts of hadrons with baryon number 1/3, spin 1/2, and fractional electric charge), together with leptons, to be taken as the basic ingredients of the micro-structure of nature. The reality of quarks, however, was established some years later, without which no one (not even Gell-Mann himself) would take the quark model seriously as an ontological basis for theorizing about the physical world. The crucial steps in establishing the reality of quarks were taken in 1969 when deep inelastic scattering experiments were performed at SLAC to probe the short-distance structure of hadrons. The observed Bjorken scaling suggested that hadrons consisted of free point-like constituents or partons (see section 10.2), some of which, when the experimental data were analyzed in terms of the operator product of currents, turned out to be charged and have baryon number 1/3 and spin 1/2. That is, they looked like quarks. Further data also showed that they were consistent with the electric charge assignment to quarks.[1] These preliminary but crucial developments certainly had convinced some pioneering physicists of the reality of quarks, and encouraged them to use the quark model for conceptualizing the subatomic world.

The concept of gauge coupling was suggested by Yang and Mills, and further developed by Utiyama, Schwinger, Glashow, and many others (see section 9.1). Complementary to this concept of gauge symmetry were the various concepts of symmetry breaking: the concept of spontaneous breaking developed by Heisenberg, Nambu, Goldstone, Anderson, and many others (section 10.1), and the concept of anomalous breaking by Adler, by Bell and Jackiw, and by many others (section 8.7). These concepts played an indispensable role in establishing the consistency of the framework and in model building. But the most important and most profound ingredient of the conceptual framework was a consequence of the anomalous breakdown of scale invariance, which was embodied in the concept of the renormalization group (see section 8.8). The idea of the renormalization group assumes a specific type of causal connection between the structures of physical interactions at different energy scales, without which no idea of running couplings, and hence no ideas of asymptotic freedom and grand unification, would be possible. Nor would a rigorous proof of renormalizability be possible either.

A proof of the consistency of the non-Abelian gauge theory

As far as the massive vector theory is concerned, the introduction of the Anderson–Higgs mechanism was an important step in the right direction, although not a decisive one. It was important because with the Anderson–Higgs mechanism, the massiveness of gauge bosons, which was crucial for saving phenomena in the area of weak interactions, could be made compatible with an exact rather than approximate or partial dynamic symmetry. And this made the proof of renormalizability much easier. However, the recognition of this advantage, combined with a widespread conviction that a Yang–Mills theory with an unbroken symmetry was renormalizable, produced a naive belief that a Higgs-type massive Yang–Mills theory would be renormalizable a priori.

The belief was naive because the conviction was based solely on the naive power-counting argument, and thus was unjustifiable. Here we have to remember that the relationship between gauge invariance and renormalizability is much subtler than was thought to be the case in the 1950s and 1960s by Yang and Mills, by Glashow, by Komar and Salam, by Weinberg, and by many others (see section 10.3). For a theory to be renormalizable, gauge invariance is neither sufficient, as in the case of gravity, nor necessary, as in the case of the neutral vector meson theory, although it certainly places severe constraints on model building. Thus a proof of the renormalizability of non-Abelian gauge theories was a great challenge to theoreticians. It required serious and difficult investigations, which were carried out by a group of theoreticians, from Lee and Yang, through Feynman, DeWitt, and Faddeev and Popov, to Veltman and 't Hooft. They quantized the theory in a consistent way with the introduction of a complex system of non-physical degrees of freedom required by accepted physical principles. They derived Feynman rules and Ward identities, invented the renormalizable gauges, and proved unitarity. Finally, they invented a gauge-invariant regularization scheme. Without these investigations, no proof of the renormalizability of non-Abelian gauge theories would be possible, and all the convictions and conjectures of an a priori kind, based solely on the symmetry argument and naive power-counting argument, would be groundless and empty.

Model building: the electroweak theory

Early attempts at building models for describing fundamental interactions within the framework of non-Abelian gauge theories can be found in section

9.3. As far as the weak interactions are concerned, two major achievements in those efforts, which were dictated by empirical data (short-range behavior demanded massive bosons, parity violation demanded chiral asymmetry), seem to be durable: (i) taking SU(2) × U(1) as the gauge group for a unified description of the weak and electromagnetic interactions, and (ii) adopting the Anderson–Higgs mechanism for mass generation.

Theoretically, however, these efforts were quite vulnerable. No consistent quantization of the models was given. Without an indication of proper Feynman rules, then, speculations about renormalizability or whether it would be spoiled by spontaneous symmetry breaking were conceptually empty. For this reason, the attempts, as Veltman commented in his influential review (1973), 'were unconvincing, and this line of research did not bear further fruits until recently'. That Veltman's comment was not a biased one was corroborated by Coleman (1979), D. Sullivan, D. Koester, D. H. White, and K. Kern (1980), and by Pais, who noticed that the model proposed by Weinberg 'was rarely quoted in the literature during 1967–9. Rapporteur talks at the biannual Rochester conferences in Vienna (1968) and Kiev (1970), as always devoted to what is currently fashionable, did not even mention SU(2) × U(1)' (Pais, 1986).[2]

The great change came, as Pais observed, when the Utrecht proof of renormalizabilty was presented at the Amsterdam conference, in a session organized by Veltman, in June 1971. In addition to a proof of renormalizability of the massive Yang–Mills theory, 't Hooft (1971b, c) also proposed three models. One of 't Hooft's models was identical to the forgotten one, proposed previously by Weinberg (1967b) and Salam (1968), which, apart from the Higgs mechanism, was equivalent to the model proposed by Glashow (1961).

The heart of this (later to be) celebrated model is its gauge group, which was first suggested by Glashow. Constrained phenomenologically by charged currents and the non-conservation of parity in weak interactions, the weak isospin group SU(2)$_T$ was suggested as responsible for the weak coupling. With respect to the weak coupling, the left-handed lepton fields were assigned as doublets

$$L_{e(\mu)} = [(1 - \gamma_5)/2] \begin{vmatrix} \nu_{e(\mu)} \\ e(\mu) \end{vmatrix},$$

the right-handed lepton fields were assigned as singlets ($R_{e(\mu)} = [(1 + \gamma_5)/2]$ $e(\mu)$), while the gauge fields were assigned as triplets ($W_\mu^\pm$ and W_μ^0). To unify weak and electromagnetic interactions, a new gauge group, the 'weak hypercharge' group U(1)$_Y$, associated with a vector field B_μ was introduced, whose corresponding quantum number satisfied the relation $Q = T_3 + Y/2$, where Q is the electric charge, T_3 is the third component of weak isospin T, and T and Y commute. In this way the combined interactions could be described by the gauge bosons of a new gauge group given by the product $SU(2)_T \times U(1)_Y$, whose masses could be acquired through the Anderson–Higgs mechanism without spoiling the symmetry.

Heuristically, the relation $Q = T_3 + Y/2$ can be understood as a statement that the photon field A_μ is some linear combination of W^0_μ and B_μ. Then the orthogonal combination Z_μ represents a neutral, massive spin-1 particle:

$$A_\mu = \sin \theta_W W^0_\mu + \cos \theta_W B_\mu$$

and

$$Z_\mu = \cos \theta_W W^0_\mu - \sin \theta_W B_\mu$$

where θ_W is the Weinberg angle ($e = g \sin \theta_W = g' \cos \theta_W$, g and g' are the coupling constants of W_μ and B_μ respectively). Thus the unification with a gauge group as a product of disconnected groups is realized through a mixing of gauge fields.

The next step in model building was dictated by the anomaly argument. In 't Hooft's model with chiral currents, the inevitable chiral anomalies would certainly spoil its renormalizability, as was indicated first by William Bardeen at the Orsay conference of January 1972, when 't Hooft thought that the anomalies were harmless.[3] Inspired by the arguments at the Orsay conference, C. Bouchiat, J. Illiopoulos and Ph. Meyer (1972), and also Gross and Jackiw (1972), suggested that the anomalies would be absent if an appropriate amount of various quarks were included in the model. In particular, they pointed out, the anomalies would cancel between quarks and leptons and renormalizability would not be spoiled if the model included a charmed quark, in addition to the three old quarks suggested by Gell-Mann and Zweig.[4] In terms of model building, therefore, the crucial point raised by the anomaly argument was that a consistent electroweak theory would be possible only when quarks were included, in a parallel way to the leptons, in the model.

Characteristic of the electroweak theory was the existence of the neutral massive gauge boson Z_μ, which was supposed to couple to a neutral, parity-violating weak current. The prediction of the neutral current was confirmed in 1973.[5] Another construction of the electroweak theory, the charmed quark, was also supported by the experimental discoveries of the J/ψ and ψ' particles,[6] which were identified as charmonium, and of the D particles,[7] which were identified as charmed particles. With the discovery of the W and Z particles in 1983,[8] it was thought that the electroweak theory was experimentally confirmed and constituted the core of the standard model.

Quantum chromodynamics (QCD)[9]

The model building in the realm of the strong interactions was directly dictated by the empirical discovery of Bjorken scaling in deep inelastic scattering experiments, which substantiated the quark model (and demanded the existence of neutral gluons as part of hadrons), required a theory of inter-quark forces as the true source of the strong nuclear forces with the latter only as a residue of the former. In the search for a framework to entertain scaling, it became clear that only asymptotically free theories can explain scaling, and only non-Abelian gauge theories could be asymptotically free (see section 10.2). Thus, physicists were forced to construct models for the strong forces within the framework of non-Abelian gauge theories, in which the basic entities were not hadrons but quarks and gluons.

The crucial step in their construction was to find an appropriate gauge group. In reporting their (re)discovery of asymptotic freedom, Gross and Wilczek (1973a) proposed an SU(3) color gauge group. The proposal was strongly suggested by empirical data. To see this, a few words about the color quantum number in the quark model are in order. Initially, the color quantum number was introduced as a response to the dilemma faced by the early (static SU(6)[10]) quark model: the lowest-lying baryons, as bound states of quarks, belong to the totally symmetrical representation of the SU(6) group, and the quarks were expected to be in S states (that is, symmetric in orbital variables). This was in direct conflict with the assumption that quarks have spin 1/2 and thus have to obey Fermi statistics. Moo-Young Han and Nambu (1965) suggested that the dilemma could be solved by introducing a new additional three-valued quantum number, color, with respect to which the baryons would be totally antisymmetric.

Han and Nambu further associated a new symmetry group with the new quantum number, a color SU(3) group or SU(3)$_c$ (in addition to the old flavor SU(3) group), and suggested a Yang–Mills-type theory for 'the superstrong interactions' among quarks, which were replaced then by SU(3)$_c$ triplets:

We introduce . . . eight gauge vector fields which behave . . . as an octet in SU(3)$_c$, but as a singlet in SU(3) . . . .. The superstrong interactions for forming baryons and mesons have the symmetry SU(3)$_c$ . . . . The lowest [baryon and meson] mass levels will be SU(3)$_c$ singlets.

(Han and Nambu, 1965)

The paper was widely read and quoted, but only the idea of three quark triplets was taken seriously as an effective means for building consistent models of hadrons, with 'profound' ideas about a color gauge theory (the SU(3)$_c$ gauge group and its eight color-charged gauge vector bosons, color-charged quarks, and color-neutral hadrons) being ignored. The fact that these 'profound' ideas were not appreciated until the coming of QCD is quite understandable, because the experimental and theoretical contexts for its appreciation, scaling, the renormalization group, and asymptotic freedom, were absent when these ideas were presented to the physics community in the mid-1960s.

The idea of the color quantum number was strongly reinforced, in the context of the anomaly argument, by experimental constraints in the early 1970s. In 1972, it was noticed that for the electroweak theory to be anomaly-free and renormalizable, not only was the inclusion of a quark sector with a charmed quark into the theory necessary, but the quarks had to be three colored, so that the anomalies produced by the lepton sector could be canceled by the three-colored quark sector.[11] In 1973, an old argument, that the coefficient of the chiral anomaly was given exactly by the lowest-order perturbation theory (the triangle graph),[12] and the relation between the $\pi^0 \rightarrow 2\gamma$ decay rate from the global chiral anomalies and the sum of the squares of the charges of the elementary constituents of the hadrons[12a] were exploited fruitfully by Bardeen, Harold Fritzsch, and Gell-Mann (1973) as implying that the observed decay rate would be explained in this theoretical context only when a factor of three provided by the color quantum number was included in the fractionally charged quark model.

Well aware of the growing arguments for the color quantum number, Gross and Wilczek proposed that the neutral constituents of the nucleon, gluons, might carry a color quantum number[13] and act as the color gauge bosons:

One particularly appealing model is based on three triplets of fermions, with Gell-Mann's SU(3) × SU(3) as a global symmetry and a SU(3) 'color' gauge group to provide the strong interactions. That is, the generators of the strong interaction gauge group commute with ordinary SU(3) × SU(3) currents and mix quarks with the same isospin and hypercharge but different 'color'. In such a model the vector mesons are (flavor) neutral, and the structure of the operator product expansion of electromagnetic or weak currents is essentially that of the free quark model (up to calculable logarithmic corrections).

(Gross and Wilczek, 1973a)

The next step in their construction of a model for the strong interactions was to choose between an exact and a broken gauge symmetry. One might expect that an unbroken symmetry would imply the existence of massless, strongly coupled vector mesons as in the case of the original Yang–Mills theory. But Gross and Wilczek argued that in asymptotically free theories

there may be little connection between the 'free' Lagrangian (of quarks and gluons) and the spectrum of (hadron) states . . .. The infrared behaviour of Green's functions (of quark and gluon fields) in this case is determined by the strong-coupling limit of the theory. It may be very well that this infrared behaviour is such so as to suppress all but color singlet states, and that the colored gauge fields as well as the quarks could be 'seen' in the large-Euclidean momentum region but never produced as real asymptotic states.

(Gross and Wilczek, 1973b)

The key argument in the Gross–Wilczek proposal was asymptotic freedom, which differentiated it from the otherwise similar Han–Nambu proposal. In terms of meeting empirical constraints, asymptotic freedom explained scaling at short distances and offered a heuristic mechanism for confinement at large distances. Furthermore, it also provided a theoretical basis for perturbation theory to be trustworthy at short distances. Thus by the end of 1973 a calculable model for the superstrong interactions, perturbative QCD, was in sight, although it was not quite relevant to the original strong nuclear interactions.

It is instructive to notice that such a consistent and plausible theory 'was awarded honorable mention – no less, no more' at the London conference on high energy physics (July 1974) (Pais, 1986). This is another example of the recognition of any scientific discoveries, ideas, models, and theories being highly context dependent.[14]

A few months later, with the discovery of charm, the situation changed dramatically. Taking the newly discovered J/ψ particle as the bound 'charmed quark–anti-charmed quark' system, or charmonium, entails that the mass of the charmed quark is large, and thus its Compton wavelength is small. Since binding holds quarks at distances even smaller than this length, it suggests that the subject involves the short-distance behavior of quarks and gluons, and can be analyzed by perturbative QCD. The fact that charmonium can be reasonably described as the hydrogen atom of QCD with a confining potential[15] created a new context, in which QCD was quickly accepted by the high-energy physics community.

From 1971, when the standard model was prompted by 't Hooft's work to rise from the horizon, to 1975, when it overwhelmed physicists' conception of fundamental interactions, it took only four years for the weak interactions to be unified with electromagnetism and described by a renormalizable theory, and for strong-interaction dynamics to transit from a state of chaos to being described by a plausible, workable model.

11.2 Further extensions

Successful as it was, however, the standard model still contained too many arbitrary parameters. In addition to the weak coupling constant, there were

Weinberg angle mixing gauge couplings, Cabibbo angles mixing quark flavors, and mass parameters of leptons, quarks, gauge bosons, and the Higgs particle. Thus, in terms of unification, one of the basic motivations for pursuing the gauge field programme (GFP), the situation was far from satisfactory.

Meanwhile, the available theoretical apparatuses, most important among which were the renormalization group and symmetry breaking, had provided powerful stimulation to extend the standard model and carry GFP to its logical conclusion. In particular, the calculation of the variation of the strong and electroweak couplings with energy, made by Howard Georgi, Helen Quinn, and Weinberg (1974) using the renormalization group equations, showed that these couplings became equal in strength somewhere around 10^{14} to 10^{16} GeV. This merger, together with the fact that both theories had a similar non-Abelian gauge structure, provided a strong impetus to the search for further unification, in which natural laws with differing invariance properties, and symmetric laws and asymmetric physical states, all emerge from a higher symmetry, which may characterize physics under the conditions present in the early universe, which passed through a sequence of phase transitions as the temperature decreased while the universe expanded, until it reached the state now described by the standard model.

In such a grand unification based on the gauge principle, the internal attributes of matter fields (fermions) possess a dynamical manifestation: the gauge fields are coupled to themselves and also, in a unique and natural way, to the conserved currents of the matter fields that carry these attributes. Thus matter, in the form of quantum fermion fields, finally acquires the status of substance in its fully fledged sense. Not only are matter and force fields as two entities unified, but matter itself becomes active, with its activity manifested in the gauge couplings.

However, such a grand unification can only be achieved when the groups $SU(2)_T \times U(1)_Y$ and $SU(3)_C$, whose gauge quanta are responsible for the corresponding electroweak and strong forces, originate from the breaking of a larger simple group, whose gauge quanta are responsible for the grand unified forces. In addition to the original $SU(4) \times SU(4)$ model for the grand unification proposed by Jogesh Pati and Salam (1973a, b) and the $SU(5)$ model proposed by Georgi and Glashow (1974), there occurred several other models.[16] All of them are based on the same essential idea and have a number of features in common.

The essential idea of grand unified theories (GUTs) was summarized by J. C. Taylor (1976) in the general form of hierarchical symmetry breaking: an underlying large local gauge symmetry of all interactions is broken down in a succession of steps, giving a hierarchy of broken symmetries. Start with a gauge group G and Higgs field ϕ. Suppose that the vacuum expectation value $F = F^{(0)} + \epsilon F^{(1)} + \epsilon^{(2)} F^{(2)} + \ldots$, where ϵ is some small parameter. Let the little group of $F^{(s)}$ be $G^{(s)}$ ($s = 0, 1, 2, \ldots$), with $G \supset G^{(0)} \supset G^{(1)} \supset G^{(2)}$ etc. Then G is strongly broken down to $G^{(0)}$, with very heavy vector mesons generated in the process (masses of order $gF^{(0)}$), $G^{(0)}$ is less strongly broken down to $G^{(1)}$ with less heavy vector mesons, and so on. One of the advantages of unifying three forces into one is that the number of arbitrary parameters can be greatly reduced. The Weinberg angle, for example, can be determined in the SU(5) model to be about $\sin^2 \theta_w = 0.20$.[17] The observable effects of the super-heavy vector mesons are supposed to be very small at ordinary energies.

One of the common features of GUTs is that quarks and leptons become amalgamated into the same multiplet. The local gauge invariance then leads to new types of gauge field with new properties. In the original SU(5) model, for example, there were twenty-four unified force fields. Twelve quanta of these fields were already known: the photon, the two W bosons, the Z boson, and eight gluons. The remaining twelve were new, and were given the collective name X. The X bosons were needed to maintain the larger gauge symmetry that mixed the quarks with the leptons. They can therefore change quarks into leptons or vice versa. That is, the X bosons provided a mechanism for baryon number non-conservation. One of its direct consequences was proton decay, which offered cosmologists a new perspective on the old problem of the cosmic baryon excess, namely, the observed matter–antimatter asymmetry might have developed from symmetric beginnings.

GUTs had many other implications for cosmology because these models directed physicists' attention to the early universe as the supreme high-energy laboratory in which particles with mass $M_x \sim 10^{15}$ GeV could be created. As a framework for understanding the structure and evolution of the universe, GUTs, although they were theoretically self-consistent, were unable to generate a sufficient baryon asymmetry to be consistent with the observed data, or give the massive neutrino solution to the universe's 'missing mass' problem. More complicated rival models based on larger and larger unifying groups were proposed, offering new possibilities: new heavy quarks and leptons and lots of massive Higgs bosons, etc. Since these particles were inaccessible experimentally, no choice could be made between rival models. Whether this fact has betrayed the possibility that GFP has been carried beyond its empirical basis is, however, not without controversy.

No matter how successful and comprehensive GUTs were, they did not exhaust the potential of GFP. Soon after GUTs entered onto the stage and GFP became a new orthodoxy in fundamental physics after the November revolution of 1974, more ambitious efforts were made to extend GFP to gravity. Before discussing this extension, however, an examination of physicists' attitudes toward the universality of the gauge principle is in order.

The claim that the gauge principle is universally applicable was challenged by some physicists when external symmetry was involved. There are two questions: (i) Are the gravitational interactions dictated by the principle of symmetry, that is, by the principle of general covariance? (ii) Can we gauge an external symmetry to get a quantum of gravity?

Concerning the first question, Einstein's original claim that general covariance, together with the equivalence principle, led to general relativity was criticized by Kretschmann as early as 1917 (see section 4.2). It was also challenged later on by Cartan, Wheeler and Michael Friedman, among others.[18] These critics argued that the principle of general covariance was of no physical content whatever since all non-covariant theories could always be recast in a covariant form. The only thing we should do, they claimed, was to replace ordinary derivatives by covariant derivatives, adding an affine connection Γ^k_{ij} into the theory and saying that there existed coordinate systems in which the components of the affine connection Γ^k_{ij} happened to be equal to zero, and the original non-covariant formalism was valid.

But some apologists argued that the gravitational interaction was actually dictated by a local external symmetry, because in the reformulation of non-covariant theories, the critics had to admit that the affine connection Γ^k_{ij} must vanish. But this was equivalent to assuming that spacetime was flat and the external symmetry was global. True general covariance, however, was a local symmetry, the only type of symmetry that could be satisfied by the curved spacetime in which gravity could be accommodated. Thus the supposed general covariance of these reformulated theories was spurious.

The second question was also not without controversy. But many attempts were made, different symmetry groups being tried by committed followers of the gauge principle, to find a quantum theory of gravity.[19] Among these, work by F. W. Hehl and his collaborators proved that the Poincaré group, i.e. the group of Lorentz transformations and translations, could lead to the most plausible gauge theory of gravity, provided it was interpreted actively.

The pity was that all earlier attempts to cast gravity in the form of a gauge field failed to renormalize their models.[20] The situation was reminiscent of the weak force before it was unified with electromagnetism. Both gravity and the weak force were non-renormalizable. In the case of the weak force, however, once an appropriate gauge symmetry $(SU(2) \times U(1))$ was found, the infinities in the resulting unified electroweak theory fell away. Guided by this lesson, theorists began to search for a more powerful group than the existing ones of external symmetry, which would render gravity renormalizable.

Part of the difficulty of unifying gravity with other forces is that the respective theories simply have too little in common, rather than that they are in conflict. The symmetries adopted for gauging gravity are different in kind from those for gauging other forces. They are external symmetries, while all the other gauge symmetries are internal and not associated with any coordinate transformation in spacetime. In the mid-1960s, some efforts were made to achieve a union of an external and an internal symmetry group.[21] It was proved, however, that under a number of rather general conditions, a unified group (G) of the Poincaré group (P) and an internal symmetry group (T) can only be achieved in the trivial sense of a direct product: $G = T \otimes P$. A number of no-go theorems were derived, which forbade a symmetry group G to contain T and P in a non-trivial way (i.e., not as a direct product) so that the imposition of G would allow for the possibility of a unified description of particles with different internal symmetries.[22]

Such no-go theorems can be circumvented, at least partly, by introducing supersymmetry.[23] While an internal symmetry only relates particles with the same spin, supersymmetry relates fermions to bosons by incorporating non-commutative numbers as an essential element. As a consequence, the super-symmetry supermultiplet consists of a series of different internal symmetry multiplets, although supersymmetry itself is just a special kind of external symmetry (see below), and not a genuine union of an external and an internal symmetry.[24]

To see the last statement we only have to remember that in QFT boson fields have dimension 1 and fermion fields dimension 3/2.[25] It is not difficult to see that two supersymmetry transformations that relate bosons to fermions lead to a gap of one unit of dimension. The only dimensional object, which is different from fields and available to fill the gap, is the derivative. Thus, in any globally supersymmetric model, we can always find, purely on dimensional grounds, a derivative appearing in a double transformation relation. Mathematically, therefore, global supersymmetry resembles taking the square root of the translation operator. That is why it is taken as an enlargement of the Poincaré group (the so-called 'super-Poincaré group') rather than an internal symmetry.

In a local theory the translation operator differs from point to point. This is precisely the notion of a general coordinate transformation which leads some physicists to expect that gravity may be present. Indeed, guided by the requirement of local supersymmetry invariance and using 'Noether's method', some physicists obtained the massless spin-3/2 (gravitino) field gauging super-symmetry and the massless spin-2 (graviton) field gauging spacetime symmetry. Thus the local gauge theory of supersymmetry implies a local gauge theory of gravity. This is why a local supersymmetry theory is called supergravity.

In simple supergravity, the number of supersymmetry generators N is one. If N is larger than one, then the theory is called extended supergravity. One of the special features of the extended supergravity theories is that they have, in addition to the spacetime symmetries related to gravity, a global $U(N)$ internal symmetry, which relates all the particles of the same spin and hence has nothing to do with supersymmetry. It was shown that the internal symmetry could be made local so that the non-gravitational interactions could be incorporated. Some physicists were greatly excited by the discovery in the $N = 8$ model of an extra local SU(8) symmetry, hoping that the local SU(8) group could produce the spin-1 and spin-1/2 bound states that were needed for grand unified theories.[26] In spite of all these developments, however, the relation between supersymmetry and the extra internal symmetry remained unclear. Thus the unification of external and internal symmetries was still out of reach, and the relationship between gravitational and other forces remained vague, within the context of supergravity.

Is supersymmetry a powerful enough symmetry to render supergravity renormalizable? The preliminary results were encouraging. In extended supergravity, it was shown that the S-matrix in the first- and second-order quantum corrections cancels owing to the symmetry between fermions (gravitinos) and bosons (gravitons).[27] That is, the new diagrams involving gravitinos canceled the divergences caused by gravitons. Despite early optimism, however, physicists' enthusiasm for supergravity quickly waned. Since the mid-1980s, with the advent of superstring theories, few physicists believe that the problem of quantum gravity can be resolved within the framework of supergravity.

11.3 GFP as a synthesis of GP and QFP

In part I we examined the geometrical programme (GP), according to which interactions are realized through continuous classical fields that are inseparably correlated to geometrical structures of spacetime, such as the metric, the affine connection, and curvature. This concept of geometrizing interactions led to a deep understanding of gravitation, but had no success elsewhere in physics. Indeed, theories of electromagnetic and weak and strong nuclear interactions took a totally different direction, in the late 1920s, with the advent of quantum electrodynamics initiating the quantum field programme for fundamental interactions (QFP), which we examined in part II. According to QFP, interactions are realized through quantum fields, which underlie the local couplings and propagations of field quanta, but have nothing to do with the geometry of spacetime.

The conceptual difference between the two programmes is so deep that we have to say that there occurs an ontological shift between the two. Here we

come across a typical conceptual revolution: two successive research programmes have different ontological commitments, and based on these they
have different mechanisms for conveying fundamental interactions. The historical fact notwithstanding, I would argue that an intimate connection between
quantum interactions and the geometrical structures of spacetime can still be
found, an argument that is far from endorsing the popular thesis of incommensurability.

Logically, there can be three approaches to establishing such a connection,
which have in fact been explored by physicists in the last four decades. First,
one can try to define the quantum field on a curved spacetime manifold rather
than on the flat Minkowskian one.[28] In this case, since the curved spacetime
manifold is a fixed, classical background structure within which the theory is to
be formulated, the only difference of the new formulation from conventional
QFT lies in effects that the geometrical and topological structures of the non-
Minkowskian manifold have on systems that are otherwise quantized in a
conventional way. Yet such a fixed view of the background structure is in direct
conflict with the true spirit of GTR, according to which the structures of
spacetime are themselves dynamical entities. More pertinent to our discussion,
however, is the fact that interactions in this formulation are not geometrical in
nature.

Second, one can take the structures of spacetime as quantum variables rather
than a fixed background and subject them to the dynamical law of quantum
mechanics. In the 1960s, the canonical approach to quantizing gravity along
this line led to the famous Wheeler–DeWitt equation.[29] Yet, in addition to the
failure to be a renormalizable formulation, the split into three spatial dimensions and one time dimension required by canonical quantization is contrary to
the spirit of relativity.[30] Moreover, there is an interpretation problem concerning the meaning of equal-time commutation relations. These relations are well
defined for matter fields only on a fixed spacetime geometry. But once the
geometrical structures are quantized and obey the uncertainty principle, it
would make no sense to say that two points are spacelike separated. In the
1970s when the path integral approach to quantization became popular,
Hawking, on the basis of the path integral over all Riemannian four-metrics on
the four-manifold M and the sum over every M with a certain required
boundary, formulated his 'Euclidean' approach to quantum gravity (1979) and,
when the manifold has only one single, connected three-boundary, obtained a
'creation ex nihilo' theory about the universe.[31] But the formulation failed to
be renormalizable too.

In response to the repeated failure to quantize gravity, some physicists, such
as Penrose (1987), seek a way out by radically revising the structure of quantum

theory; others, such as the superstring theorists,[32] are trying totally different physical theory beyond the framework of QFT; still others, such as Chris Isham (1990), who try to see how far conventional QFT can be pushed and the extent to which the geometrical and topological ideas underlying GP can still be maintained in QFT, appeal to an extremely radical idea about the geometrical and topological structure of spacetime, that is, the idea that there are only finite points in the universe and there are only discrete topologies on finite sets of points, which can be interpreted as the structures of spacetime. Although it is not objectionable to go beyond the existing frameworks of quantum theory, quantum field theory, or spacetime theory, these responses have not yet resulted in any sustainable physical theories. For this reason, the second approach to synthesizing GP and QFP has not succeeded.

Third, and this is the central topic of this section, one can try to explore the geometrical implications of gauge theory. Although the framework of gauge theories is successful only in domains other than gravity, a geometrical interpretation of it still embodies a synthesis of GP and QFP in a proper sense. This is so because GFP itself is a sub-programme of QFP.

To pursue the synthesis thesis, however, we need only to focus on the essential element that differentiates GFP from QFP in general, that is, focus on the gauge principle.

Gauge invariance requires the introduction of gauge potentials to compensate the additional changes of internal degrees of freedom at different spacetime points. This mimics GTR, where taking the group of all admissible coordinate transformations as the symmetry group requires the introduction of the gravitational potential[33] to compensate the additional changes of spatio-temporal degrees of freedom. Gravitational potentials in GTR are correlated with a definite kind of geometrical structure, that is, the linear connection in four-dimensional curved spacetime. Gauge potentials can also be interpreted as being correlated with a certain type of structure, that is, the connection on a fiber bundle which is a generalization of spacetime. The deep similarity in theoretical structures between GTR and gauge theories allows the possibility that gauge theory is geometrical in nature. This impression is deepened by the fact that GTR itself can be viewed as a special case of gauge theory. Thus Yang, one of the founders of GFP, has repeatedly talked about the 'geometrical nature of gauge fields' and advocated the 'geometrization of physics' through GFP (Yang, 1974, 1977, 1980, 1983; Wu and Yang, 1975). He has even claimed that the gauge field is exactly the geometrical structure that Einstein was striving to find to give rise to his unified field theory (1980).

The starting point of this kind of claim is, of course, the geometrization of gravity, which is well established at least in a weak sense (see section 5.2). Making Poincaré symmetry local, which is equivalent to the requirement of general covariance, removes the flatness of spacetime and requires the introduction of non-trivial geometrical structures, such as connections and curvatures that are correlated with gravity.

For other fundamental forces that can be described by gauge potentials, it is easy to find that the theoretical structures of the corresponding theories are exactly parallel to those of gravity. There are internal symmetry spaces: phase space, for electromagnetism, which looks like a circle; isospace, which looks like the interior of a three-dimensional sphere; color space for strong interactions; etc. The internal space defined at each spacetime point is called a fiber, and the union of this internal space with spacetime is called fiber bundle space. If the internal space directions of a physical system at different spacetime points are assumed to be different, then the local gauge symmetries remove the 'flatness' of the fiber-bundle space and require the introduction of gauge potentials, which are responsible for the gauge interactions, to connect internal directions at different spacetime points. Since the role the gauge potentials play in fiber bundle space in gauge theory is exactly the same as the role the affine connection plays in curved spacetime in general relativity, Yang's geometrization thesis seems to be justifiable.

A critic might reply, however, that this is not a genuine geometrization because the geometrical structure in gauge theories is only defined on the fiber bundle space, which is only a kind of mathematical structure (a trivial product of spacetime and internal space, two factor spaces having nothing to do with each other), unlike real spacetime. Thus the so-called geometrization can only be taken as meaningless rhetoric.

The difference in opinion seems to be only terminological. Yet a closer examination reveals that it is deeper than this and has bearings on the way of perceiving fundamental physics.

To see this, let us start with a simple question: what is geometry? Before special relativity, the only physical geometry was Euclidean geometry in three-dimensional space. Other geometries, though logically possible, were viewed as only a kind of mental fiction. With special relativity came Minkowskian geometry in four-dimensional spacetime. No one today would deny that this is a genuine geometry. An important reason for this belief is that the four-dimensional spacetime on which the Minkowskian geometry is defined is not just a trivial product of three-dimensional space and one-dimensional time. Rather, the Lorentz rotation mixes the spatial and temporal indices. Riemannian geometry in four-dimensional spacetime is different again from

Minkowskian geometry. Is this also a genuine geometry, or just a kind of mathematical trick for describing gravitation against the background of Minkowskian geometry?

One of the lessons general relativity teaches us is that the gravitational interactions are of necessity reflected in the geometrical character of spacetime. Even if we start with Minkowskian geometry and reformulate general relativity in QFT, in which gravity is described by a massless spin-2 field, the resultant theory always involves a curved metric, not the originally postulated flat metric.[34] The relation between flat and curved metrics here is quite similar to the relation between bare and dressed charges in QFT. So Riemannian geometry should be viewed as a genuine geometry of spacetime. Or, equivalently, we should recognize that the character of the genuine geometry of spacetime is actually affected by gravity. This is precisely what Riemann speculated in his famous inaugural lecture and what Einstein insisted throughout his life.

Then what is the situation when internal symmetries (or equivalently, when interactions other than gravitation) are involved? To properly assess the situation, a few words on the further evolution of ideas of geometry after general relativity are in order.

In 1921, Eddington introduced an idea about 'the geometry of the world-structure, which is the common basis of space and time and things'. In this broadest sense, the Eddingtonian geometry contains elements correlated both with matter and with interaction mechanisms of material systems. Such a geometry was explored by Einstein and Schrödinger in their unified field theories, and by Kaluza and Klein in their five-dimensional relativity (see section 5.3). As we shall see, interest in the Eddingtonian geometry has revived in the last two decades in the fiber bundle version of gauge theory,[35] and in modern Kaluza–Klein theory.[36]

In the original Yang–Mills theory, the so-called differential formalism, internal degrees of freedom, though generalized from the simple phase of complex numbers to the generators of a Lie group, had nothing to do with spacetime. Yang and Mills merely generalized quantum electrodynamics, knowing nothing of its geometrical meaning. But in 1967 a contact point between internal and external degrees of freedom emerged. T. T. Wu and Yang in their (1967) paper suggested a solution to isospin gauge field equations: $b_{ia} = \Sigma_{ia\tau} f(r)/r$, where $\alpha = 1, 2, 3$, designates isospin index, i and τ are spacetime indices, and r is the length of the three-vector (x_1, x_2, x_3). This solution explicitly mixes isospin and space indices. Conceptually, this was a significant development. If the Minkowskian geometry can be regarded as a genuine geometry because it is that in which spatial and temporal indices are mixed by Lorentz rotation, then the Wu–Yang solution provided a wider basis for a new genuine geometry.

A further development in this direction was the monopole solution of an SU(2) gauge theory, suggested independently by 't Hooft and by Polyakov in 1974. In this case, the monopole solution relates directions in physical space with those in internal space. The same type of correlation happens also in the instanton solutions of gauge theory (cf. section 10.4).

The geometrical implication of gauge theory became clearer when an integral formalism for gauge fields was suggested by Yang (1974). The essential point of the integral formalism is the introduction of the concept of a non-integrable phase factor, which is defined as $\phi_{A(A+dx)} = I + b\mu_\mu^k(x)X_k\,dx^\mu$, where $b_\mu^k(x)$ are the components of a gauge potential, and X_k are the generators of the gauge group. With the help of this concept, Yang found that (i) the parallel displacement concept of Levi-Civita was a special case of a non-integrable phase factor with the gauge group GL(4); (ii) linear connection was a special type of gauge potential; and (iii) Riemannian curvature was a special case of gauge field strength. Although Yang's specific GL(4) gauge theory of gravity was criticized by R. Pavelle (1975) for differing from Einstein's theory and being in possible conflict with observations, these general ideas of Yang's are widely accepted. This work was impressive in showing that the concept of gauge fields is deeply geometrical. This impression was strengthened by the work of Wu and Yang (1975) on the global formulation of gauge fields. Here the global geometrical connotations of gauge fields, including those implied by the Aharanov–Bohm effect and the non-Abelian monopoles, were explored and formulated in terms of fiber bundle concepts.[37]

Given the above developments culminating in the fiber-bundle version of gauge theory, it is clear that parallel to the external geometry associated with spatio-temporal degrees of freedom, there is an internal geometry associated with internal degrees of freedom. The two types of geometry can be put in the same mathematical framework. However, despite the impressive mathematical similarity and more impressive possibility of blending external with internal degrees of freedom, the real picture of the unification of the two types of geometry, which is correlated with the unification of gravitational and other fundamental interactions, is still quite vague.

The picture becomes much clearer in modern Kaluza–Klein theories. In the original Kaluza–Klein theory, an extra dimension of space, compactified in low-energy experiments, was grafted onto the known four-dimensional space-time in order to accommodate electromagnetism. In one of the revitalized Kaluza–Klein theories, the number of extra space dimensions becomes seven, the number of symmetry operations embodied in grand unified theories and extended $N = 8$ supergravity being taken into account. It is supposed that the seven extra space dimensions are compactified in low energy as a hypersphere. The seven-sphere contains many additional symmetries which are intended to model the underlying gauge symmetries of the force fields. This means that the internal symmetries are the manifestation of the geometrical symmetries associated with the extra compactified space dimensions, and that all kinds of geometries associated with internal symmetries are genuine space geometries, namely the geometries associated with extra space dimensions.

In this case, however, the question of what is the relation between internal and external geometries seems to be empty. But in a deeper sense the question remains profound. What is the real issue when we talk about the geometrization of gauge field theory? The geometrization thesis makes sense only when the geometrical structures in four-dimensional spacetime are actually correlated to the gauge interactions other than gravity or supergravity, or, equivalently, if they are mixed with the geometrical structures associated with extra dimensions. We shall not know whether this is the case or not until the hypothetical compactification scale, the Planck scale $T = 10^{38}$ GeV or so, is reached. So it will remain an open question until then. Nevertheless, the modern Kaluza–Klein theory does open the door to establishing the correlation between non-gravitational gauge potentials and the geometrical structures in four-dimensional spacetime via the geometrical structures in extra dimensions. Within this theoretical context, then, the geometrization thesis, which is related to, though not identified with, the unification of gravity with other gauge interactions, is in principle testable, and cannot be accused of being non-falsifiable or irrelevant to the future development of fundamental physics.

It is interesting to note that, in the context of a geometrical interpretation of GFP, what differentiates GFP from QFP is exactly what assimilates GFP to GP. Thus despite the radical difference between GFP and GP, as radical as the difference between QFP and GP, a deep inherent connection between them emerges. Both of them are field theories. Both of them are guided by symmetry principles and thus relate interactions to geometrical structures. In the transition from GP to QFP, of course, some underlying ideas are transformed: classical fields are replaced by quantum fields, spatio-temporal symmetries are generalized to gauge symmetries incorporating both spatio-temporal and internal symmetries, and so on. All these transformations are accomplished within a rival research programme, namely QFP. So it seems reasonable to regard GFP as a synthesis of GP and QFP if we express GFP as a research programme according to which interactions are realized through quantized gauge fields which are inseparably correlated to some generalized geometrical structures.

11.4 Stagnation and a new direction: effective field theories

After a short period of optimism in the late 1970s and early 1980s, GFP, as Weinberg pointed out, 'had failed to make further progress in explaining or predicting the properties of elementary particles beyond the progress that was already well in hand by the early 1970s'. Conceptually, he realized that 'we could understand its successes in the low-energy range, up to a TeV or so, without having to believe in quantum field theory as a fundamental theory' (1986a, b). But its difficulties, even in its most sophisticated form of the standard model, became more and more conspicuous from the mid-1980s on: the explanation of low-energy pion–nucleon interactions by QCD seems almost unattainable. Even the self-consistency of the electroweak theory itself seems also to be in a dubious situation, owing to the difficulties related with such issues as quark confinement in a four-dimensional spacetime, the existence of the Higgs particles, the generation problems, etc. Moreover, the unification of the electroweak with the strong interactions has been attacked but without success, let alone the quantization of gravity and its unification with other interactions.

In the meantime, it became much clearer to the most perspicacious theorists, such as Weinberg (1979), from the late 1970s that the conceptual foundations of QFT had undergone some radical transformations during the last four decades, as a result both of attempts to solve conceptual anomalies within the theory, and of fruitful interactions between QFT and statistical physics. Implicit assumptions concerning such concepts as regularization, the cutoff, dimensionality, symmetry, and renormalizability were clarified, and the original understanding of these concepts was transformed. New concepts, such as 'symmetry breakings' (either spontaneous or anomalous), 'renormalization group', 'decoupling of high-energy processes from low energy phenomena', 'sensible non-renormalizable theories', and 'effective field theories', were developed, drawing heavily on the dramatic progress in statistical physics. At the heart of these transformations was the emergence of the new concept of 'broken scale invariance' and the related renormalization group approach.

It was Weinberg (1978) who had first assimilated the physical insights, mainly developed by Fisher, Kadanoff, and Wilson in the context of critical phenomena (see section 8.8), such as the existence of the fixed-point solutions of renormalization group equations and the conditions for trajectories in coupling-constant space passing through fixed points, needed to explain or even replace the principle of renormalizability with a more fundamental guiding principle called 'asymptotic safety'. Yet this programme was soon overshadowed by another programme, that of 'effective field theory' (EFT), initiated also by Weinberg. At the beginning, EFT was less ambitious than asymptotically safe theories because it still took renormalizability as its guiding principle. Eventually, however, it led to a new understanding of renormalization and a serious challenge to the fundamentality of renormalizability, thus clarifying the theoretical structure of QFT and its ontological basis and, most importantly, introducing a radical shift of outlook in fundamental physics.

Foundational transformations

Cutoff

As we noted in section 7.6, the renormalization procedure consists essentially of two steps. First, for a given theory, such as QED, an algorithm is specified for an unambiguous separation of the ascertainable low-energy processes from the high-energy processes that are not known, the latter being describable only by new theories. Second, the incorporation of the effects of the neglected high-energy processes on the low-energy physics described by the theory is accomplished by redefining a finite number of parameters of the theory. The redefined parameters are not calculable from the theory, but can be determined by experiments (Schwinger, 1948a, b). The implicit requirements for the incorporation and redefinition to be possible will be examined below. Here our focus is on the separation.

For Schwinger (1948b), and Tomonaga (1946), who directly separated the infinite terms by contact transformations, the unknown contributions were simply represented by divergent terms with proper gauge and Lorentz transformation properties. There was, however, no clue whatsoever in their formulations as to where the boundary separating the knowable from the unknowable energy region lies. It is buried and hidden somewhere in the divergent integrals. Thus the incorporation and redefinition can only be viewed as a species of essentially formalistic manipulations of divergent quantities, with an extremely tenuous logical justification.[38]

Feynman, Pauli and Villars, and most other physicists, took an approach that was different from that taken by Schwinger and Tomonaga. They temporarily modified the theory with the help of a regularization procedure so as to make the integrals finite. In the momentum cutoff regularization scheme introduced by Feynman (1948c), and by Pauli and Villars (1949), the boundary line separating the knowable region from the unknown one was clearly indicated by the momentum cutoff introduced.[39] Below the cutoff, the theory was supposed to be trustworthy, and the integrals for the higher-order corrections could be justifiably manipulated and calculated. The unknown high-energy processes that occur above the cutoff were excluded from consideration as they have to be. Up to this point, Feynman's scheme seemed superior to Schwinger's in implementing the basic ideas of renormalization, which were first clearly stated by Schwinger. It also seemed to be more respectable logically and mathematically.

However, the following difficult question must be answered by various regularization schemes: How are the effects of the excluded high-energy processes upon the low-energy phenomena taken into account? This question is specific to local field theories and is unavoidable within that framework. Feynman's solution, which became the ruling orthodoxy, was to take the cutoff to infinity at the end of the calculation. In this way, all high-energy processes were taken into consideration, and their effects upon the low-energy phenomena could be incorporated by redefining the parameters that appeared in the specification of the theory's Lagrangian in the manner of Schwinger. The price for accomplishing this was that the cutoff could no longer be taken as the threshold energy, at which the theory stops being valid and new theories are required for a correct physical description. Otherwise a serious conceptual anomaly would arise: taking the cutoff to infinity would mean that the theory is trustworthy everywhere and high-energy processes are not unknowable. This is in direct contradiction with the basic idea of renormalization, and the divergent integrals that result when the cutoff is taken to infinity clearly indicate that this is not the case.[40]

The implications of taking the cutoff to infinity are significant. First of all, the boundary line separating knowable from unknowable regions becomes buried and hidden. It also changes the status of the cutoff from a tentative threshold energy to a purely formalistic device, and thus reduces the Feynman–Pauli–Villars scheme to Schwinger's purely formalistic one. In this case, one can take Feynman's momentum cutoff regularization, which replaced Schwinger's canonical transformations, as a more efficient formalistic algorithm for manipulating the divergent quantities. Or, equivalently, one can view Schwinger's direct identification of the divergent integrals as combining Feynman's two steps of introducing a finite cutoff, and then taking it to infinity. More significantly, the step of taking the cutoff to infinity also reinforces a prevailing formalistic claim that physics should be cutoff independent, and thus all explicit reference to the cutoff should be removed upon redefining the parameters. The claim seems compelling because the step has indeed deprived the cutoff-dependent quantities of any physical meaning. Conversely, the claim in turn allows one to take a formalistic interpretation of the cutoff, and forces its removal from physics.

What if the cutoff is taken seriously and interpreted realistically as the threshold energy for new physics? Then the orthodox formalistic scheme collapses and the entire perspective changes: the cutoff cannot be taken to infinity, and the obverse side of this same coin is that the physics cannot be claimed to be cutoff independent. In fact, important advances, since the mid-1970s, in understanding the physics and philosophy of renormalization have come from such a realist interpretation.[41] The strands of physical reasoning that led to this position were intertwined. Thus to clarify the conceptual situation centered around this issue, they have to be disentangled.

To begin with, let us examine the reason why it is possible to take a realist position on the cutoff. As we noted above, the motivation for taking the cutoff to infinity is to take into account the high-energy effects on low-energy phenomena, which are excluded by introducing a finite cutoff. If we can find other ways of retaining these effects while keeping the cutoff finite, then there is no compelling reason for taking the cutoff to infinity. In fact, the realist position has become attractive since the late 1970s precisely because theorists have gradually come to realize that the high-energy effects can be retained without taking the cutoff to infinity. The objective can be achieved by adding a finite number of new non-renormalizable interactions that have the same symmetries as the original Lagrangian, combined with a redefinition of the parameters of the theory.[42] It should be noted that the introduction of non-renormalizable interactions causes no difficulty, because the theory has a finite cutoff.

There is a price to be paid for taking the realist position. First, the formalism becomes more complicated by adding new compensating interactions. Yet there are only a finite number of new interactions that need be added because these are subject to various constraints. Moreover, the realist position is conceptually simpler than the formalistic one. Thus the cost is not very high. Second, the realist formalism is supposed to be valid only up to the cutoff energy. Since any experiment can only probe a limited range of energies, this limitation of the realist formalism has actually not caused any real loss in accuracy. Thus the apparent cost is illusionary.

The next question is how to articulate the physical realization of the cutoff so that ways can be found to determine its energy scale. The cutoff in realist theory is no longer a formalistic device or an arbitrary parameter, but acquires physical significance as the embodiment of the hierarchical structure of QFT, and as a boundary separating energy regions which are separately describable by different sets of parameters and different physical laws (interactions) with different symmetries. The discovery of spontaneous symmetry breaking (SSB) and the decoupling theorems (see below) suggests that the value of the cutoff is connected with the masses of heavy bosons which are associated with SSB. Since the symmetry breaking makes the otherwise negligible non-renormaliz-able interactions[43] detectable, owing to the absence of all other interactions that are forbidden by the symmetry, the energy scale of the cutoff can be established by measuring the strength of the non-renormalizable interactions in a theory.

The above preliminary arguments have shown that a realist conception of the cutoff is not an untenable position. However, a convincing proof of its viabil-ity is possible only when this conception is integrated into a new conceptual network that provides new foundations for understanding renormalizability, non-renormalizable interactions, and QFT in general. Let us turn to other strands in this network.

Symmetry and symmetry breaking

In the traditional procedure, after separating the invalid (divergent) parts from the valid (finite) parts of the solutions, the high-energy effects are absorbed into the modified parameters of the theory. For this amalgamation to be possible, however, the structure of the amplitudes that simulate the unknowable high-energy dynamics has to be same as that of the amplitudes responsible for the low-energy processes. Otherwise the multiplicative renormalizations would be impossible. To insure the required structural similarity, a crucial assumption has to be made and is in fact implicitly built into the very scheme of multiplicative renormalization. This is the assumption that the high-energy dynamics is constrained by the same symmetries as those that constrain the low-energy dynamics. Since the solutions of a theory constitute a represent-ation of the symmetry group of the theory, if different symmetries were displayed by the dynamics in different energy regions, this would imply different group-theoretical constraints and thus different structures for the solutions in different pieces of the dynamics. If this were the case, then the renormalizability of the theory would be spoiled.

In the case of QED, the renormalizability is secured by the somewhat mysterious universality of the U(1) gauge symmetry. With the discovery of symmetry breaking (SSB in the early 1960s and anomalous symmetry breaking (ASB) in the late 1960s), however, the situation became more and more complicated,[44] and demanded that the above general consideration about the relationship between symmetry and renormalizability had to be refined.

Consider SSB first. In condensed matter and statistical physics, SSB is a statement concerning the properties of the solutions of a dynamical system, namely, that some asymmetric configurations are energetically more stable than symmetric ones. Essentially, SSB is concerned with the low-energy behavior of the solutions and asserts that some low-energy solutions exhibit less symmetry than the symmetry exhibited by the Lagrangian of the system, while others possess the full symmetry of the system. Traced to its foundation, SSB is an inherent property of the system because the existence and the specification of the asymmetric solutions are completely determined by the dynamics and parameters of the system. They are connected to the hierarchical structure of the solutions, which, in statistical physics, is manifested in the phenomena of continuous (second-order) phase transitions.

In QFT, SSB makes physical sense only in gauge theories when continuous symmetries are involved. Otherwise, one of its mathematical predictions (the existence of massless Goldstone bosons) would contradict physical observations. Within the framework of gauge theories, all the statements concerning SSB mentioned above are valid. In addition, there is another important assertion that is of relevance to our discussion. In gauge theories, in contradistinction to the case of explicit symmetry breaking, diverse low-energy phenomena can be accommodated into a hierarchy with the help of SSB, without spoiling the renormalizability of the theory (see section 10.3). The reason for this is that SSB affects the structure of physics only at energies lower than the scale at which the symmetry is broken, and thus does not affect the renormalizability of a theory, which is essentially a statement of the high-energy behavior of the theory.

Generally speaking, ASB is the breakdown of a classical symmetry caused by quantum mechanical effects (see section 8.7). It is possible that some symmetries that the system possessed in its classical formulation may disappear in its quantized version, because the latter may introduce some symmetry-violating processes. In QFT these arise because of loop corrections and are related to the renormalization procedure and the absence of an invariant regulator.

ASB plays an important role in QFT. In particular, the desire to safeguard a symmetry from being anomalously broken can place a very strong constraint on model building (see section 11.1). If the symmetries concerned are local, such as gauge symmetries and general covariance, then the occurrence of ASB is fatal because the renormalizability of the theory is spoiled. If the symmetries concerned are global, then the occurrence of ASB is harmless or even desirable, as in the case of global γ_5 invariance for explaining $\pi^0 \rightarrow \gamma\gamma$ decay, or in the case of scale invariance in QCD with massless quarks for obtaining massive hadrons as bound states. The most profound implication of scale anomaly is undoubtedly the concept of the renormalization group, which, however, we have reviewed in section 8.8 already.

The decoupling theorem and effective field theories (EFT)

According to the renormalization group approach, different renormalization prescriptions only lead to different parameterizations of a theory. An important application of this freedom in choosing a convenient prescription is expressed in the decoupling theorem, first formulated by Symanzik in (1973), and then by T. Appelquist and J. Carazzone in (1975). Based on the power-counting argument and concerned with renormalizable theories in which some fields have masses much larger than the others, the theorem asserts that in these theories a renormalization prescription can be found such that the heavy particles can be shown to decouple from the low-energy physics, except for producing renormalization effects and corrections that are suppressed by a power of the experimental momentum divided by the heavy mass.

An important corollary to this theorem is that low-energy processes are describable by an EFT that incorporates only those particles which are actually important at the energy being studied. That is, there is no need to solve the complete theory describing all light and heavy particles (Weinberg, 1980b).

An EFT can be obtained by deleting all heavy fields from the complete renormalizable theory and suitably redefining, with the help of the renormalization group equations, the coupling constants, masses, and the scale of the Green's functions. Clearly, a description of physical processes by an EFT is context dependent. It is delimited by the experimental energy available, and is thus able to keep close track of the experimental situation. The context dependence of EFT is embodied in effective cutoffs which are represented by heavy masses associated with SSB. Thus with the decoupling theorem and the concept of EFT emerges a hierarchical conception of nature offered by QFT, which explains why a physical description at any one level is stable and undisturbable by whatever happens at higher energies, and thus justifies the use of such descriptions.

There seems to be an apparent contradiction between the idea underlying the concept of renormalization group and the idea underlying the concept of EFT. While the former is predicated on the absence of a characteristic scale in the system under consideration, the latter takes seriously the mass scales of heavy particles as physical cutoffs or characteristic scales separating different validity domains of EFTs. The contradiction disappears immediately, however, if one remembers that heavy particles still make contributions to the renormalization effects in EFT. Thus the mass scales of heavy particles in an EFT should only be taken to be pseudo-characteristic rather than genuinely characteristic scales. The existence of such pseudo-characteristic scales reflects a hierarchical ordering of couplings at different energy scales, but it does not change the essential feature of systems described by QFT, namely, the absence of characteristic scales and the coupling of fluctuations at various energy scales (see section 8.8). While some couplings between fluctuations at high- and low-energy scales exist universally and manifest themselves in the renormalization effects in low-energy physics, others are suppressed and reveal no observable clues in low-energy physics.

The above assertion that the decoupling is not absolute is reinforced by the important observation that the influence of heavy particles on the low-energy physics is directly detectable in some circumstances. If there are processes (e.g. those of weak interactions) that are exactly forbidden by symmetries (e.g. parity, strangeness conservation, etc.) in the absence of the heavy particles that are involved in symmetry-breaking interactions leading to these processes (e.g. W and Z bosons in the weak interactions), then the influence of the heavy particles on the low-energy phenomena is observable, although, owing to the decoupling theorem, suppressed by a power of energy divided by the heavy mass. Typically, these effects can be described by an effective non-renormalizable theory (e.g. Fermi's theory of weak interactions), which, as a low-energy approximation to the renormalizable theory (e.g. the electroweak theory), possesses a physical cutoff or characteristic energy scale set by the heavy particles (e.g. 300 GeV for Fermi's theory). When the experimental energy approaches the cutoff energy, the non-renormalizable theory becomes inapplicable, and new physics appears that requires for its description either a renormalizable theory or a new effective theory with a higher cutoff energy. The first choice represents the orthodoxy. The second choice presents a serious challenge to the fundamentality of the principle of renormalizability, and is gaining momentum and popularity at the present time.

A challenge to renormalizability

The concept of EFT clarifies how QFT at different scales takes different forms, and allows two different ways of looking at the situation. First, if a renormalizable theory at high energy is available, then the effective theory at any lower energy can be obtained in a systematic way by integrating out the heavy fields of the theory. Thus the renormalizable electroweak theory and QCD, understood as effective theories at low energy of some grand unified theory, have lost their presumed status of being fundamental theories. Another possibility, also compatible with this way of looking at the situation, is to assume that there exists a tower of effective theories that contain non-renormalizable interactions, each with fewer numbers of particles and with more small non-renormalizable interaction terms than the last. When the physical cutoff (heavy particle mass

M) is much larger than the experimental energy E, the effective theory is approximately renormalizable because the non-renormalizable terms are suppressed by a power of E/M.

The second way corresponds more closely to what high-energy theorists actually do in their investigations. Since nobody knows what the renormalizable theory at the unattainable higher energies is, or even whether it exists at all, they have to probe the accessible low-energies first, and design representations that fit this energy range. They extend the theory to higher energies only when it becomes relevant to the understanding of physics. This process of practice is embodied in the concept of an endless tower of theories,[45] in which each theory is a particular response to a particular experimental situation, and none can ultimately be regarded as the fundamental theory. According to this conception, the requirement of renormalizability is replaced by a condition on the non-renormalizable interactions in EFT: all non-renormalizable interactions in an effective theory describing physics at a scale m must be produced by heavy particles with a mass scale M ($\gg m$), and are thus suppressed by powers of m/M. Furthermore, in the renormalizable effective theory including the heavy particles with mass M, these non-renormalizable interactions must disappear.

These clarifications, together with the renormalization group equations, have helped physicists to come to a new understanding of renormalization. As Gross (1985) puts it, 'renormalization is an expression of the variation of the structure of physical interactions with changes in the scale of the phenomena being probed'. Notice that this new understanding is very different from the old one, which focused exclusively on the high-energy behavior and on ways of circumventing the divergences. It shows a more general concern with the finite variations of the various physical interactions with finite changes of energy scales, and thus provides enough leeway for considering non-renormalizable interactions.

A significant change in the attitude of physicists with respect to what should be taken as guiding principles in theory construction has taken place in recent years in the context of EFT. For many years, renormalizability had been taken as a necessary requirement for a theory to be acceptable. With the realization that experiments can probe only a limited range of energies, EFT appears to many physicists a natural framework for analyzing experimental results. Since non-renormalizable interactions occur quite naturally within this framework, there is no a priori reason to exclude them when constructing theoretical models to describe currently accessible physics.

In addition to being congenial and compatible with the new understanding of renormalization, taking non-renormalizable interactions seriously is also supported by some other arguments. First, non-renormalizable theories are malleable enough to accommodate experiments and observations, especially in the area of gravity. Second, they possess predictive power and are able to improve this power by taking higher and higher cutoffs. Third, because of their phenomenological nature, they are conceptually simpler than the renormalizable theories, which, as stressed by Schwinger, involve physically extraneous speculations about the dynamic structure of the physical particles. The fourth supportive argument comes from constructive theorists who, since the mid-1970s, have helped to understand the structure of non-renormalizable theories, and to discover conditions in which a non-renormalizable theory can make sense (see section 8.4).

The traditional argument against non-renormalizable theories is that they are undefinable at energies higher than their physical cutoff. So let us look at the high-energy behavior of non-renormalizable interactions, which many physicists have regarded as one of the most fundamental questions in QFT.

In the initial framework of EFT, non-renormalizable theories were only taken as auxiliary devices. When the experimentally available energy approaches their cutoffs and new physics begins to appear, they become incorrect and have to be replaced by renormalizable theories. In the framework of Weinberg's asymptotically safe theories, non-renormalizable theories have acquired a more fundamental status. Nevertheless, they still share a common feature with EFT, namely, all the discussion of them is based on taking the cutoff to infinity, and thus fall into the category of formalistic interpretations of the cutoff.

However, if we take the idea underlying EFT to its logical conclusion, then a radical change in outlook takes place, a new perspective appears, a new interpretation of QFT can be developed, and a new theoretical structure of QFT waits to be explored. Thoroughgoing advocates of EFT, such as Georgi (1989b) and Lepage (1989) argue that when the experimentally available energy approaches the cutoff of a non-renormalizable effective theory, it can always be replaced by another non-renormalizable effective theory with a much higher cutoff. In this way, the high-energy behavior of non-renormalizable interactions above the cutoff can be properly taken care of by (i) the variation of the renormalization effects, caused by the change of the cutoff and calculable by the renormalization group equations, and (ii) additional non-renormalizable counter-terms.[46]

Thus, at any stage of development, the cutoff is always finite and can be given a realist interpretation. In addition to the finite cutoff, two new ingredients, which are absent or forbidden in the traditional structure of QFT, become legitimate and indispensable in the theoretical structure of EFT. These are (i) the variations of renormalization effects with specific changes of cutoff, and (ii) the non-renormalizable counter-terms which are legitimatized by the introduction of finite cutoffs.

Some difficulties with the new conception of renormalization should be mentioned. First, its starting assumption is that the renormalization group equations have fixed-point solutions. But there is no guarantee that such fixed points exist in general. So the ground for the whole edifice is not so solid. Second, EFT is justified by the decoupling theorem. The theorem, however, encounters a number of complicated difficulties when symmetry breaking, whether spontaneous or anomalous, is involved, which is certainly the case in contemporary field-theoretical models.[47] Third, the decoupling argument has not addressed the smallness assumption, namely that the divergences incorporated in the renormalization effects (which also exist in the case of decoupling) are actually small effects. Finally, the long-standing difficulty of local field theories, first raised by Landau, namely the zero-charge argument, remains to be addressed.

Atomism and pluralism

Various models developed within the framework of QFT to describe the subatomic world remain atomistic in nature. The particles described by the fields appearing in Lagrangians are to be regarded as the elementary constituents of the world. The EFT approach, in some sense, extends the atomistic paradigm further because within this framework the domain under investigation is given a more discernible and a more sharply defined hierarchical structure. The hierarchy is delimited by mass scales associated with a chain of SSB and is justified by the decoupling theorem.

The decoupling theorem does not reject the general idea of causal connections between different hierarchical levels. In fact, such connections are assumed to exist – most noticeably, they manifest themselves in the renormalization effects of a high-energy process upon low-energy phenomena – and to be describable by the renormalization group equations; they are thus built into the very conceptual basis of the theorem. It is the attempt to give the connections universal significance and the stipulation of their direct relevance to scientific inquiry that is rejected. More precisely, what is to be rejected is the suggestion that it is possible simply by means of this kind of connection to infer the complexity and the novelty that emerge at the lower-energy scales from the simplicity at higher-energy scales without any empirical input. The necessity, as required by the decoupling theorem and EFT, of an empirical input into the theoretical ontologies applicable at the lower energy scales, to which the ontologies at the higher energy scales have no direct relevance in scientific investigations, is fostering a particular representation of the physical world. In this picture the latter can be considered layered into quasi-autono-mous domains, each layer having its own ontology and associated 'fundamen-tal' laws. The ontology and dynamics of each layer are quasi-stable, almost immune to whatever happen in other layers. Thus the name quasi-autonomous domains.

An examination of the EFT-based hierarchical structure from a metaphysical perspective yields two seemingly contradictory implications. On the one hand, the structure seems to lend support to the possibility of interpreting physical phenomena in a reductionist or even reconstructionist way, at least to the extent SSB works. Most efforts expended in the past two decades by the mainstream of the high-energy physics community, from the standard model to superstring theories, can be regarded as exploring this potential. In a weak sense, therefore, such a hierarchical structure still falls into the category of atomism.

On the other hand, taking the decoupling theorem and EFT seriously would endorse the existence of objective emergent properties, which entails a pluralist view of possible theoretical ontologies.[48] This in turn has set an intrinsic limit to the reductionist methodology. Thus the development of QFT as an atomistic-reductionist pursuit, dictated by its inner logic, has reached a critical point, at which, ironically, its own reductionist foundations have been somewhat under-mined.

It is the strong anti-reductionist commitment concerning the relationship between different levels that differentiates the pluralist version of atomism nurtured by EFT from the crude version of atomism adopted by conventional QFT, the constitutive components of which are reductionism and reconstruc-tionism. In addition, the emphasis on empirical inputs, which are historically contingent, also sharply contrasts the hierarchical-pluralist version of atomism with the Neoplatonist mathematical atomism. That in the traditional pursuit of quantum field theorists Neoplatonism is always implicitly assumed is testified by the prevailing conviction that ahistorical mathematical entities should be taken to be the ontological foundation of their investigations, from which em-pirical phenomena can be deduced.

Three approaches to the foundations of QFT

The foundational transformations and related conceptual developments re-viewed in this section have provided fertile soil for the acceptance and the further development of Schwinger's insightful ideas, which were advanced in his criticism of the operator formulation of QFT and renormalization and detailed in the presentation of his source theory. Schwinger's views strongly influenced Weinberg in his work on the phenomenological Lagrangian approach to chiral dynamics and on EFT. We can easily find three features shared by Schwinger's source theory and EFT: first, their denial of being fundamental theories; second, their flexibility in being able to incorporate new particles and new interactions into existing schemes; and third, the capacity of each of them to consider non-renormalizable interactions.

However, a fundamental difference exists between the two schemes. EFT is a local operator field theory and contains no characteristic scale, and thus has to deal with the contributions from fluctuations at arbitrarily high energy. The behavior of local couplings among local fields at various momentum scales can be traced by the renormalization group equations, which often, though not always, possess a fixed point. If this is the case, the local field theory is calculable, and is able to make effective predictions. Thus for EFT the notion of an effective local operator field is acceptable. By contrast, such a notion in Schwinger's theory is totally rejected. Schwinger's theory is a thoroughly phenomenological one, in which the numerical field, unlike the operator field, is only responsible for the one-particle excitation at low energy. There is therefore no question of renormalization in Schwinger's theory. By contrast, in the formulation of EFT, renormalization, as we noticed above, has taken on a more and more sophisticated form, and has become an ever more powerful calculational tool.

If we take EFT seriously as suggesting a new world picture, a new conception of the foundations of QFT, then what some of the conceptual difficulties mentioned above represent are unlikely to be normal puzzles that can be solved by the established methodology. What is required in dealing with these conceptual problems, it seems, is a drastic change of our conception of fundamental physics itself, a change from aiming at a fundamental theory (as the foundation of physics) to having effective theories valid at various energy scales.

Many theorists have rejected this interpretation of EFT. For Gross (1985) and Weinberg (1995), EFTs are only the low-energy approximations of a deeper theory, and can be obtained from it in a systematic way. An interesting point, however, is worth noticing. Although they believe that within the reductionist methodology, ways out of conceptual difficulties facing high-energy physics will be found sooner or later, with the help of more sophisticated mathematics or novel physical ideas, or simply by digging deeper and deeper into the layers of subquark physics, both of them have lost their confidence in QFT as the foundation of physics, and conceive a deeper theory or a final theory not as a field theory, but as a string theory, although the latter at this stage cannot be properly taken as a physical theory.

A question raised thereby, with profound implications and worthy of exploration, is this: from the string theorists' point of view, what kind of defect in the foundations of QFT deprives QFT of its status as the foundation of physics? For some more 'conservative' mathematical physicists, such as Wightman (1992) and Jaffe (1995), and many other physicists, this question is non-existent because, they believe, by more and more mathematical elaborations, a consistent formulation of QFT, most likely in the form of gauge theory, can be established and continue to serve as the foundation of physics.

Thus, at the present, there are essentially three approaches to the questions of the foundations of QFT, of QFT as the foundation of physics, and of the reasons why QFT can no longer be taken as the foundation of physics. In addition to physical investigations, the assessment of the three approaches also requires a philosophical clarification about reductionism and emergentism, which, however, is beyond the scope of this volume.[49]

Notes

1. For a consistent picture of hadrons, some neutral parts (gluons, which, some time later, were identified as gauge bosons for interquark forces) were also required. For theoretical analyses, see Callan and Gross (1968), Gross and Llewelyn-Smith (1969), and Bjorken and Paschos (1969, 1970). For experimental reports, see Bloom *et al.* (1969) and Breidenbach *et al.* (1969).
2. The rapporteur talks mentioned by Pais can be found in the *Proceedings of 1968 International Conference on High Energy Physics* (Vienna), and the *Proceedings of 1970 International Conference on High Energy Physics* (Kiev).
3. See Veltman (1992).
4. The fourth quark was first introduced by Y. Hara (1964) to achieve a fundamental lepton–baryon symmetry with respect to the weak and electromagnetic interactions, and to avoid the undesirable neutral strangeness-changing currents. It became known as the charmed quark owing to the paper by Bjorken and Glashow (1964) and was made famous by the paper of Glashow, Illiopolos and Maiani (1970).
5. Hasert *et al.* (1973a, b).
6. Aubert *et al.* (1974), Augustin *et al.* (1974), and Abrams (1974).
7. Goldhaber *et al.* (1976), and Peruzzi *et al.* (1976).
8. Arnison *et al.* (1983a, b) and Bagnaia *et al.* (1983).
9. The name QCD first appeared in Marciano and Pagels (1978) and was attributed to Gell-Mann.
10. The SU(6) group contains an SU(3) subgroup transforming quark species (or flavor) for fixed spin and an SU(2) subgroup transforming quark spin for fixed species (or flavor) because quarks were supposed to form flavor SU(3) triplets and spin doublets.
11. Bouchiat, Illiopoulos, and Meyer (1972); and Gross and Jackiw (1972).
12. Adler and Bardeen (1969).
12a. See Adler (1969).
13. They noticed that since the deep inelastic experiments indicated that the charged constituents of the nucleon were quarks, the neutral gluons had to be flavor neutral. This was not an argument for their color non-neutrality, but the gluons would otherwise be dynamically totally idle.
14. Other examples in the realm of QCD are provided by the Han–Nambu proposal and

't Hooft's discovery of the asymptotic freedom of non-Abelian gauge theories; in the realm of the electroweak theory they are provided by the Glashow–Weinberg–Salam model.

15. Appelquist and Politzer (1975), Harrington, Park, and Yildiz (1975), Eichten, Gottfried, Kinoshita, Koght, Lane, and Yan (1975).

16. See, for example, Fritzsch and Minkowski (1975); Gürsey and Sikivie (1976); Gell-Mann, Ramond and Slansky (1978).

17. Weinberg (1980a).

18. See Misner, Thorne, and Wheeler (1973) and Friedman (1983).

19. See, for example, Utiyama (1956) and Yang (1974).

20. Cf. note 49 of chapter 10.

21. By an analogy with SU(4) symmetry in nuclear physics proposed by Wigner in (1937), which is a union of spin rotations and isospin rotations, B. Sakita (1964) and F. Gürsey and L. A. Radicati (1964) suggested a non-relativistic static SU(6) symmetry for the quark model which contained an internal symmetry group SU(3) and the spin SU(2) group as its subgroups. Since spin is one of the two conserved quantities of the Poincaré group (the other is mass), this was regarded as a kind of union of an external and an internal symmetry.

22. See, for example, McGlinn (1964), Coleman (1965), and O'Raifeartaigh (1965). For a review, see Pais (1966).

23. The idea of supersymmetry as a global symmetry first appeared in work by Ramond (1971a, b), and by Neveu and Schwarz (1971a, b) on the dual model in the tradition of the S-matrix theory. The idea was then extended by Wess and Zumino to QFT (1974). One of its most surprising properties was that two supersymmetry transformations led to a spacetime translation.

24. Cf. van Nieuwenhuizen (1981).

25. The convention comes from the demand that the action be dimensionless (in units $h = c = 1$), together with another convention that boson fields have two derivatives in the action, while fermion fields have only one.

26. For example, Stephen Hawking in his inaugural lecture (1980) took $N = 8$ supergravity as the culmination of GFP, or even of theoretical physics itself, since it can in principle explain everything, all forces and all particles, in the physical world.

27. The infinities cancel only in extended supergravity theories because only there can all particles be transformed into gravitons and all the diagrams be reduced to those having only gravitons, which can be shown to have a finite sum.

28. See, for example, Friedlander (1976).

29. Wheeler (1964b), DeWitt (1967c).

30. If one rejects the idea of external time and tries to identify time with some function of the metric variables and then reinterpret the dynamics with respect to this 'internal time', then one finds that in practice it is impossible to make such a precise identification. Cf. Isham (1990).

31. Hartle and Hawking (1983).

32. Green, Schwarz, and Witten (1987).

33. For the sake of comparison, we call Christoffel symbols of the second kind $\Gamma^{\sigma}_{\mu\nu}$ gravitational potentials. $\Gamma^{\sigma}_{\mu\nu}$ are expressed in terms of $g_{\mu\nu}$, which are loosely called gravitational fields, and their first derivatives.

34. This was shown by Stanley Deser (1970).

35. Yang (1974), Wu and Yang (1975), and Daniel and Viallet (1980).

36. For modern Kaluza–Klein theory, see Hosotani (1983). Superstring theorists have also become interested in Eddingtonian geometry, but superstring theories are beyond the scope of this volume.

37. A fiber bundle over R^4 is a generalization of a product space of spacetime and an internal space, which allows for possible twisting in the bundle space and therefore gives rise to a non-trivial fusion of spacetime with internal space. Mathematically, the external and internal indices can be blended together by certain transition functions whose role in gauge

theory is played by the generalized phase transformations. For techniques, see Daniel and Viallet (1980).

38. For Dirac's criticism, see his (1969a, b).

39. As to other regularization schemes, the same claim can be made for the lattice cutoff scheme, which is essentially equivalent, but not for dimensional regularization, which is more formalistic and not relevant to the point discussed here.

40. Lepage (1989) asserts without further explanation that 'it now appears likely that this last step [taking the cutoff to infinity] is also a wrong step in the nonperturbative analysis of many theories, including QED'.

41. Polchinski (1984) and Lepage (1989).

42. See Wilson (1983), Symanzik (1983), Polchinski (1984), and especially Lepage (1989).

43. The non-renormalizable interactions simulate the low-energy evidence of the inaccessible high-energy dynamics and are thus suppressed by a power of the experimental energy divided by the mass of the heavy boson.

44. Explicit symmetry breaking, e.g., adding non-gauge-invariant mass terms to a pure Yang–Mills theory, is irrelevant to our discussion here.

45. That the tower of EFTs would be endless is entailed by the local operator formulation of QFT. See section 8.1.

46. See Lepage (1989).

47. There are extensive discussions on this topic. See, for example, Veltman (1977) and Collins, Wilczek, and Zee (1978). The main argument is this. The non-renormalizable effects arising from heavy particles will be detectable if there is a process that is forbidden in the absence of the heavy particles.

48. Karl Popper (1970) argues convincingly about the implications of the emergence viewpoint for pluralism in theoretical ontology.

49. Cao and Schweber (1993).

12

Ontological synthesis and scientific realism

The historical study of 20th century field theories in the preceding chapters provides an adequate testing ground for models of how science develops. On this basis I shall argue in this chapter that one of the possible ways of achieving conceptual revolutions is what I shall call 'ontological synthesis', and thus propose an argument for a certain kind of scientific realism and for the rationality of scientific growth.

12.1 Two views on how science develops

There are many views in contemporary philosophy of science concerning the question of how science develops. I shall consider in particular two of them. According to the first view, science evolves through the progressive incorporation of past results in present theories, or in short, science is a continuing progression. Such a 'growth by incorporation' view was taken by the empiricist philosopher Ernest Nagel. Nagel took for granted that knowledge tended to accumulate and claimed that 'the phenomenon of a relatively autonomous theory becoming absorbed by, or reduced to, some inclusive theory is an undeniable and recurrent feature of the history of modern science' (1961). Thus he spoke of stable content and continuity in the growth of science, and took this stable content as a common measure for comparing scientific theories. The idea of commensurability was taken to be the basis for a rational comparison of scientific theories.

A more sophisticated version of the 'growth by incorporation' view was proposed by Wilfrid Sellars (1965) and Heinz R. Post (1971). Post, in particular, appealed to what was called 'the general correspondence principle', according to which 'any acceptable new theory L should account for the success of its predecessor (S) by "degenerating" into that theory (S) under those conditions under which S has been well confirmed by test' (*ibid.*). Thus

355

Post argued that 'science may well be converging ultimately to a unique truth', and that 'the progress of science appears to be linear' (*ibid.*).

The advocates of the opposite view, which emerged in the late 1950s, rejected the idea of 'superimposing' a 'false continuity' on the history of science. For example, N. R. Hanson (1958) suggested that a conceptual revolution in science was analogous to a gestalt shift in which the relevant facts came to be viewed in a new way. Stephen Toulmin (1961) also pointed out that drastic conceptual changes often accompanied the replacement of one inclusive theory by another (1961). In their view, the replacement of one theory by another is often by revolutionary overthrow.

The best-known advocates of this radical view are Thomas Kuhn and Paul Feyerabend. Their position can be characterized by the incommensurability thesis. According to this thesis, successive and competing theories within the same domain speak different theoretical languages. Thus these theories can neither be strictly compared to nor translated into each other. The languages of different theories are the linguistic counterparts of the different worlds we may conceive. We can pass from one world to another by a Gestalt switch, but not by any continuous process.[1] In this regard, Kuhn and Feyerabend have some followers. For example, Larry Laudan (1981) denied the possibility of ontological progress and claimed that 'changing ontologies or conceptual frameworks make it impossible to capture many of the central theoretical laws and mechanisms postulated by the earlier theory', let alone the elements of earlier ontology. Concerning the history of modern physics, Andrew Pickering (1984) asserted that the new physics after 1974 (roughly corresponding to the gauge field programme in this volume) and old physics (roughly corresponding to the quantum field programme, but including S-matrix theory as well) were incommensurable in the most radical sense: they belonged to different worlds and had no common set of phenomena.

The very word 'incommensurable' points to various problems at different levels: topic, meaning, reference, and ontology. But in every case the thesis results in the same claim that there can only be 'instrumental progress' in the history of science, which is not expressed in progressively truer and truer theoretical propositions. The relativist and anti-realist implications of the claim are straightforward. If we come to see the world differently with each paradigm shift (in Kuhn's term), then the truth value of theoretical statements would be theory bound and the aim of discovering the 'deep truth' about the world, in the correspondence sense, should be abandoned. Moreover, there would be no accumulation of true theoretical propositions corresponding to the world, nor even progress towards the truth.

A moderate version of the second view was suggested by Sellars, Hesse, and

Laudan. The incommensurability of the language of local laws was rejected, on the basis of certain views of meaning (see section 12.2). Progress in science was admitted on the basis of the following arguments. First, they argued, science as knowledge is not primarily grandiose theory, but rather a corpus of low-level laws and particular descriptions. Second, these laws and descriptions persist through theoretical revolutions because science has the function of learning from experience. Third, changes or revolutions in scientific theories often turn on conceptual issues rather than on questions of empirical support, thus leaving intact many of the low-level laws.[2] Science thus exhibits progress in the accumulation of particular truths and local regularities. But this kind of 'instrumental progress' is different from the progress in general laws and theoretical ontologies, since, according to Hesse (1974, 1985), the local and particular success of science does not require there to be strictly true general laws and theoretical ontologies.

Thus a significant distinction is made between local and ontological progress. While local progress is admitted, ontological progress is denied. Hesse claims that 'there is no reason to suppose that these theoretical ontologies will show stability or convergence, or even at any given time that there will ever be just one uncontroversially "best" ontology' (1985). This conforms to Kuhn's original position about theories: 'I can see [in the systems of Aristotle, Newton, and Einstein] no coherent direction of ontological development' (1970).

The arguments against the concept of progress in theoretical ontologies are quite strong. First, the Duhem–Quine thesis of the underdetermination of theory by evidence seems to make the concept of true theoretical ontology itself dubious. Second, historical records seem to suggest that there is no permanent existence of theoretical ontologies. Third, it has been argued, ontological progress can only be defined if it is in principle capable of reaching an ideally true and general theory, in which theoretical entities 'converge' to true entities; yet 'this would require the absence in future science of the kind of conceptual revolutions that have been adequately documented in the history of science up to now' (Hesse, 1981; see also her 1985).

There are many controversial issues in modern philosophy of science involved in these arguments, and I do not want to venture a solution to the problem of ontological progress. Rather, my aim is quite modest: to draw a lesson from the historical study in the preceding chapters and to argue for a certain kind of ontological progress, and thus for a certain kind of scientific realism and for the rationality of scientific growth. To this end, however, a brief outline of my own framework for discussing ontological progress is required. Before giving such an outline, it may be helpful to make some comments on existing frameworks that can be adopted to meet the challenge of the incommensurability thesis.

12.2 Frameworks opposed to the incommensurability thesis

In order to combat the incommensurability thesis, Peter Achinstein (1968) suggested the concept of 'semantically relevant properties', which were restricted to perceptual ones but were responsible for a physical object being recognized as a member of the ontological class. On the basis of this concept, it could be argued that certain perceptual properties, or conditions 'semantically relevant' for a theoretical term X, could be known independently of a theory, and have a particularly intimate connection with the meaning of the term X. In this framework, therefore, a theoretical term (or a universal predicate) specifying a particular ontological class could have the same meaning in two theories, and the meaning of a term used in a theory could be known without knowing the theory.

Quite similarly, in discussing the meaning-variance and theory-ladenness of descriptive terms, Hesse (1974) suggested an interpretation of meaning-stability in terms of 'intentional reference'. In Hesse's network model of universals, it is the concept of intentional reference that makes it possible to provide a perceptual basis for ascribing a predicate to a physical object, or provide a predicate with empirical meaning. In virtue of perceptually recognizable similarities and differences between objects, many theoretical assertions about observable properties and processes and their analogies, and some approximate forms of empirical laws, are translatable from theory to theory, and thus may be both approximately stable and cumulative. Here, although a predicate may be one that specifies a particular class of objects in the theoretical ontology, it is still based on perceptual recognition of similarity. The intentional reference is therefore an empirical means of recognizing observables, and relates to unobservables only insofar as they occur in a theory that makes predictions about observables.

Within both of the above frameworks, therefore, it is difficult to find room for the continuity of an unobservable ontology. Another framework, which seems able to accommodate this kind of continuity, is Hilary Putnam's causal theory of reference.[3] According to Putnam, a scientific term (e.g. water) is not synonymous with a description (of water), but refers to objects having properties whose occurrence is connected by a causal chain of the appropriate type to a statement involving the term. This theory can be broadened to cover unobservable terms in science (e.g. electron) with the help of 'the principle of benefit of the doubt'. The principle says that reasonable reformulations of a description that was used to specify a referent for an unobservable term but failed to refer (e.g. Bohr's description of the electron), are always accepted. Such reformulations (e.g. the description of the electron in quantum mechanics) make those earlier descriptions refer to classes of objects with somewhat the same roles which do exist from the standpoint of the later theory (Putnam, 1978). Thus we can talk about cross-theoretical reference of unobservable terms and interpret meaning-stability in terms of stability of this reference. In contrast to the radical view that an unobservable term cannot have the same referent in different theories, this causal theory of reference allows us to talk about different theories of the same class of objects specified by an unobservable term (*ibid.*).

Putnam (1978) further argued that if there was no convergence in the sense that earlier theories are limiting cases of later theories, then 'benefit of the doubt' would always turn out to be unreasonable. It would be unreasonable because then theoretical terms like phlogiston would have to be taken to refer, and it would be impossible to take theoretical terms as preserving their references across most changes of theory. In addition, Putnam argued, a belief in convergence would lead to a methodology of preserving the mechanisms of earlier theories as often as possible. And this would restrict the class of candidate theories and thereby increase the chance of success.

The methodological implication of this restriction will be discussed in sections 12.5 and 12.6. Here I just want to make a comment on Putnam's causal theory of reference, which is relevant to the framework I adopt for accommodating ontological progress. Putnam (1978) is right in observing that what are relevant for specifying the referent of a theoretical term are not just monadic properties (charge, mass, etc.) of the hypothetic referent, but also effects explained and roles played by the hypothetic referent. Even if we can interpret the monadic properties of theoretical entities as perceptual by their analogy to those of observable objects, other properties that are hinted at by the 'effects' and 'roles' are, in Putnam's view, not directly perceptual. In my view, these properties are associated with what I shall call 'structural properties' or properties of properties and relations of entities. These structural properties, though not directly perceptual, are nevertheless crucial for specifying the reference of theoretical terms, and thus play a decisive role in determining the meaning of theoretical terms.

12.3 Structural properties of ontologies

Before outlining my own framework for describing ontological progress through conceptual revolutions, a question should first be answered: why is there any need for ontology in science if science is primarily a corpus of empirical laws? The answer is this. Most practicing scientists believe that empirical laws have only 'local' validity, while ontology endows science with unifying power. As a model of the basic structure of the world, ontology is believed to be the carrier of the general mechanism which underlies the discovered empirical laws. Thus as a basis on which a unifying conceptual framework of a scientific theory can be established, ontology is theoretically much more fundamental than individual empirical laws.

The theoretical terms of an ontology are not merely useful intellectual fictions but in some sense referential. It was noticed by Moritz Schlick (1918), Bertrand Russell (1927), Rudolf Carnap (1929), and Grover Maxwell (1971), and I agree, that the statements about an ontology refer to underlying particular entities and their intrinsic properties mainly through the reference to the structural characteristics of these entities.[4] With these statements, however, we are still quite ignorant concerning what exactly these entities are and what exactly these intrinsic and structural properties are. We can nevertheless at least know that there are entities and there are intrinsic and structural properties. Here intrinsic properties (or relations) are first-order properties (or relations) of individual entities, and are direct referents of monadic (or polyadic) predicate symbols; and structural properties are properties of properties and relations: they are referents of a higher logical type.

Among the properties that are relevant for something to be such and such a particular entity are the so-called essential properties. These properties are defined in a theory, and their descriptions may change from theory to theory. Once a former essential property is found to be explainable by new essential properties, it stops being essential. As a consequence of this theory dependence, what a particular entity essentially is will never be finally settled as long as science continues to develop. Thus a theoretical ontology in the sense of entity cannot be regarded as a true replica but only as a model of physical reality, based on the analogies suggested by what we know, ultimately by observation.

What such a model actually provides us with, therefore, is not literally true descriptions of the underlying entities themselves, but rather, by analogy, the assertions of the observable structural relations carried by the hypothetical entity. In fact, a stronger case can be argued that any ontological characterization of a system is always and exclusively structural in nature. That is, part of what an ontology is is mainly specified by the established structural relations of the underlying entities. Thus structural properties and relations are part of what an ontology is. Moreover, they are the only part of the ontology that is accessible to scientific investigators through the causal chains that relate the structural assertions to the hypothetical entities. Although structural assertions will also be modified when the theory changes and new structural properties are discovered, these properties, like observables, are both approximately stable and cumulative because they are translatable from theory to theory in virtue of their recognizable identities.

Some examples of structural properties that are crucial to my discussion in this chapter include external and internal symmetries, geometrizability, and quantizability. External symmetries (e.g. Lorentz symmetry) satisfied by laws of physical objects are obviously structural in nature. In fact, the foundational role of transformation groups and the idea of invariants with respect to transformation groups as a description of the structural features of collections of objects were realized and advocated by Poincaré from the late 19th century (1895, 1902), and were subsequently built into the collective consciousness of mathematical physicists, such as Einstein, Dirac, Wigner, Yang, and Gell-Mann.[5] Internal symmetries (e.g. isospin symmetry) are connected with, but not exhausted by, intrinsic properties of physical objects (conserved quantities, such as charge, isospin charge, etc.) through Noether's theorem. As symmetries in abstract internal spaces, however, they are also higher-order properties. Geometrizability is a structural property that is isomorphic to the structural characteristics of a spacetime manifold or its extension. Quantizability is a structural property of a continuous plenum, which is connected with a mechanism by which the discrete can be created from, or annihilated into, the continuous.

Thus statements about an ontology include terms which signify observables (defined in the network sense) and logical items. That is, the potential reference of an ontology is specified by its observable structural properties. For example, the reference of a particle ontology is specified in part by the inclusion of such structural properties as 'physical objects have isolable constituents', and 'these constituents have a certain theoretically characterizable autonomy'. The reference of a field ontology is specified by, for example, the inclusion of such structural properties as 'the superimposability between different portions of the entity' and 'the impossibility of individualizing the entity'.

Among the structural properties of a theoretical ontology are guiding principles, such as the principles of geometrization, quantization, and gauge invariance. These principles underlie and unify empirical laws (relations); thus their continuity can be justified by the continuity of empirical laws. In what follows I shall restrict my discussion to these guiding principles.

In scientific theories, the structural properties that have been discovered are supposed to be carried by theoretical entities. With the accumulation of structural properties, old theoretical entities will inevitably be replaced by new ones, together with a change of the ontological character of the whole theory. However, as far as older structural properties are concerned, the change of ontology only means the change of their functions and places in the whole body of structural properties. Thus the replacement of theoretical entities should be properly regarded as an analogical extension of the theoretical ontology, which is caused by the accumulation of structural properties, rather than a revolutionary overthrow.

12.4 Conceptual revolutions via ontological synthesis

The remarks I made in the last section have provided a framework that can be used for the discussion of ontological progress. By ontological progress I mean the accumulation and extension of the structural relations exhibited by successive theories of the world. The source of unifying power of a theory lies precisely in the fact that its ontology in the sense of entities is a metaphor and capable of what may be called a metaphorical extension, and in the sense of structural properties is stable and cumulative in nature.[6]

At first glance the geometrical programme (GP) and the quantum field programme (QFP) are so different in describing fundamental interactions that they might be taken as a paradigm case of the incommensurability thesis. In the former, the representation of the physical reality of the agent transmitting interactions is geometrical in nature, continuous in spacetime; while in the latter, the corresponding representation is discrete quanta. In the former, general coordinate covariance is a guiding principle; while in the latter, no continuous spatio-temporal description for the behavior of quanta is even possible, and special coordinate systems are assumed without apology. The contrast between the two programmes is so sharp that no one would deny that a conceptual revolution had indeed happened. Who then dares to say that the geometrical structures of the old ontology can be incorporated in the quanta of the new ontology? Where can we see the coherent direction of ontological development?

There is indeed a conceptual revolution. Yet its implications for an ontology shift are a subject for debate. The whole picture will be quite different from that provided by Kuhn and Feyerabend if we explore the ontological bases and latent possibilities of the two programmes at a deeper level, with the advantage of hindsight.

The historical study in the preceding chapters can be summarized as follows. All versions of 20th century field theories have originated from classical electrodynamics (CED). CED is a theory of the substantial electromagnetic fields, and has the Lorentz group as its symmetry group (see section 3.3).[7]

The gravitational field theory is a direct descendent of CED: Einstein developed his general theory of relativity (GTR) when Lorentz invariance (a global external symmetry) was generalized to general covariance (a local symmetry), together with the introduction of the gravitational field through the principle of equivalence (see sections 3.4 and 4.2). GTR initiated GP, in which the substantial fields are inseparably correlated with the geometrical structures of spacetime, through which interactions are mediated. The basic ontology of GP, according to its weak version, is also substantial fields: spacetime together with its geometrical structures has no existence on its own, but only as a structural quality of the fields (see sections 5.1 and 5.2). Here the ontological continuity between GP and CED is obvious.

Quantum electrodynamics (QED) is another direct descendent of CED. When the classical electromagnetic fields were replaced by the quantized ones, and, by analogy with the boson fields, substantial fermion fields were introduced, QED was available for the physics community (see section 7.3). QED initiated QFP, in which fields manifest themselves in the form of discrete quanta by which interactions are transmitted. The basic ontology in this programme is also some kind of substantial field (see section 7.3).

It is true that the substantial fields in QFP, the quantum fields, are radically different in their structural properties from the substantial fields in CED, the classical fields, and this ontological difference makes them belong to different paradigms. Yet the quantum fields and the classical fields still share such hard-core structural properties as 'the superimposability of its different portions' and 'the impossibility of individualization'. Generally speaking, on the basis of the hard-core structural properties shared by any two theories with seemingly different ontologies, we can always establish an ontological correspondence between the two theories,[8] and thus make the referential continuity of the ontologies, in the sense of structural properties, in two theories discernible. In particular, we can find a kind of continuity in ontological change between CED and QED, that is, a referential continuity between the quantum field and classical field in their structural properties, with the help of Bohr's correspondence principle.

The gauge theory is a direct descendent of QED. What Yang and Mills did was merely replace the local U(1) phase symmetry of QED by a local SU(2) isospin symmetry that was supposed to be respected by the strong interactions. The Yang–Mills theory initiated GFP, in which the forms of fundamental interactions are fixed by the requirement of gauge invariance. The claim of ontological continuity between GFP and QFP faces no challenge since the gauge fields, like the 'material fields', are also to be quantized. But what about the relationship between the ontologies of GFP and GP?

The ontological continuity between these two programmes, like that between QED and CED, lies in the referential continuity between the quantized gauge field and the classical geometrical field in their structural properties. Not only can a theory in GP, such as GTR or its extension or variations (at least in principle), be put in a quantized form, which is very similar in its mathematical structure to a theory in GFP (see sections 9.3, 11.2, and 11.3); but also theories in GFP can be given a kind of geometrical interpretation (see section 11.3). In this way the ontologies of the two programmes can be shown to share structural properties or even essential features. The ontological continuity between GFP and GP is manifested in modern Kaluza–Klein theories in such a particularly straightforward way (see section 11.3) that we can claim without hesitation that GFP is a direct descendent of GP.

It is interesting to note that the intimate link between the ontologies of QFP and GP became discernible only after GFP took the stage as a synthesis of the two (see section 11.3). This fact suggests to us that the concept of synthesis is of some help in recognizing the referential continuity of theoretical ontologies across conceptual revolutions.

Here a 'synthesis' of scientific ideas does not mean a combination of previous ideas or principles in a 'synthetic stew', but rather refers to a highly selective combination, presupposing transformations of previous scientific ideas (concepts, principles, mechanisms, hypotheses, etc.). Some useful elements in each of the previous ideas are chosen and the rest rejected. A synthesis becomes possible only when the selected ideas are transformed and become essentially new and different ideas. For example, Newton synthesized Kepler's concept of inertia (as a property of matter which would bring bodies to rest whenever the force producing their motion ceased to act) with Descartes' concept of inertial motion (as a law of nature rather than a law of motion, namely that everything is always preserved in the same state), and formed his own concept of inertia as a property of matter which would keep bodies in whatever state they were in, whether a state of rest or a state of uniform motion in a straight line. In this synthesis, both of the previous concepts were transformed.

We can extend the concept of synthesis to the discussion of ontology and find that an ontological synthesis also presupposes transformation. As a matter of fact, the birth of GP was the result of synthesizing the idea of general covariance and the principle of equivalence, while the idea of general covariance originated in, but differed from, the idea of Lorentz invariance; the birth of QFP was the result of synthesizing CED and the quantum principle, while the idea of field quantization originated in, but ontologically differed from, the idea of quantization of atomic motions (see sections 6.1, 6.2, 7.2, 7.3); and the birth of GFP was the result of synthesizing QFP and the gauge principle, while the gauge principle originated in, but differed from, minimal electromagnetic coupling (see section 9.2). In all three cases, the previous principles were transformed into new forms, and only then became useful for ontological synthesis.

Moreover, as a general feature, ontological synthesis often makes a substance (as a primary entity) into an epiphenomenon (or a derivative entity), and thus accompanies a change of basic ontology. For example, in GP the Newtonian gravitational potential is regarded as a manifestation of (a component of) the metric field; in QFP, the classical field is regarded as an epiphenomenon of the quantum field; in the standard model of GFP, the quantum electromagnetic field is regarded as an epiphenomenon of the quantized gauge fields (see section 11.1). Thus the definite links between the three programmes, as summarized in this section, suggest that ontological synthesis is one of the possible ways of achieving conceptual revolutions. The result of the synthesis is the birth of a new research programme, which is based on a new basic ontology. On this view, a direct incorporation, as suggested by Nagel (1961) and Post (1971), of the old ontology of a prerevolutionary programme into the new ontology of the postrevolutionary programme is very unlikely.

On the other hand, the discovered structural relations of the world (e.g. external and internal symmetries, geometrization, quantization, etc.) embodied in the old ontology persist, in a limiting sense, across the conceptual revolutions. A conceptual revolution achieved through transformations and synthesis is by no means an absolute negation, but may be regarded as an 'Aufhebung' ('sublation') in the Hegelian sense, a sense of overcoming, changing while preserving. Thus science exhibits progress not only in the form of the accumulation of empirical laws, but also, and even more significantly, in the form of the conceptual revolutions. What science aims at are fruitful metaphors and ever more detailed structures of the world. We move nearer to this goal after a conceptual revolution because with a revolution the empirical laws are better unified by a new ontology than by the old one.

It is in this sense of preserving and accumulating the structures of the world that we claim a coherent direction for ontological developments towards the true structures of the world. Notice that the expression 'the true structures of the world' here should be understood as 'the true structures of a definite, though always enlargeable, domain of investigations'. Since there is no reason to assume that the structural properties of the world can only be divided into finite levels, and no theory would be able to capture an infinite number of structural properties, the concept of an ultimate true ontology, which will provide us with the whole of the true structures of the world, is obviously meaningless. Accordingly, the words 'better' and 'nearer' used in this paragraph have only a comparative implication between older and newer theories, without an absolutist connotation.

12.5 Conceptual revolutions and scientific realism

In this section I shall turn to the question of how the lesson drawn in the last section bears on the argument for scientific realism. According to the accepted view,[9] scientific realism assumes that (i) the entities and their structural properties postulated by progressively acceptable theories actually exist, although no claim is made for the truth of every particular description of the entities given by a theory; (ii) theories aim at true propositions corresponding to the world; and (iii) the history of at least the mature sciences[10] shows progressive approximation to a true, in the correspondence sense, account of a domain under investigation, progressive both in terms of 'low-level' laws and also, or even more importantly, in terms of ontology (in the sense of local but expansible structural relations), which underlies the 'low-level' laws. Point (iii) is most relevant to the historical study in this volume.

Scientific realism requires ontological continuity across conceptual revolutions. Otherwise the theoretical ontology in a paradigm would not have any existential claim on the next paradigm, and then realism would simply be false. Anti-realists reject point (iii) as far as ontology is concerned. For them the existence of conceptual revolutions implies that, historically, theoretical sciences have been radically discontinuous and gives a strong argument against realism as defined above. Now that successful theories have been rejected by theories postulating radically different ontologies, in which nothing but formerly recognized low-level laws have survived, scientific realism of the sort defined seems not to be a tenable position.

The situation was clearly summarized by Putnam (1978) in his famous discussion on 'meta-induction'. If conceptual revolutions entail that the history of science is absolutely discontinuous, then 'eventually the following meta-induction becomes compelling: just as no term used in the science of more than 50 years ago referred, so it will turn out that no term used now refers' (*ibid.*), or that scientific realism is a false doctrine.

The interpretation of the nature of conceptual revolutions is so crucial to scientific realism that realists have to find some arguments to block the disastrous 'meta-induction'. One argument for realism is provided by Putnam in his causal theory of reference. Logically, the causal theory of reference with 'the principle of benefit of the doubt' makes it possible to describe scientific revolutions as involving referential continuity for the theoretical terms. It is obvious that with such a referential continuity, the disastrous 'meta-induction' can be successfully blocked, and a unilinear view of scientific growth, namely the uniquely convergent realism adopted by Nagel and Post, can be more or less justified.

But this kind of rebuttal of the anti-realist argument is both over-simplistic and too abstract. It is over-simplistic because the structure of scientific progress is much richer than a unilinear form of continuity suggests, although the concept of referential continuity can be used as a basis for a realistic conception of the progressiveness of conceptual revolutions. It is too abstract because the logical argumentation involved explains neither the emergence of conceptual revolutions, nor the development of theories, in particular concerning the postulation of new unobservable entities. Thus it does not convincingly support realism with a historical analysis of actual scientific developments.

Another argument for realism is provided by the network model of universals (Hesse, 1974). Logically, the network model of universals makes it possible to describe scientific revolutions as involving analogical continuity for theoretical terms. It is obvious that with such an idea of analogical continuity, together with the understanding that the reference of a theoretical term must be derived from observables and understood intentionally, the disastrous 'meta-induction' can also be successfully blocked. On the basis of such an argument, a weak realism of 'instrumental' progress can be justified (*ibid.*).

My approach to this issue purports to advance beyond both the uniquely convergent realism and the weak realism by introducing the concept of ontological synthesis. The concept is directly drawn from the historical analysis of 20th century field theories, and thus cannot be accused of being irrelevant to actual scientific practice. Also, it can serve to explain complex forms of progress across conceptual revolutions. The growth and progress of science do not necessarily present themselves in the unilinear form of continuity and accumulation. The concept of ontological synthesis as a dialectical form of continuity and accumulation of the discovered world structures is more powerful in explaining the mechanism of conceptual revolutions and the patterns of scientific progress. In fact, continuity, accumulation, and unification in science are rarely realized in a straightforward form, but often in the form of synthesis-via-transformations, as summarized in the last section.

In conclusion, we can say that the concept of ontological synthesis-via-transformations as a possible way of achieving conceptual revolutions, represents a reconciliation between change and conservation in scientific growth. It is therefore suitable for capturing essential continuity through discontinuous appearance in the history of science. On the basis of such a concept, a stronger argument for scientific realism may be developed.

12.6 Conceptual revolutions and scientific rationality

The notion of scientific rationality[11] is closely connected with the notion of truth. Only characterizations of scientific progress which take the central aim of science to be evolving towards truer and truer assertions about the empirical laws and structural properties of the world allow us to represent science as a rational activity. In this sense realism can be regarded as a foundation, though not necessarily the only foundation, for scientific rationality. On the other hand, scientific reasoning plays a crucial role in settling which theory is getting at the truer description of the world. Thus it is also necessary for evaluating theories, in terms of their truth claims, and for making a decisive choice among competing theories.[12]

A threat to scientific rationality comes from the concept of the revolutionary paradigm shift. With this concept, scientific revolutions are compared to religious conversions or to the phenomenon of a Gestalt switch. This comparison, in fact, becomes the major argument for the incommensurability thesis and against the notions of truth and ontological progress. In this sense, the idea of scientific revolutions calls scientific rationality into question, and produces a crisis of scientific rationality. One reason why this is a serious worry is that eventually rational theory evaluations and theory choosing would become impossible. Yet science continues to develop theories. This is particularly true of fundamental physics, where significant developments have often taken the form of a conceptual revolution, involving a change of ontology, but without obvious irrationality. Thus a philosophy of science based on the idea of incommensurability would become irrelevant to actual scientific activities.

In order to defend scientific rationality against the threat, therefore, one has to argue that conceptual revolutions are progressive in the sense that theories become truer and truer descriptions of the domain under investigation. It is precisely at this juncture that I find that the concept of ontological synthesis as a way of achieving conceptual revolutions is crucial to the argument. Here the progressiveness is defined in terms of the extension of structural relations, which underlies the continuity, but not unique convergence, of theoretical structures (ontologies, mechanisms, principles, etc.).

The rational position based on the concept of ontological synthesis entails that the future direction of fundamental research suggested by the synthetic view of scientific progress is different from that suggested by the incommensurability view. For some adherents of the latter view, the direction of scientific research is decided mainly by external, such as social or psychological, factors and has little to do with intellectual factors. For others, intellectual considerations, though important, play an essentially unpredictable role since the evolution of science has a catastrophic character. By contrast, the synthetic view requires that the internal mechanisms of the earlier theory must be preserved as far as possible.

Another significant implication of this position is that scientific growth is not unilinear but dialectical. New practice always produces new data since aspects of the world may actually be infinite in number. From new data new ideas emerge, often with transformations of existing ideas. Thus a new synthesis, and hence a new conceptual revolution, is always needed. Scientific progress in this sense is therefore not incompatible with future conceptional revolutions. The revolution is perpetual. Convergence to a fixed truth is incompatible with the synthetic view of progress.

There is another difference between the synthetic view and the unilinear view. According to the latter, the existing successful theory must be the model for future developments. For example, Einstein's general theory of relativity was expected to be able to lead to an understanding of electromagnetism and matter. As is well known, the expectation proved false. In fact, the abandonment of earlier models in the history of science has repeatedly happened. The 'mechanical philosophy' that was popular in the 17th and 18th centuries was abandoned later on; the 'electromagnetic world view' that was popular at the end of the 19th century was also abandoned in the 20th century. In the 1920s, shortly after Dirac published his relativistic wave equations for the electron, Born claimed that 'physics, as we know it, will be over in six months'.[13] All these expectations proved false. Unfortunately, the same thing has happened again in recent years. This time the successful theory to act as the model is the standard model of GFP.[14] What will be the fate of GFP? In fact, GFP is far from being a complete theoretical framework, even if simply because the Higgs mechanism and Yukawa coupling cannot be accounted for by the gauge principle.[15]

In contrast, the suggestion for future research arising from the synthetic view is totally different in kind. Scientists are advised to keep their minds open to all kinds of possibilities, since a new synthesis beyond the existing conceptual framework is always possible. In this spirit, future developments of field theories may not exclusively come from research within GFP, and try to incorporate the Anderson–Higgs mechanism and Yukawa coupling into the programme. It is quite possible that they may be stimulated by some use of the ideas and techniques developed, for example, in the S-matrix theory, whose underlying ideas, such as those of ontology and of the nature of force, are radically different from those in GFP.

In conclusion, scientific rationality lies in the intention to obtain approximate knowledge of the real structures of the physical world; and on the synthetic view of scientific growth, one way of realizing this rationality is through conceptual revolutions.

Notes

1. Kuhn (1962) and Feyerabend (1962).
2. See, for example, Sellars (1961, 1963, 1965); Hesse (1974, 1981, 1985); Laudan (1981).
3. Putnam (1975, 1978, 1981); see also Kripke (1972).
4. For further discussions of the structuralist view of ontology, see Chihara (1973), Resnik (1981, 1982), Giedymin (1982), and Demopoulos and Friedman (1985).
5. See, for example, Einstein (1949), Dirac (1977), Wigner (1949, 1964a, b), Yang (1980), and Gell-Mann (1987).
6. See also McMullin (1982, 1984) and Boyd (1979, 1984).
7. Here the symmetry group of a theory is to be understood as the covariance group of the standard formulation of the theory, which is a system of differential equations for the dynamic objects of the theory alone, see Friedman (1983).
8. This, of course, is not without controversy. There is an extensive literature dealing with this issue from different positions on theory change. See, for example, Feyerabend (1962), Sneed (1971, 1979), Post (1971), Popper (1974), Kuhn (1976), Stegmüller (1976, 1979), Krajewski (1977), Spector (1978), Yoshida (1977, 1981), and Moulines (1984). For recent contributions on this issue, see the survey by Balzer, Pearce and Schmidt (1984).
9. See Putnam (1975) and Hacking (1983).
10. A mature science contains one or more mature theories, which are characterized by a consistent mathematical structure, by a delineated validity domain (empirical support), and by horizontal (in terms of its relations with other theories in different branches) and vertical (in terms of its relations with other mature theories) coherence. All the following discussion concerning scientific realism makes sense only in terms of mature sciences. For more discussion of mature sciences, see Rohrlich and Hardin (1983).
11. The discussion of rationality in this section is restricted to the notion of scientific rationality, excluding other forms of it, such as practical or esthetic considerations. It should be pointed out that Kuhn, who is criticized in this section, has developed, in recent years (1990, 1991, 1993), a very profound notion of practical rationality in scientific enterprise.
12. On the issue of theory choice, I assume a strongly internalist position, according to which ultimately empirical and logical considerations play a crucial role in making choice. It is too complicated, however, and also not very relevant, to defend this position against attacks by social constructivists in this short philosophical discussion. See, however, Cao (1993).
13. Quoted by Steven Hawking (1980).
14. See, for example, Hawking (1980) and Glashow (1980).
15. Further discussions on the problems with GFP can be found in section 11.4.

Appendices

Appendix 1. The rise of intrinsic, local, and dynamic geometry

A1.1 Intrinsic geometry

Until Gauss's work on differential geometry (1827), surfaces had been studied as figures in three-dimensional Euclidean space. But Gauss showed that the geometry of a surface could be studied by concentrating on the surface itself. A surface S is a set of points with two degrees of freedom, so any point r of S can be expressed in terms of two parameters u_1 and u_2. We can get expressions: $dr = (\partial r/\partial u^1)\, du^1 + (\partial r/\partial u^2)\, du^2 = r_i\, du^i$ ($r_i \equiv \partial r/\partial u^i$, with the summation convention over $i = 1,\ 2$), and $ds^2 = dr^2 = r_i r_j\, du^i\, du^j = g_{ij}\, du^i\, du^j$. Gauss had made the observation that the properties of a surface, such as the elements of arc length, the angles between two curves on a surface, and the so-called Gaussian curvature of a surface, depended only on g_{ij}, and this had many implications. If we introduce the u^1 and u^2 coordinates, which come from the parametric representation $x = x(u^1,\ u^2)$, $y = y(u^1,\ u^2)$, $z = z(u^1,\ u^2)$, of the surface in three-dimensional space, and use the g_{ij} determined thereby, then we obtain the Euclidean properties of that surface. But we could start with the surface, introduce the two families of parametric curves u^1 and u^2, and get the expression for ds^2 in terms of g_{ij} as functions of u^1 and u^2. Thus the surface has a geometry determined by the g_{ij}.

The geometry is intrinsic to the surface and has no connection with the surrounding space. This suggests that the surface can be considered a space in itself. What kind of geometry does the surface possess if it is regarded as a space in itself? If we take the 'straight lines' on that surface to be the geodesics (the shortest line connecting two points on the surface), then the geometry may be non-Euclidean. Therefore what Gauss's work implied is that there are non-Euclidean geometries at least on surfaces when they are regarded as space in themselves.

Guided by Gauss's intrinsic geometry of surfaces in Euclidean space, Riemann developed an intrinsic geometry for a wider class of spaces (1854). He preferred to treat n-dimensional geometry, even though the three-dimensional case was clearly the important one. He spoke of n-dimensional space as a manifold. A point in a manifold of n-dimensions was represented by assigning special values to n variable parameters: x_1, x_2, ..., x_n, and the aggregate of all such possible points constituted the n-dimensional manifold itself. Like Gauss's intrinsic geometry of surfaces, Riemann's geometrical properties of the manifold were defined in terms of quantities determinable on the manifold itself, and there was no need to think of the manifold as lying in some higher-dimensional manifold.

A1.2 Local geometry

According to Riemann,

it is upon the exactness with which we follow phenomena into the infinitely small that our knowledge of their causal relations essentially depends. The progress of recent centuries in the knowledge of mechanics depends almost entirely on the exactness of the construction which has become possible through the invention of the infinitesimal calculus.

(Riemann, 1854)

So the 'question about the measure relations of space in the infinitely small' was of paramount importance *(ibid.)*.

In contrast with the finite geometry of Euclid, Riemann's geometry, as essentially a geometry of infinitely near points, conformed to the Leibnizian idea of the continuity principle, according to which no law of interactions can be formulated by action at a distance. So it can be compared with Faraday's field conception of electromagnetic phenomena, or with Riemann's own ether field theory of electromagnetism, gravity, and light (cf. section 2.3). Hermann Weyl had characterized this situation as follows: 'The principle of gaining knowledge of the external world from the behavior of its infinitesimal parts is the mainspring of the theory of knowledge in infinitesimal physics as in Riemann's geometry' (1918a).

A1.3 Dynamical geometry

Riemann pointed out that if bodies were dependent on position, 'we cannot draw conclusions from metric relations of the great to those of the infinitely small'. In this case,

it seems that the empirical notions on which the metrical determinations of space are founded, the notions of a solid body and of a ray of light, cease to be valid for the infinitely small. We are therefore quite at liberty to suppose that the metric relations of space in the infinitely small do not conform to the hypotheses of (Euclidean) geometry; and we ought in fact to suppose it if we can thereby obtain a simpler explanation of phenomena.

(Riemann, 1854)

This suggested that the geometry of physical space, as a special kind of manifold, cannot be derived only from pure geometric notions about manifold. The properties that distinguished physical space from other triply extended manifolds were to be obtained only from experience, that is, by introducing measuring instruments or by having a theory of the ether force, etc. Riemann continued:

The question of the validity of the hypotheses of geometry in the infinitely small is bound up with the question of the ground of the metric relations of space, . . . in a continuous manifold . . . we must seek the ground of its metric relations outside it, in binding forces which act upon it . . . .. This leads us into the domain of another science, of physics.

(Ibid.)

This paragraph shows clearly that Riemann rejected the notion that the metric structure of space was fixed and inherently independent of the physical phenomena for which it served as a background. On the contrary, he asserted that space in itself was nothing more than a three-dimensional manifold devoid of all form: it acquired a definite form only through the material content filling it and determining its metric relations. Here the material content was described by his ether theory (see section 2.3). In view of the fact that the disposition of matter in the world changes, the metric groundform will alter in the course of time. Riemann's anticipation of such a dependence of the metric on physical data later provided a justification for avoiding the notion of absolute space, whose metric is independent of physical forces. For example, more than sixty years later, Einstein took Riemann's empirical conception of geometry presented here as an important justification for his general theory of relativity (see section 4.4).

Riemann's idea of associating matter with space in order to determine what was true of physical space was further developed by William Clifford. For Riemann, matter was the efficient cause of spatial structure: for Clifford, matter and its motion were manifestations of the varying curvature of space. As Clifford put it: 'Slight variation of the curvature may occur from point to point, and themselves vary with time . . . . We might even go so far as to assign to their variation of the curvature of space "what really happens in that phenomenon which we term the motion of matter"' (in Newman (ed.), 1946, p. 202). In 1876, Clifford published a paper "On the space-theory of matter', in which he wrote,

I hold in fact (1) That small portions of space are in fact of a nature analogous to little hills on a surface which is on the average flat; namely, that the ordinary laws of geometry are not valid in them. (2) That this property of being curved or distorted is continually being passed on from one portion of space to another after the manner of a wave. (3) That this variation of the curvature of space is what really happens in that phenomenon which we call the motion of matter, whether ponderable or ethereal. (4) That in the physical world nothing else takes place but this variation, subject (possibly) to the laws of continuity.

(1876)

It is clear that all these ideas were strongly influenced by Riemann's ether field theory, though Riemann's ether was renamed by Clifford space.

A1.4 Invariants

In his 1861 paper, Riemann formulated the general question of when a metric $ds^2 = g_{ij} dx^i dx^j$ can be transformed by the equations $x_i = x_i(y_1, \ldots, y_n)$, into a given metric $ds'^2 = h_{ij} dy^i dy^j$. The understanding was that ds would equal ds' so that the geometries of the two spaces would be the same except for the choice of coordinates (1861b). This theme was reconsidered and amplified by Elwin Christoffel in his two 1869 papers, in which the Christoffel symbols were introduced. Christoffel showed that for the μ-ply differential form $G_\mu = \Sigma(i_1, \ldots, i_\mu)\partial_1 x_{i1} \ldots \partial_\mu x_{i\mu}$, the relations $(\alpha_1, \ldots, \alpha_\mu) = \Sigma(i_1, \ldots, i_\mu)\partial x_{i1}/\partial y_{\alpha 1} \ldots \partial x_{i\mu}/\partial y_{\alpha\mu}$ were necessary and sufficient for $G_\mu = G'_\mu \equiv \Sigma(\alpha_1, \ldots, \alpha_\mu)\partial_1 y_{\alpha 1} \ldots \partial_\mu y_{\alpha\mu}$, where $(\mu + 1)$-index symbols can be derived, by the procedure that Gregorio Ricci and Tullio Levi-Civita later called 'covariant differentiation', from a μ-index symbol $(i_1, \ldots, i_\mu)$, which was defined in terms of the g_{rs}.

The above study implied that for the very same manifold different coordinate representations could be obtained. However, the geometrical properties of the manifold had to be independent of the particular coordinate system used to represent it. Analytically these geometrical properties would be represented by invariants. The invariants of interest in Riemannian geometry involve the fundamental quadratic form ds^2. From this Christoffel derived higher-order differential forms, his G_4 and G_μ, which were also invariants. Moreover, he showed how from G_μ one could derive $G_{\mu+1}$ which was also an invariant.

The concept of invariant was extended by Felix Klein (1872). For any set S and its transformation group G, if for every $x \in S$ and $f \in G$, whenever x has the property Q, $f(x)$ has Q, we say that the group G preserves Q. We may say likewise that G preserves a relation or a function defined on S^n. Any property, relation, etc., preserved by G is said to be G-invariant. Klein used these ideas to define and clarify the notion of geometry. Let S be on an n-dimensional manifold and let G be a group of transformations of S. By adjoining G to S, Klein defined a geometry on S, which consisted in the theory of G invariants. So a geometry was determined not by the particular nature of the elements of the manifold on which it was defined, but by the structure of the group of transformations that defined it.

Not all of geometry can be incorporated into Klein's scheme. Riemann's geometry of the manifold will not fit into it. If S is a manifold of non-constant curvature, it may happen that arc length is preserved by no group of transformations of S other than the trivial one which consists of the identity alone. But this trivial group cannot be said to characterize anything, let alone Riemann's geometry of the manifold S.

A1.5 Tensor calculus

A new approach to differential invariants was initiated by Ricci. Ricci was influenced by Luigi Bianchi whose work had followed Christoffel's. Ricci and his pupil Levi-Civita worked out their approach and a comprehensive notation for the subject, which they called the absolute differential calculus (1901).

Ricci's idea was that instead of concentrating on the invariant differential form $G_\mu = \Sigma (i_1, \ldots, i_\mu)\partial_1 x_{i1} \ldots \partial_\mu x_{i\mu}$, it would be sufficient and more expeditious to treat the set of n^μ components $(i_1, \ldots, i_\mu)$. He called this set a (covariant or contravariant) tensor provided that they transformed under change of coordinates in accordance with certain (covariant or contravariant) rules. The physical and geometrical significance which a tensor possessed in one coordinate system was preserved by the transformation so that it obtained again in the second coordinate system. In their 1901 paper Ricci and Levi-Civita showed how physical laws can be expressed in tensor form so as to render them independent of the coordinate system. Thus tensor analysis can be used, as Einstein did in his formulation of GTR, to express the mathematical invariance of physical laws that held for all frames of reference represented by corresponding co-ordinate systems.

Operations on tensors include addition, multiplication, covariant differentiation, and contraction. By contraction Ricci obtained what is now called the Ricci tensor or the Einstein tensor from the Riemann–Christoffel tensor. The components R_{jl} are $\sum_{k=1}^{n} R_{jlk}^{k}$. This tensor for $n = 4$ was used by Einstein to express the curvature of his spacetime Riemannian geometry.

Appendix 2. Homotopy classes and homotopy groups

A2.1 Homotopy classes

Let f_0 and f_1 be two continuous mappings from a topological space X into another one Y. They are said to be homotopic if they are continuously deformable into each other. That is, if and only if there exists a continuous deformation of maps $F(x, t)$, $0 \leqslant t \leqslant 1$, such that $F(x, 0) = f_0(x)$ and $F(x, 1) = f_1(x)$. The function $F(x, t)$ is called the homotopy. All mappings of X into Y can be devided into homotopy classes. Two mappings are in the same class if they are homotopic.

A2.2 Homotopy groups

A group structure can be defined on the set of homotopy classes. The simplest case is this. Let S^1 be a unit circle parameterized by the angle θ, with θ and $\theta + 2\pi$ identified. Then the group of homotopy classes of mappings from S^1 into the manifold of a Lie group G is called the first homotopy group of G, and is denoted by $\Pi_1(G)$. In case G is the group U(1) represented by a set of unimodular complex numbers $u = \exp(i\alpha)$, which is topologically equivalent to S^1, the elements of $\Pi_1(U(1))$ (or the continuous functions from S^1 to S^1), $\alpha(\theta) = \exp[i(N\theta + a)]$, form a homotopy class for different values of a and a fixed integer N. One can think of $\alpha(\theta)$ as a mapping of a circle into another circle such that N points of the first circle are mapped into one point of the second circle (winding N times around the second circle). For this reason the integer N is called the winding number and each homotopy class is characterized by its winding number, which takes the form $N = -i\int_0^{2\pi} (d\theta/2\pi)[(1/\alpha(\theta))(d\alpha/d\theta)]$. Thus, $\Pi_1(U(1)) = \Pi_1(S^1) = Z$, where Z denotes an additive group of integers. The mappings of any winding number can be obtained by taking powers of $\alpha^{(1)}(\theta) = \exp(i\theta)$.

This discussion can be generalized by taking $X = S^n$ (the n-dimensional sphere). The classes of mappings $S^n \rightarrow S^m$ form a group called the nth homotopy group of S^m and designated by $\Pi_n(S^m)$. Similar to $\Pi_1(S^1) = Z$, we have $\Pi_n(S^n) = Z$. That is, mappings $S^n \rightarrow S^n$ are also classified by the number of times one n-sphere covers the other. Ordinary space R^3 with all points at infinity identified is equivalent to S^3. Since any element M in the SU(2) group can be written as $M = a + i\mathbf{b} \cdot \boldsymbol{\tau}$, where the τs are Pauli matrices and where a and $\mathbf{b}$ satisfy $a^2 + \mathbf{b}^2 = 1$, the manifold of the SU(2) group elements is also topologically equivalent to S^3. Thus $\Pi_3(SU(2)) = \Pi_3(S^3) = Z$. Since any continuous mappings of S^3 into an arbitrary group SU(N) can be continuously deformed into a mapping into the SU(2) subgroup of SU(N), the same result holds for compact non-Abelian gauge groups, including SU(N), in general. That is, Π_3 (compact non-Abelian gauge groups) $= Z$.

The winding number for gauge transformations providing the mapping $S^3 \rightarrow G$ can be shown to be given by $N = (1/24\pi^2)\int d\mathbf{r} \ \epsilon^{ijk} \operatorname{tr}[U^{-1}(\mathbf{r})\partial_i U(\mathbf{r})U^{-1}(\mathbf{r})\partial_j U(\mathbf{r})U^{-1}(\mathbf{r})\partial_k U(\mathbf{r})]$, which is determined by the large distance properties of the gauge potential.

Only when the group G is U(1) is every mapping of S^3 into U(1) continuously deformable into the constant map, corresponding to $N = 0$. In this case, the gauge transformations are called homotopically trivial. All homotopicaly non-trivial gauge transformations which are not deformable to the identity are called finite or large, while the trivial ones are called infinitesimal or small.

Bibliography

Abbreviations

AJP	*American Journal of Physics*
AP	*Annals of Physics (New York)*
CMP	*Communications in Mathematical Physics*
DAN	*Doklady Akademii Nauk SSSR*
EA	The Einstein Archives in the Hebrew University of Jerusalem
JETP	*Soviet Physics, Journal of Experimental and Theoretical Physics*
JMP	*Journal of Mathematical Physics (New York)*
NC	*Nuovo Cimento*
NP	*Nuclear Physics*
PL	*Physics Letters*
PR	*Physical Review*
PRL	*Physical Review Letters*
PRS	*Proceedings of the Royal Society of London*
PTP	*Progress of Theoretical Physics*
RMP	*Reviews of Modern Physics*
ZP	*Zeitschrift für Physik*

Abers, E., Zachariasan, F. and Zemach, C. (1963). 'Origin of internal symmetries', *PR*, **132**: 1831–1836.

Abrams, G. S. *et al.* (1974). 'Discovery of a second narrow resonance', *PRL*, **33**: 1453–1454.

Achinstein, P. (1968). *Concepts of Science: A Philosophical Analysis* (The Johns Hopkins University Press, Baltimore).

Adler, S. L. (1964). 'Tests of the conserved vector current and partially conserved axial-vector current hypotheses in high-energy neutrino reactions', *PR*, **B135**: 963–966.

Adler, S. L. (1965). 'Sum-rules for axial-vector coupling-constant renormalization in beta decay', *PR*, **B140**: 736–747.

Adler, S. L. and Dashen, R. F. (1968). *Current Algebra and Applications to Particle Physics* (Benjamin, New York).

Adler, S. L. (1969). 'Axial-vector vertex in spinor electrodynamics', *PR*, **177**: 2426–2438.

Adler, S. L. and Bardeen, W. (1969). 'Absence of higher-order corrections in the anomalous axial-vector divergence equation', *PR*, **182**: 1517–1532.

Adler, S. L. and Wu-Ki Tung (1969). 'Breakdown of asymptotic sum rules in perturbation theory', *PRL*, **22**: 978–981.

Adler, S. L. (1970). 'π^0 decay', in *High-Energy Physics and Nuclear Structure*, ed. Devons, S. (Plenum, New York), 647–655.

Adler, S. L. (1991). Taped interview at Princeton, 5 Dec. 1991.

Aharonov, Y. and Bohm, D. (1959). 'Significance of electromagnetic potentials in quantum theory', *PR*, **115**: 485–491.

Ambarzumian, V. and Iwanenko, D. (1930). 'Unobservable electrons and β-rays', *Compt. Rend. Acad. Sci. Paris*, **190**: 582–584.

Anderson, P. W. (1952). 'An approximate quantum theory of the antiferromagnetic ground state', *PR*, **86**: 694–701.

Anderson, P. W. (1958a). 'Coherent excited states in the theory of superconductivity: Gauge invariance and the Meissner effect', *PR*, **110**: 827–835.

Anderson, P. W. (1958b). 'Random phase approximation in the theory of super-conductivity', *PR*, **112**: 1900–1916.

Anderson, P. W. (1963). 'Plasmons, gauge invariance, and mass', *PR*, **130**: 439–442.

Appelquist, T. and Carazzone, J. (1975). 'Infrared singularities and massive fields', *PR*, **D11**: 2856–2861.

Appelquist, T. and Politzer, H. D. (1975). 'Heavy quarks and e^+e^- annihilation', *PRL*, **34**: 43–45.

Arnison, G. *et al.* (1983a). 'Experimental observation of isolated large transverse energy electrons with asociated missing energy at $\sqrt{s} = 540$ Gev', *PL*, **122B**: 103–116.

Arnison, G. *et al.* (1983b). 'Experimental observation of lepton pairs of invariant mass around 95 GeV/c^2 at the CERN SPS collider', *PL*, **126B**: 398–410.

Arnowitt, R., Friedman, M. H. and Nath, P. (1968). 'Hard meson analysis of photon decays of π^0, η and vector mesons', *PL*, **27B**: 657–659.

Aubert, J. J. *et al.* (1974). 'Observation of a heavy particle J', *PRL*, **33**: 1404–1405.

Augustin, J.-E. *et al.* (1974). 'Discovery of a narrow resonance in e^+e^- annihilation', *PRL*, **33**: 1406–1407.

Bagnaia, P. *et al.* (1983). 'Evidence for $Z^0 \rightarrow e^+e^-$ at the CERN p$^+$p$^-$ collider', *PL*, **129B**: 130–140.

Baker, M. and Glashow, S. L. (1962). 'Spontaneous breakdown of elementary particle symmetries', *PR*, **128**: 2462–2471.

Balzer, W., Pearce, D. A. and Schmidt, H.-J. (1984). *Reduction in Science* (Reidel, Dordrecht).

Bardeen, J. (1955). 'Theory of the Meissner effect in superconductors', *PR*, **97**: 1724–1725.

Bardeen, J. (1957). 'Gauge invariance and the energy gap model of superconductivity', *NC*, **5**: 1766–1768.

Bardeen, J., Cooper, L. N. and Schrieffer, J. R. (1957). 'Theory of superconductivity', *PR*, **108**: 1175–1204.

Bardeen, W. (1969). 'Anomalous Ward identities in spinor field theories', *PR*, **184**: 1848–1859.

Bardeen, W., Fritzsch, H. and Gell-Mann, M. (1973). 'Light cone current algebra, π^0 decay and e^+e^- annihilation', in *Scale and Conformal Symmetry in Hadron Physics*, ed. Gatto, R. (Wiley, New York), 139–153.

Bardeen, W. (1985). 'Gauge anomalies, gravitational anomalies, and superstrings', talk presented at the INS International Symposium, Tokyo, August 1985; Fermilab preprint: Conf. 85/110-T.

Barnes, B. (1977). *Interests and the Growth of Knowledge* (Routledge and Kegan Paul, London).

Belavin, A. A., Polyakov, A. M., Schwartz, A. and Tyupkin, Y. (1975). 'Pseudoparticle solutions of the Yang–Mills equations', *PL*, **59B**: 85–87.

Bell, J. S. (1967a). 'Equal-time commutator in a solvable model', *NC*, **47A**: 616–625.

Bell, J. S. (1967b). 'Current algebra and gauge invariance', *NC*, **50A**: 129–134.

Bell, J. S. and Berman, S. M. (1967). 'On current algebra and CVC in pion beta-decay', *NC*, **47A**: 807–810.

Bell, J. S. and Jackiw, R. (1969). 'A PCAC puzzle: $\pi^0 \to \gamma\gamma$ in the σ-model', *NC*, **60A**: 47–61.

Bell, J. S. and van Royen, R. P. (1968). 'Pion mass difference and current algebra', *PL*, **25B**: 354–356.

Bell, J. S. and Sutherland, D. G. (1968). 'Current algebra and $\eta \to 3\pi$', *NP*, **B4**: 315–325.

Bernstein, J., (1968). *Elementary Particles and Their Currents* (Freeman, San Francisco).

Bernstein, J., Fubini, S., Gell-Mann, M. and Thirring, W. (1960). 'On the decay rate of the charged pion', *NC*, **17**: 758–766.

Bethe, H. A. (1947). 'The electromagnetic shift of energy levels', *PR*, **72**: 339-341.

Bjorken, J. D. and Glashow, S. L. (1964). 'Elementary particles and SU(4)', *PL*, **11**: 255–257.

Bjorken, J. D. (1966). 'Applications of the chiral U(6) $\otimes$ U(6) algebra of current densities', *PR*, **148**: 1467–1478.

Bjorken, J. D. (1969). 'Asymptotic sum rules at infinite momentum', *PR*, **179**: 1547–1553.

Bjorken, J. D. and Paschos, E. A. (1969). 'Inelastic electron–proton and γ–proton scattering and the structure of the nucleon', *PR*, **185**: 1975–1982.

Bjorken, J. D. and Paschos, E. A. (1970). 'High-energy inelastic neutrino–nucleon interactions', *PR*, **D1**: 3151–3160.

Blankenbecler, R., Cook, L. F. and Goldberger, M. L. (1962). 'Is the photon an elementary particle?' *PR*, **8**: 463–465.

Blankenbecler, R., Coon, D. D. and Roy, S. M. (1967). 'S-matrix approach to internal symmetry', *PR*, **156**: 1624–1636.

Blatt, J. M., Matsubara, T. and May, R.M. (1959). 'Gauge invariance in the theory of superconductivity', *PTP*, **21**: 745– 757.

Bloom, E. D. *et al.* (1969). 'High energy inelastic e–p scattering at 6° and 10°', *PRL*, **23**: 930–934.

Bloor, D. (1976). *Knowledge and Social Imagery* (Routledge and Kegan Paul, London).

Bludman, S. A. (1958). 'On the universal Fermi interaction', *NC*, **9**: 433–444.

Bogoliubov, N. N. (1958). 'A new method in the theory of superconductivity', *JETP*, **34** (7): 41–46, 51–55.

Bohr, N. (1912). In *On the Constitution of Atoms and Molecules*, ed. Rosenfeld, L. (Benjamin, New York, 1963), p.xxxii.

Bohr, N. (1913a, b, c). 'On the constitution of atoms and molecules. I, II, III', *Philos. Mag.*, **26**: 1–25, 476–502, 857–875.

Bohr, N. (1918). 'On the quantum theory of line-spectra', *Kgl. Dan. Vid. Selsk. Skr. Nat-Mat. Afd.* series 8, vol. 4, number 1, Part I-III (Høst, Copenhagen).

Bohr, N., Kramers, H. A. and Slater, J. C. (1924). 'The quantum theory of radiation', *Philos. Mag.*, **47**: 785–802.

Bohr, N. (1927). 'Atomic theory and wave mechanics', *Nature*, **119**: 262.

Bohr, N. (1928). 'The quantum postulate and the recent development of atomic theory' (a version of the Como lecture given in 1927), *Nature*, **121**: 580–590.

Bohr, N. (1930). 'Philosophical aspects of atomic theory', *Nature*, **125**: 958.

Boltzmann, L. (1888). Quoted from R. S. Cohen's 'Dialectical materialism and Carnap's logical empiricism' in *The Philosophy of Rudolf Carnap*, ed. Schilpp, P. A. (Open Court, LaSalle, 1963), 109.

Bopp, F. (1940). 'Eine lineare theorie des elektrons', *Ann. Phys.* **38**: 345–384.

Born, M. (1924). 'Über Quantenmechanik', *ZP*, **26**: 379–395.

Born, M. and Jordan, P. (1925). 'Zur Quantenmechanik', *ZP*, **34**: 858–888.

Born, M., Heisenberg, W. and Jordan, P. (1926). 'Zur Quantenmechnik. II', *ZP*, **35**: 557–615.

Born, M. (1926). 'Zur Quantenmechanik der Stossvorgange', *ZP*, **37**: 863–867.

Born, M. (1949). *Natural Philosophy of Cause and Chance* (Oxford University Press, Oxford).

Born, M. (1956). *Physics in My Generation* (Pergamon, London).

Bose, S. N. (1924). 'Plancks Gesetz und Lichtquantenhypothese', *ZP*, **26**: 178–181.

Bouchiat, C., Iliopoulos, J. and Meyer, Ph. (1972). 'An anomaly-free version of Weinberg's model', *PL*, **38B**: 519–523.

Boulware, D. (1970). 'Renormalizability of massive non-Abelian gauge fields: a functional integral approach', *AP*, **56**: 140–171.

Boyd, R. N. (1979). 'Metaphor and theory change', in *Metaphor and Thought*, ed. Ortony, A. (Cambridge University Press, Cambridge).

Boyd, R. N. (1983). 'On the current status of scientific realism', *Erkenntnis*, **19**: 45–90.

Brandt, R. A. (1967). 'Derivation of renormalized relativistic perturbation theory from finite local field equations', *AP*, **44**: 221–265

Brandt, R. A. and Orzalesi, C. A. (1967). 'Equal-time commutator and zero-energy theorem in the Lee model', *PR*, **162**: 1747–1750.

Brans, C. and Dicke, R. H. (1961). 'Mach's principle and a relativistic theory of gravitation', *PR*, **124**: 925–935.

Brans, C. (1962). 'Mach's principle and the locally measured gravitational constant in general relativity', *PR*, **125**: 388–396.

Braunbeck, W. and Weinmann, E. (1938). 'Die Rutherford-Streuung mit Berücksichtigung der Austrahlung', *ZP*, **110**: 360–372.

Breidenbach, M. *et al.* (1969). 'Observed behavior of highly inelastic electron–proton scattering', *PRL*, **23**: 935–999.

Bridgeman, P. W. (1927). *The Logic of Modern Physics* (Macmillan, New York).

Bromberg, J. (1976). 'The concept of particle creation before and after quantum mechanics', *Historical Studies in the Physical Sciences*, **7**: 161–191.

Brown, H. R. and Harré, R. (1988). *Philosophical Foundations of Quantum Field Theory* (Clarendon, Oxford).

Brown, L. M. and Hoddeson, L. (eds.) (1983). *The Birth of Particle Physics* (Cambridge University Press, Cambridge).

Brown, L. M. and Rechenberg, H. (1988). 'Landau's work on quantum field theory and high energy physics (1930–1961)', Max Planck Institute, Preprint, MPI-PAE/Pth 42/88 (July 1988).

Brown, L. M., Dresden, M. and Hoddeson, L. (eds.) (1989). *Pions to Quarks* (Cambridge University Press, Cambridge).

Brown, L. M. and Cao, T. Y. (1991). 'Spontaneous breakdown of symmetry: its rediscovery and integration into quantum field theory', *Historical Studies in the Physical and Biological Sciences* **21**: 211–235.

Bucella, F., Veneziano, G., Gatto, R. and Okubo, S. (1966). 'Necessity of additional unitary-antisymmetric q-number terms in the commutators of spatial current components', *PR*, **149**: 1268– 1272.

Buckingham, M. J. (1957). 'A note on the energy gap model of superconductivity', *NC*, **5**: 1763–1765.

Burtt, E. A. (1932). *The Metaphysical Foundations of Modern Physical Science* (Doubleday, Garden City). (First edition in 1924, revised version first appeared in 1932.)

Cabibbo, N. (1963). 'Unitary symmetry and leptonic decays', *PRL*, **10**: 531–533.

Cabrera, B. (1982). 'First results from a superconductive detector for moving magnetic monopoles', *PRL*, **48**: 1378–1381.

Callan, C. and Gross, D. (1968). 'Crucial test of a theory of currents', *PRL*, **21**: 311–313.

Callan, C. and Gross, D. (1969). 'High-energy electroproduction and the constitution of the electric current', *PRL*, **22**: 156–159.

Callan, C. (1970). 'Bjorken scale invariance in scalar field theory', *PR*, **D2**: 1541–1547.

Callan, C., Coleman, S. and Jackiw, R. (1970). 'A new improved energy-momentum tensor', *AP*, **59**: 42–73.

Callan, C. (1972). 'Bjorken scale invariance and asymptotic behavior', *PR*, **D5**: 3202–3210.

Callan, C. and Gross, D. (1973). 'Bjorken scaling in quantum field theory', *PR*, **D8**: 4383–4394.

Callan, C., Dashen, R. and Gross, D. (1976). 'The structure of the vacuum', *PL*, **63B**: 334–340.

Carmeli, M. (1976). 'Modified gravitational Lagrangian', *PR*, **D14**: 1727.

Carmeli, M. (1977). 'SL_{2c} conservation laws of general relativity', *NC*, **18**: 17–20.

Cantor, G. N. and Hodge, M. J. S. (eds.) (1981). *Conceptions of Ether* (Cambridge University Press, Cambridge).

Cao, T. Y. (1991). 'The Reggeization program 1962–1982: attempts at reconciling quantum field theory with S-matrix theory', *Archive for History of Exact Sciences*, **41**: 239–283.

Cao, T. Y. (1993). 'What is meant by social constructivism? – A critical exposition', a talk given at the Dibner Institute, MIT, on 19 October 1993.

Cao, T. Y. and Schweber, S. S. (1993). 'The conceptual foundations and the philosophical aspects of renormalization theory', *Synthese*, **97**: 33–108.

Capps, R. H. (1963). 'Prediction of an interaction symmetry from dispersion relations, *PRL*, **10**: 312–314.

Capra, F. (1979). 'Quark physics without quarks: a review of recent developments in S-matrix theory', *AJP*, **47**: 11–23.

Capra, F. (1985). 'Bootstrap physics: a conversation with Geoffrey Chew', in *A Passion for Physics: Essays in Honour of Geoffrey Chew*, eds. De Tar, C., Finkelstein, J. and Tan, C. I. (Taylor and Francis, Philadelphia), 247–286.

Carnap, R. (1929). *Der Logisches Aufbau der Welt* (Schlachtensee Weltkreis-Verlag, Berlin).

Carnap, R. (1956). *Meaning and Necessity*, enlarged edition (University of Chicago Press, Chicago).

Cartan, E. (1922). 'Sur une géneralisation de la notion de courbure de Riemann et les éspaces à torsion', *Compt. Rend. Acad. Sci. Paris*, **174**: 593–595.

Case, K. M. (1949). 'Equivalence theorems for meson–nucleon coupling', *PR*, **76**: 1–14.

Cassidy, D. C. and Rechenberg, H. (1985). 'Biographical data, Werner Heisenberg (1901–1976)', in W. Heisenberg, *Collected Works*, eds. Blum, W. *et al.*, Series A, part I (Berlin), 1–14.

Castillejo, L., Dalitz, R. H. and Dyson, F. J. (1956). 'Low's scattering equation for the charged and neutral scalar theories', *PR*, **101**: 453–458.

Cauchy, A. L. (1828). *Exercise de Mathématiques*, **3**: 160.

Cauchy, A. L. (1830). 'Mémoire sur la théorie de la lumière', *Mém. de l'Acad.*, **10**: 293–316.

Chadwick, J. (1914). 'Intensitätsvertieilung im magnetischen Spektrum der β-strahlen von Radium B + C', *Ber. Deut. Phys. Gens.* **12**: 383–391.

Chadwick, J. (1932). 'Possible existence of a neutron', *Nature*, **129**: 312.

Chambers, R. G. (1960). 'Shift of an electron interference pattern by enclosed magnetic flux', *PRL*. **5**: 3–5.

Chandler, C. (1968). 'Causality in S-matrix theory', *PR*. **174**: 1749–1758.

Chandler, C., and Stapp, H. P. (1969). 'Macroscopic causality and properties of scattering amplitudes', *JMP*. **10**: 826–859.

Chandrasekhar, S. (1931). 'The maximum mass of ideal white dwarfs', *Astrophys. J.*, **74**: 81.

Chandrasekhar, S. (1934). 'Stellar configurations with degenerate cores', *Observatory*, **57**: 373–377.

Chandrasekhar, S. (1935). *Observatory*, **58**: 38.

Chanowitz, M. S., Furman, M. A. and Hinchliffe, Z. (1978). 'Weak interactions of ultra heavy fermions', *PR*, **B78**: 285–289.

Charap, J. M. and Fubini, S. (1959). 'The field theoretic definition of the nuclear potential-I', *NC*, **14**: 540–559.

Chew, G. F. (1953a). 'Pion–nucleon scattering when the coupling is weak and extended', *PR*, **89**: 591–593.

Chew, G. F. (1953b). 'A new theoretical approach to the pion-nucleaon interaction', *PR*, **89**: 904.

Chew, G. F. and Low, F. E. (1956). 'Effective-range approach to the low-energy p-wave pion–nucleon interaction', *PR*, **101**: 1570–1579.

Chew, G. F., Goldberger, M. L., Low, F. E. and Nambu, Y. (1957a). 'Application of dispersion relations to low-energy meson-nucleon scattering', *PR*, **106**: 1337–1344.

Chew, G. F., Goldberger, M. L., Low, F. E. and Nambu, Y. (1957b). 'Relativistic dispersion relation approach to photomeson production', *PR*, **106**: 1345–1355.

Chew, G. F. and Frautschi, S. C. (1960). 'Unified approach to high- and low-energy strong interactions on the basis of the Mandelstam representation', *PRL*, **5**: 580–583.

Chew, G. F. and Mandelstam, S. (1960). 'Theory of low energy pion-pion interaction', *PR*, **119**: 467–477.

Chew, G. F. (1961). *S-Matrix Theory of Strong Interactions* (Benjamin, New York).

Chew, G. F. and Frautschi, S. C. (1961a). 'Dynamical theory for strong interactions at low momentum transfer but arbitrary energies', *PR*, **123**: 1478–1486.

Chew, G. F. and Frautschi, S. C. (1961b). 'Potential scattering as opposed to scattering associated with independent particles in the S-matrix theory of strong inter-actions', *PR*, **124**: 264–268.

Chew, G. F. and Frautschi, S. C. (1961c). 'Principle of equivalence for all strongly interacting particles within the S-matrix framework', *PRL*, **7**: 394–397.

Chew, G. F. and Frautschi, S. C. (1962). 'Regge trajectories and the principle of maximum strength for strong interactions', *PRL*, **8**: 41–44.

Chew, G. F. (1962a). 'S-matrix theory of strong interactions without elementary particles', *RMP*, **34**: 394–401.

Chew, G. F. (1962b). 'Reciprocal bootstrap relationship of the nucleon and the (3, 3) resonance', *PRL*, **9**: 233–235.

Chew, G. F. (1962c). 'Strong interaction theory without elementary particles', in *Proceedings of the 1962 International Conference on High Energy Physics at CERN*, ed. Prentki, J. (CERN, Geneva), 525–530.

Chew, G. F. (1989). 'Particles as S-matrix poles: hadron democracy', in *Pions to Quarks: Particle Physics in the 1950s*, eds. Brown, L. M., Dresden, M. and Hoddeson, L. (Cambridge University Press, Cambridge), 600–607.

Chihara, C. (1973). *Ontology and the Vicious Circle Principle* (Cornell University Press, Ithaca).

Cho, Y. M. (1975). 'Higher-dimensional unifications of gravitation and gauge theories', *JMP*, **16**: 2029–2035.

Cho, Y. M. (1976). 'Einstein Lagrangian as the translational Yang-Mills Lagrangian', *PR*, **D14**: 2521–2525.

Christoffel, E. B. (1869a). 'Ueber die Transformation der homogenen Differential-ausdrücke zweiten Grades', *J. r. angew. Math.* **70**: 46–70.

Christoffel, E. B. (1869b). 'Ueber ein die Transformation hamogen Differential-ausdrücke zweiten Grades betreffendes Theorem', *J. r. angew. Math.* **70**: 241–245.

Clifford, W. K. (1876). 'On the space-theory of matter', in *Mathematical Papers*, ed. Tucker, R. (Macmillan, London).

Coleman, S. (1965). 'Trouble with relativistic SU(6)', *PR*, **B138**: 1262–1267.

Coleman, S. and Mandula, J. (1967). 'All possible symmetries of the S matrix', *PR*, **B159**: 1251–1256.

Coleman, S. and Jackiw, R. (1971). 'Why dilatation generators do not generate dilatations', *AP*, **67**: 552–598.

Coleman, S. and Gross, D. (1973). 'Price of asymptotic freedom', *PRL*, **31**: 851–854.

Coleman, S. and Weinberg, E. (1973). 'Radiative corrections as the origin of sponta-neous symmetry breaking', *PR*, **D7**: 1888–1910.

Coleman, S. (1977). 'The use of instantons', a talk later published in *The Ways in Subnuclear Physics*, ed. Zichichi, A. (Plenum, New York, 1979).

Coleman, S. (1979). 'The 1979 Nobel Prize in physics', *Science*, **206**: 1290–1292.

Coleman, S. (1985). *Aspects of Symmetry* (Cambridge University Press, Cambridge).

Collins, C. B. and Hawking, S. W. (1973a). 'The rotation and distortion of the universe' *Monthly Notices of the Royal Astronomical Society*, **162**: 307–320.

Collins, C. B. and Hawking, S. W. (1973b). 'Why is the universe isotropic?', *Astrophys. J.* **180**: 317–334.

Collins, J. C., Wilczek and Zee, A. (1978). 'Low-energy manifestations of heavy particles: application to the neutral current', *PR*. **D18**: 242–247.

Collins, J. C. (1984). *Renormalization* (Cambridge University Press, Cambridge).

Collins, P. D. B. (1977). *An Introduction to Regge Theory and High Energy Physics* (Cambridge University Press, Cambridge).

Compton, A. (1923a). 'Total reflection of X-rays', *Philos. Mag.* **45**: 1121–1131.

Compton, A. (1923b). 'Quantum theory of the scattering of X-rays by light elements', *PR*, **21**: 483–502.

Coster, J. and Stapp, H. P. (1969). 'Physical-region discontinuity equations for many-particle scattering amplitudes. I', *JMP*, **10**: 371–396.

Coster, J. and Stapp, H. P. (1970a). 'Physical-region discontinuity equations for many-particle scattering amplitudes. II', *JMP*, **11**: 1441–1463.

Coster, J. and Stapp, H. P. (1970b). 'Physical-region discontinuity equations', *JMP*, **11**: 2743–2763.

Creutz, M. (1981). 'Roulette wheels and quark confinement', *Comments on Nuclear and Particle Physics*, **10**: 163–173.

Crewther, R., Divecchia, P., Veneziano, G. and Witten, E. (1979). 'Chiral estimate of the electric dipole moment of the neutron in quantum chromodynamics', *PL*, **88B**: 123–127.

Curie, P. (1894). 'Sur la symetrie dans les phenomenes physiques, symetrie d'un champ electrique et d'un champ magnetique', *J. Phys. (Paris)*, **3**: 393–415.

Cushing, J. T. (1986). 'The importance of Heisenberg's S-matrix program for the theoretical high-energy physics of the 1950's', *Centaurus*, **29**: 110–149.

Cushing, J. T. (1990). *Theory Construction and Selection in Modern Physics: The S-Matrix Theory* (Cambridge University Press, Cambridge).

Cutkosky, R. E. (1960). 'Singularities and discontinuities of Feynman amplitudes', *JMP*, **1**: 429–433.

Cutkosky, R. E. (1963a). 'A model of baryon states', *AP*, **23**: 415–438.

Cutkosky, R. E. (1963b). 'A mechanism for the induction of symmetries among the strong interactions', *PR*, **131**: 1888–1890.

Cutkosky, R. E. and Tarjanne, P. (1963). 'Self-consistent derivations from unitary symmetry', *PR*, **132**: 1354–1361.

Dancoff, S. M. (1939). 'On radiative corrections for electron scattering', *PR*, **55**: 959–963.

Daniel, M, and Viallet, C. M. (1980). 'The geometrical setting of gauge theories of the Yang–Mills type', *RMP*, **52**: 175–197.

de Broglie, L. (1923a). 'Ondes et quanta', *Compt. Rend. Acad. Sci. Paris*, **177**: 507–510.

de Broglie, L. (1923b). 'Quanta de lumiere, diffraction et interferences', *Compt. Rend. Acad. Sci. Paris*, **177**: 548–550.

de Broglie, L. (1926). 'The new undulatory mechanics', *Compt. Rend. Acad. Sci. Paris*, **183**: 272–274.

de Broglie, L. (1927a). 'Possibility of relating interference and diffraction phenomena to the theory of light quanta', *Compt. Rend. Acad. Sci. Paris*, **183**: 447–448.

de Broglie, L. (1927b). 'La mécanique ondulatoire et la structure atomique de la matiére et du rayonnement', *J. Phys. Radium* **8**: 225–241.

de Broglie, L. (1960). *Non-linear Wave Mechanics: A Causal Interpretation*, trans. Knobel, A. J. and Miller, J. C. (Elsevier, Amsterdam).

de Broglie, L. (1962). *New Perspectives in Physics*, trans. Pomerans, A. J. (Oliver and Boyd, Edinburgh).

Debye, P. (1910a). Letter to Sommerfeld, 2 March 1910.

Debye, P. (1910b). 'Der Wahrscheinlichkeitsbegriff in der Theorie der Strahlung', *Ann. Phys.*, **33**: 1427–1434.

Debye, P. (1923). 'Zerstreuung von Röntgenstrahlen und quantentheorie', *Phys. Z.*, **24**: 161–166.

Demopoulos, W. and Friedman, M. (1985). 'Critical notice: Bertrand Russell's *The Analysis of Matter*: its historical context and contemporary interest', *Philos. Sci.*, **52**: 621–639.

Deser, S. (1970). 'Self-interaction and gauge invariance', *General Relativity and Gravitation*, **1**: 9.

de Sitter, W. (1916a). 'On the relativity of rotation in Einstein's theory', *Proceedings of the Section of Science (Koninklijke Akademie van Wetenschappen te Amsterdam)*, **19**: 527–532.

de Sitter, W. (1916b). 'On Einstein's theory of gravitation and its astronomical consequences. I, II', *Monthly Notices of the Royal Astronomical Society*, **76**: 699–738; **77**: 155–183.

de Sitter, W. (1917a). 'On the relativity of inertia: remarks concerning Einstein's latest hypothesis', *Proceedings of the Section of Science (Koninklijke Akademie van Wetenschappen te Amsterdam)*, **19**: 1217–1225.

de Sitter, W. (1917b). 'On the curvature of space', *Proceedings of the Section of Science (Koninklijke Akademie van Wetenschappen te Amsterdam)*, **20**: 229–242.

de Sitter, W. (1917c). 'On Einstein's theory of gravitation and its astronomical consequences. Third paper', *Monthly Notices of the Royal Astronomical Society*, **78**: 3–28.

de Sitter, W. (1917d). 'Further remarks on the solutions of the field equations of Einstein's theory of gravitation', *Proceedings of the Section of Science (Koninklijke Akademie van Wetenschappen te Amsterdam)*, **20**: 1309–1312.

de Sitter, W. (1917e). Letter to Einstein, 1 April 1917, EA: 20–551.

de Sitter, W. (1920). Letter to Einstein, 4 November 1920, EA: 20–571.

de Sitter, W. (1931). 'Contributions to a British Association discussion on the evolution of the universe', *Nature*, **128**: 706–709.

DeWitt, B. S. (1964). 'Theory of radiative corrections for non-Abellian gauge fields', *PRL*, **12**: 742–746.

DeWitt, B. S. (1967a). 'Quantum theory of gravity. I. The canonical theory', *PR*, **160**: 1113–1148.

DeWitt, B. S. (1967b). 'Quantum theory of gravity. II. The manifestly covariant theory', *PR*, **162**: 1195–1239.

DeWitt, B. S. (1967c). 'Quantum theory of gravity. III. Applications of the covariant theory', *PR*, **162**: 1239–1256.

Dirac, P. A. M. (1925). 'The fundamental equations of quantum mechanics', *PRS*, **A109**: 642–653.

Dirac, P. A. M. (1926a). 'Quantum mechanics and a preliminary investigation of the hydrogen atom', *PRS*, **A110**: 561–579.

Dirac, P. A. M. (1926b). 'On the theory of quantum mechanics', *PRS*, **A112**: 661–677.

Dirac, P. A. M. (1927a). 'The physical interpretation of the quantum dynamics', *PRS*, **A113**: 621–641.

Dirac, P. A. M. (1927b). 'The quantum theory of emission and absorption of radiation', *PRS*, **A114**: 243–265.

Dirac, P. A. M. (1927c). 'The quantum theory of dispersion', *PRS*, **A114**: 710–728.

Dirac, P. A. M. (1928a). 'The quantum theory of the electron', *PRS*, **A117**: 610–624.

Dirac, P. A. M. (1928b). 'The quantum theory of the electron. Part II', *PRS*, **A118**: 351–361.

Dirac, P. A. M. (1930a). 'A theory of electrons and protons', *PRS*, **A126**: 360–365.

Dirac, P. A. M. (1930b). *The Principles of Quantum Mechanics* (Clarendon, Oxford).

Dirac, P. A. M. (1931). 'Quantized singularities in the electromagnetic field', *PRS*, **A133**: 60–72.

Dirac, P. A. M. (1932). 'Relativistic quantum mechanics', *PRS*, **A136**: 453–464.

Dirac, P. A. M. (1933). 'Théorie du positron', in *Rapport du Septième Conseil de Solvay Physique, Structure et Propriétés des noyaux atomiques (22–29 Oct. 1933)* (Gauthier-Villars, Paris), 203–212.

Dirac, P. A. M. (1934). 'Discussion of the infinite distribution of electrons in the theory of the positron', *Proc. Cambridge Philos. Soc.*, **30**: 150–163.

Dirac, P. A. M. (1938). 'Classical theory of radiating electrons', *PRS*, **A167**: 148–169.

Dirac, P. A. M. (1939). 'La théorie de l'électron et du champ électromagnetique', *Ann. Inst. Henri Poincaré*, **9**: 13–49.

Dirac, P. A. M. (1942). 'The physical interpretation of quantum mechanics', *PRS*, **A180**: 1–40.

Dirac, P. A. M. (1948). 'Quantum theory of localizable dynamic systems', *PR*, **73**: 1092–1103.

Dirac, P. A. M. (1951). 'Is there an aether?' *Nature*, **168**: 906–907.

Dirac, P. A. M. (1952). 'Is there an aether?' *Nature*, **169**: 146 and 702.

Dirac, P. A. M. (1963). 'The evolution of the physicist's picture of nature', *Scientific American*, **208**(5): 45–53.

Dirac, P. A. M. (1968). 'Methods in theoretical physics', in *Special Supplement of IAEA Bulletin* (IAEA, Vienna, 1969), 21–28.

Dirac, P. A. M. (1969a). 'Can equations of motion be used?', in *Coral Gables Conference on Fundamental Interactions at High Energy, Coral Gables, 22–24 Jan. 1969* (Gordon and Breach, New York), 1–18.

Dirac, P. A. M. (1969b). 'Hopes and fears', *Eureka*, **32**: 2–4.

Dirac, P. A. M. (1973a). 'Relativity and quantum mechanics', in *The Past Decades in Particle Theory*, eds. Sudarshan, C. G. and Neéman, Y. (Gordon and Breach, New York), 741–772.

Dirac, P. A. M. (1973b). 'Development of the physicist's conception of nature', in *The Physicist's Conception of Nature*, ed. Mehra, J. (Reidel, Dordrecht), 1–14.

Dirac, P. A. M. (1977). 'Recollections of an exciting era', in *History of Twentieth Century Physics*, ed. Weiner, C. (Academic Press, New York), 109–146.

Dirac, P. A. M. (1978). *Directions in Physics* (Wiley, New York)

Dirac, P. A. M. (1981). 'Does renormalization make sense?', in *Perturbative Quantum Chromodynamics*, eds. Duke, D. W. and Owen, J. F. (AIP Conference Proceedings No. 74, American Institute of Physics, New York), 129–130.

Dirac, P. A. M. (1983). 'The origin of quantum field theory', in *The Birth of Particle Physics*, eds. Brown, L. M. and Hoddeson, L. (Cambridge University Press, Cambridge), 39–55.

Dirac, P. A. M. (1984a). 'The future of atomic physics', *Int. J. Theor. Phys.*, **23**(8): 677–681.

Dirac, P. A. M. (1984b). 'The requirements of fundamental physical theory', *Eur. J. Phys.*, **5**: 65–67.

Dirac, P. A. M. (1987). 'The inadequacies of quantum field theory', in *Reminiscences about a Great Physicist: Paul Adrien Maurice Dirac*, eds. Kursunoglu, B. N. and Wigner, E. P. (Cambridge University Press, Cambridge).

Dolen, R., Horn, O. and Schmid, C. (1967). 'Prediction of Regge parameters of ρ poles from low-energy πN data', *PRL*, **19**: 402–407.

Dolen, R., Horn, O. and Schmid, C. (1968). 'Finite-energy sum rules and their applications to πN charge exchange', *PR*, **166**: 1768–1781.

Doran, B. G. (1975). 'Origins and consolidation of field theory in nineteenth century Britain', *Hist. Stud. Phys. Sci.*, **6**: 133–260.

Dorfman, J. (1930). 'Zur Frage über die magnetischen Momente der Atomkerne', *ZP*, **62**: 90–94.

Duane, W. (1923). 'The transfer in quanta of radiation momentum to matter', *Proc. Nat. Acad. Sci.*, **9**: 158–164.

Duhem, P. (1906). *The Aim and Structure of Physical Theory*, (Princeton University Press, Princeton, 1954).

Dürr, H. P., Heisenberg, W., Mitter, H., Schlieder, S. and Yamazaki, K. (1959). 'Zur theorie der elementarteilchen', *Zeitschrift für Naturforschung*, **14A**: 441–485.

Dürr, H. P. and Heisenberg, W. (1961). 'Zur theorie der "seltsamen" teilchen', *Zeitschrift für Naturforschung*, **16A**: 726–747.

Dyson, F. J. (1949a). 'The radiation theories of Tomonaga, Schwinger and Feynman', *PR*, **75**: 486–502.

Dyson, F. J. (1949b). 'The *S*-matrix in quantum electrodynamics', *PR*, **75**: 1736–1755.

Dyson, F. J. (1951). 'The renormalization method in quantum electrodynamics', *PRS*, **A207**: 395–401.

Dyson, F.J. (1952). 'Divergence of perturbation theory in quantum electrodynamics', *PR*, **85**: 631–632.

Dyson, F. J. (1965). 'Old and new fashions in field theory', *Physics Today*, **18**(6): 21–24.

Earman, J., Glymore, C. and Stachel, J. (eds.) (1977). *Foundations of Space-Time Theories* (University of Minnesota Press, Minneapolis).

Earman, J. (1979). 'Was Leibniz a relationist?' In *Studies in Metaphysics*, eds. French, P. and Wettstein, H. (University of Minnesota Press, Minneapolis).

Earman, J. and Norton, J. (1987). 'What price space-time substantivalism? The hole story', *British Journal for the Philosophy of Science*, **38**: 515–525.

Earman, J. (1989). *World-Enough and Space-Time* (MIT Press, Cambridge, MA).

Eddington, A. S. (1916). Letter to de Sitter, 13 October 1916, Leiden Observatory, quoted by Kerszberg (1989).

Eddington, A. S. (1918). *Report on the Relativity theory of Gravitation* (Fleetway, London).

Eddington, A. S. (1921). 'A generalization of Weyl's theory of the electromagnetic and gravitational fields', *PRS*, **A99**: 104–122.

Eddington, A. S. (1923). *The Mathematical Theory of Gravitation* (Cambridge University Press, Cambridge).

Eddington, A. S. (1926). *The Internal Constitution of the Stars* (Cambridge University Press, Cambridge).

Eddington, A. S. (1930). 'On the instability of Einstein's spherical world', *Monthly Notices of the Royal Astronomical Society*, **90**: 668–678.

Eddington, A. S. (1935). 'Relativistic degeneracy', *Observatory*, **58**: 37–39.

Ehrenfest, P. (1906). 'Zur Planckschen Strahlungstheorie', *Phys. Z.*, **7**: 528–532.

Ehrenfest, P. (1911). 'Welche Zuge der lichtquantenhypothese spielen in der Theorie die Warmestrahlung eine wesentliche Rolle?' *Ann. Phys.*, **36**: 91–118.

Ehrenfest, P. and Kamerling-Onnes, H. (1915). 'Simplified deduction of the formula from the theory of combinations which Planck uses as the basis for radiation theory', *Proc. Amsterdam Acad.*, **23**: 789–792.

Ehrenfest, P. (1916). 'Adiabatische invarianten und quantentheorie', *Ann. Phys.*, **51**: 327–352.

Eichten, E., Gottfried, K., Kinoshita, T., Koght, J., Lane, K. D. and Yan, T.-M. (1975). 'Spectrum of charmed quark-antiquark bound states', *PRL*, **34**: 369–372.

Einstein, A. (1905a). 'Über einen die Erzeugung und Verwandlung des lichtes betreffenden heuristischen Gesichtspunkt', *Ann. Phys.*, **17**: 132–148.

Einstein, A. (1905b). 'Die von der molekulärkinetischen Theorie der Wärme geforderte bewegung von in ruhenden Flüssigkeiten suspendierten Teilchen', *Ann. Phys.*, **17**: 549–560.

Einstein, A. (1905c). 'Zur Elektrodynamik bewegter Körper', *Ann. Phys.*, **17**: 891–921.

Einstein, A. (1905d). 'Ist die Trägheit eines Körpers von seinem Energiegehalt abhängig?', *Ann. Phys.*, **18**: 639–641.

Einstein, A. (1906a). 'Zur Theorie der Lichterzeugung und Lichtabsorption', *Ann. Phys.*, **20**: 199–206.

Einstein, A. (1906b). 'Das Prinzip von der Erhaltung der Schwerpunktsbewegung und die Trägheit der Energie', *Ann. Phys.*, **20**: 627–633.

Einstein, A. (1907a). 'Die vom Relativitätsprinzip geforderte Trägheit der Energie', *Ann. Phys.*, **23**: 371–384.

Einstein, A. (1907b). 'Über das Relativitätsprinzip und aus demselben gezogenen Folgerungen', *Jahrb. Radioakt. Elektron.*, **4**: 411–462.

Einstein, A. (1909a). 'Zum gegenwärtigen Stand des Strhlungsproblems', *Phys. Z.*, **10**: 185–193.

Einstein, A. (1909b). 'Über die Entwicklung unserer Anschauungen über das wesen und die Konstitution der Strahlung', *Phys. Z.*, **10**: 817–825.

Einstein, A. (1909c). Letter to Lorentz, 23 May 1909.

Einstein, A. (1911). 'Über den Einfluss der Schwerkraft aus die Ausbreitung des Lichtes', *Ann. Phys.* **35**: 898–908.

Einstein, A. (1912a). 'Lichtgeschwindigkeit und statik des Gravitationsfeldes', *Ann. Phys.* **38**: 355–369.

Einstein, A. (1912b). 'Zur Theorie des statischen Gravitationsfeldes', *Ann. Phys.* **38**: 443–458.

Einstein, A. and Grossmann, M. (1913). 'Entwurf einer verallgemeinerten Relativitätstheorie und einer Theorie der Gravitation', *Z. Math. Phys.*, **62**: 225–261.

Einstein, A. (1913a). 'Zum gegenwärtigen Stande des Gravitationsproblems', *Phys. Z.*, **14**: 1249–1266.

Einstein, A. (1913b). 'Physikalische Grundlagen einer gravitationstheorie', *Vierteljahrsschr. Naturforsch. Ges. Zürich*, **58**: 284–290.

Einstein, A. (1913c). A letter to Mach, E. 25 June 1913; quoted by Holton, G. (1973).

Einstein, A. (1914a). 'Prinzipielles zur verallgeneinerten Relativitästheorie und Gravitationstheorie', *Phys. Z.*, **15**: 176–180.

Einstein, A. (1914b). 'Die formale Grundlage der allgemeinen Relativitätstheorie', *Preussische Akad. Wiss. Sitzungsber.*: 1030–1085.

Einstein, A. (1915a). 'Zur allgemeinen Relativitätstheorie', *Preussische Akad. Wiss. Sitzungsber.*: 778–786.

Einstein, A. (1915b). 'Zur allgemeinen Relativitätstheorie (Nachtrag)', *Preussische Akad. Wiss. Sitzungsber.*: 799–801.

Einstein, A. (1915c). 'Erklärung der perihelbewegung des Merkur aus der allgemeinen Relativitätstheorie', *Preussische Akad. Wiss. Sitzungsber.*: 831–839.

Einstein, A. (1915d). 'Die Feldgleichungen der Gravitation', *Preussische Akad. Wiss. Sitzungsber.*: 844–847.

Einstein, A. (1915e). Letter to Ehrenfest, P., 26 Dec. 1915. EA: 9–363.

Einstein, A. (1916a). Letter to Besso, M., 3 Jan. 1916, in Speziali, P. (1972).

Einstein, A. (1916b). Letter to Schwarzschild, K., 9 Jan. 1916. EA: 21–516.

Einstein, A. (1916c). 'Die Grundlage der allgemeinen Relativitätstheorie', *Ann. Phys.* **49**: 769–822.

Einstein, A. (1916d). 'Ernst Mach', *Phys. Z.*, **17**: 101–104.

Einstein, A. (1917a). 'Kosmologische Betrachtungen zur allgemeinen Relativitätstheorie', *Preussische Akad. Wiss. Sitzungsber.*, 142–152.

Einstein, A. (1917b). 'Zur quantentheorie der strahlung', *Phys. Z.*, **18**: 121–128.

Einstein, A. (1917c). Letter to de Sitter, 12 March 1917. EA: 20–542.

Einstein, A. (1917d). Letter to de Sitter, 14 June 1917. EA: 20–556.

Einstein, A. (1917e). Letter to de Sitter, 8 August 1917. EA: 20–562.

Einstein, A. (1918a). 'Prinzipielles zur allgemeinen Relativitätstheorie', *Ann. Phys.* **55**: 241–244.

Einstein, A. (1918b). 'Kritischen zu einer von Herrn de Sitter gegebenen Lösung der Gravitationsgleichungen', *Preussische Akad. Wiss. Sitzungsber.*, 270–272.

Einstein, A. (1918c). 'Dialog über Einwande gegen die Relativitätstheorie', *Naturwissenschaften.* **6**: 197–702.

Einstein, A. (1918d). ' "Nachtrag" zu H. Weyl: "Gravitation und elektrizität" ', *Preussische Akad. Wiss. Sitzungsber.*, 478–480.

Einstein, A. (1919). 'Spielen Gravitationsfelder im Aufbau der materiellen Elementarteilchen eine wesentliche Rolle?', *Preussische Akad. Wiss. Sitzungsber.*, 349–356.

Einstein, A. (1920a). Äther und Relativitätstheorie (Springer, Berlin).

Einstein, A. (1920b). *Relativity* (Methuen).

Einstein, A. (1921a). 'Geometrie und Erfahrung', *Preussische Akad. Wiss. Sitzungsber.*, 123–130.

Einstein, A. (1921b). 'A brief outline of the development of the theory of relativity', *Nature*, 106: 782–784.

Einstein, A. (1922). *The Meaning of Relativity* (Princeton University Press, Princeton).

Einstein, A. (1923a). 'Zur affinen Feldtheorie', *Preussische Akad. Wiss. Sitzungsber.*, 137–140.

Einstein, A. (1923b). 'Bietet die Feldtheorie Möglichkeiten für die Lösung des Quanten problems?', *Preussische Akad. Wiss. Sitzungsber.*, 359–364.

Einstein, A. (1923c). 'Notiz zu der Arbeit von A. Friedmann "Über die Krümmung des Raumes" ', *ZP*, **16**: 228.

Einstein, A. (1924). 'Quantentheorie des einatomigen idealen Gases', *Preussische Akad. Wiss. Sitzungsber.*, 261–267.

Einstein, A. (1925a). 'Quantentheorie des einatomigen idealen Gases. Zweite Abhandlung', *Preussische Akad. Wiss. Sitzungsber.*, 3–14

Einstein, A. (1925b). 'Non-Euclidean geometry and physics', *Neue Rundschau*, **1**: 16–20.

Einstein, A. (1927). 'The meaning of Newton and influence upon the development of theoretical physics', *Naturwissenschaften*, **15**: 273–276.

Einstein, A. (1929a). 'Field: old and new', *New York Times*, 3 Feb. 1929.

Einstein, A. (1929b). 'Zur einheitlichen Feldtheorie', *Preussische Akad. Wiss. Sitzungsber.*, 2–7.

Einstein, A. (1929c). 'Professor Einstein spricht über das physikalische Raum- und Äther-Problem', *Deutsche Bergwerks-Zeitung*, 15 Dec. 1929, p. 11.

Einstein, A. (1930a). 'Raum-, Feld- und Äther-Problem in der physik', in *Gesamtbericht, Zweite Weltkraftkonferenz, Berlin, 1930*, eds. Neden, F. and Kromer, C. (VDI-Verlag, Berlin), 1–5.

Einstein, A. (1930b). Letter to Weiner, A., 18 Sept. 1930; quoted by Holton (1973).

Einstein, A. (1931). 'Zum kosmologischen problem der allgemeinen Relativitätstheorie', *Preussische Akad. Wiss. Sitzungsber.* 235–237.

Einstein, A. (1933a). *On the Method of Theoretical Physics* (Clarendon, Oxford).

Einstein, A. (1933b). *Origins of the General Theory of Relativity* (Jackson, Glasgow).

Einstein, A. (1936). 'Physics and reality', *J. Franklin Inst.* **221**: 313–347.

Einstein, A., Infeld, L. and Hoffmann, B. (1938). 'Gravitational equations and the problem of motion', *Ann. Math.*, **39**: 65–100.

Einstein, A. (1945). 'A generalization of relativistic theory of gravitation', *Ann. Math.*, **46**: 578–584.

Einstein, A. (1948a). A letter to Barnett, L. 19 June 1948, quoted by Stachel. J. in his 'Notes on the Andover conference', in Earman *et al.* (1977), ix.

Einstein, A. (1948b). 'Generalized theory of gravitation', *RMP*, **20**: 35–39.

Einstein, A. (1949). 'Autobiographical notes', in *Albert Einstein: Philosopher-Scientist*, ed. Schilpp, P. A. (The Library of Living Philosophers, Evanston), 1–95.

Einstein, A. (1950a). 'On the generalized theory of gravitation', *Scientific American*, **182**(4): 13–17.

Einstein, A. (1950b). Letter to Viscount Samuel, 13 Oct. 1950, in *In Search of Reality*, by Samuel, V. (Blackwell, 1957), p. 17.

Einstein, A. (1952a). 'Relativity and the problem of space', appendix 5 in the 15th edition of *Relativity: The Special and the General Theory* (Methuen, London, 1954), 135–157.

Einstein, A. (1952b). Preface to the fifteenth edition of *Relativity: The Special and the General Theory* (Methuen, 1954).

Einstein, A. (1952c). Letter to Seelig, C., 8 April 1952, quoted by Holton, G. (1973).

Einstein, A. (1952d). Letter to Max Born, 12 May 1952, in *The Born–Einstein Letters* (Walker, New York, 1971)

Einstein, A. (1953). Foreword to *Concepts of Space*, by Jammer, M. (Harvard University Press, Cambridge, MA).

Einstein, A. (1954a). Letter to Pirani, F., 2 Feb. 1954.

Einstein, A. (1954b). 'Relativity theory of the non-symmetrical field', appendix 2 in *The Meaning of Relativity* (Princeton University Press, Princeton, 1955), 133–166.

Ellis, G. (1989). 'The expanding universe: a history of cosmology from 1917 to 1960', in *Einstein and the History of General Relativity*, eds. Howard, D. and Stachel, J. (Birkhäuser, Boston/Basel/Berlin), 367–431.

Elsasser, W. (1925). 'Bemerkunggen zur Quantenmechanik Elektronen', *Naturwissenschaften*, **13**: 711.

Englert, F. and Brout, R. (1964). 'Broken symmetry and the mass of gauge vector mesons', *PRL*, **13**: 321–323.

Englert, F. and Brout, R. and Thiry, M. F. (1966). 'Vector mesons in presence of broken symmetry', *NC*, **43**: 244–257.

Essam, J. W. and Fisher, M. E. (1963). 'Padé approximant studies of the lattice gas and Ising ferromagnet below the critical point', *J. Chem. Phys.*, **38**: 802–812.

Euler, H. (1936). 'Über die streuung von Licht an Licht nach Diracschen Theorie', *Ann. Phys.*, **21**: 398–448.

Faddeev, L. D. and Popov, V. N. (1967). 'Feynman diagrams for the Yang–Mills field', *PL*, **25B**: 29–30.

Faraday, M. (1844). 'A speculation touching electric conduction and the nature of matter', *Philos. Mag.*, **24**: 136–144.

Feinberg, G. and Gürsey, F. (1959). 'Space-time properties and internal symmetries of strong interactions', *PR*, **114**: 1153–1170.

Feldman, D. (1949). 'On realistic field theories and the polarization of the vacuum', *PR*, **76**: 1369–1375.

Feldman, D. (1967). (ed.) *Proceedings of the Fifth Annual Eastern Theoretical Physics Conference* (Benjamin, New York).

Fermi, E. (1922). 'Sopra i fenomeni che avvengono in vicinanza di una linea oraria', *Accad. Lincei*, **311**: 184–187, 306–309.

Fermi, E. (1929). 'Sopra l'electrodinamica quantistica. I.', *Rend. Lincei*, **9**: 881–887.

Fermi, E. (1930). 'Sopra l'electrodinamica quantistica. II.', *Rend. Lincei*, **12**: 431–435.

Fermi, E. (1932). 'Quantum theory of radiation', *RMP*, **4**: 87–132.

Fermi, E. (1933). 'Tentativo di una teoria del l'emissione dei raggi β', *Ric. Sci.*, **4**(2): 491–495.

Fermi, E. (1934). 'Versuch einer Theorie der β-Strahlen. I', *ZP*, **88**: 161–171.

Feyerabend, P. (1962). 'Explanation, reduction and empiricism', in *Scientific Explanation, Space and Time*, eds. Feigl, H. and Maxwell, G. (University of Minnesota Press, Minneapolis), 28–97.

Feynman, R. P. (1948a). 'A relativistic cut-off for classical electrodynamics', *PR*, **74**: 939–946.

Feynman, R. P. (1948b). 'Relativistic cut-off for quantum electrodynamics', *PR*, **74**: 1430–1438.

Feynman, R. P. (1948c). 'Space-time approach to non-relativistic quantum mechanics', *RMP*, **20**: 367–387.

Feynman, R. P. (1949a). 'The theory of positrons', *PR*, **76**: 749–768.

Feynman, R.P. (1949b). 'The space-time approach to quantum electrodynamics', *PR*, **76**: 769–789.

Feynman, R. P. and Gell-Mann, M. (1958). 'Theory of the Fermi interaction', *PR*, **109**: 193–198.

Feynman, R. P. (1963). 'Quantum theory of gravity', *Acta Phys. Polonica*, **24**: 697–722.

Feynman, R. P. and Hibbs, A. R. (1965). *Quantum Mechanics and Path Integrals* (McGraw-Hill, New York).

Feynman, R. P. (1969). 'Very high-energy collisions of hadrons', *PRL*, **23**: 1415–1417.

Feynman, R. P. (1973). 'Partons', in *The Past Decade in Particle Theory*, eds. Sudarshan, C. G. and Ne'eman, Y. (Gordon and Breach, New York), 775.

Finkelstein, D. (1958). 'Past-future asymmetry of the gravitational field of a point particle', *PR*, **110**: 965.

Fisher, M. E. (1964). 'Correlation functions and the critical region of simple fluids', *JMP*, **5**: 944–962.

Fitzgerald, G. E. (1885). 'On a model illustrating some properties of the ether', in *The Scientific Writings* (Dublin).

Fock, V. (1926). 'Über die invariante from der Wellen- und der Bewegungs- gleichungen für einen geladenen Massenpunkt', *ZP*, **39**: 226–233.

Forman, P. (1971). 'Weimar culture, causality, and quantum theory, 1918-27: adaptation by German physicists and mathematicians to a hostile intellectual environment', *Historical Studies in the Physical Sciences*, **3**: 1–115.

Fradkin, E. S. (1956). 'Concerning some general relations of quantum electrodynamics', *JETP*, **2**: 361–363.

Fradkin, E. S. and Tyutin, I. V. (1970). 'S matrix for Yang–Mills and gravitational fields', *PR*, **D2**: 2841–2857.

Frenkel, J. (1925). 'Zur elektrodynamik punktfoermiger elektronen', *ZP*, **32**: 518–534.

Friedlander, F. G. (1976). *The Wave Equation on a Curved Space-time* (Cambridge University Press, Cambridge).

Friedman, A. (1922). 'Über die Krümmung des Raumes', *ZP*, **10**: 377–386.

Friedman, A. (1924). 'Über die Möglichkeit einer Welt mit konstant negativer Krümmung des Raumes', *ZP*, **21**: 326–332.

Friedman, M. (1983). *Foundations of Space-Time Theories* (Princeton University Press, Princeton).

Frishman Y. (1971). 'Operator products at almost light like distances', *AP*, **66**: 373–389.

Fritzsch, H., Gell-Mann, M. and Leutwyler, H. (1973). 'Advantages of the color octet gluon picture', *PL*, **47B**: 365–368.

Fritzsch, H. and Minkowski, P. (1975). 'Unified interactions of leptons and hadrons', *AP*, **93**: 193–266.

Fubini, S and Furlan, G. (1965). 'Renormalization effects for partially conserved currents', *Physics*, **1**: 229–247.

Fukuda, H. and Miyamoto, Y. (1949). 'On the γ-decay of neutral meson', *PTP*, **4**: 347–357.

Furry, W. H. and Oppenheimer, J. R. (1934). 'On the theory of the electron and positron', *PR*, **45**: 245–262.

Furry, W. H. and Neuman, M. (1949). 'Interaction of meson with electromagnetic field', *PR*, **76**: 432.

Gasiorowicz, S. G., Yennie, P. R. and Suura, H. (1959). 'Magnitude of renormalization constants', *PRL*, **2**: 513–516.

Gauss, C. F. (1827). 'Disquisitiones generales circa superficies curvas', *Comm. Soc. reg. Sci. Gött. cl. math.* **6**: 99–146.

Gauss, C. F. (1845). Letter to Weber, 19 March 1845, quoted by Whittaker (1951), 240.

Gell-Mann, M. and Low, F. E. (1954). 'Quantum electrodynamics at small distances', *PR*, **95**: 1300–1312.

Gell-Mann, M., Goldberger, M. L. and Thirring, W. (1954). 'Use of causality conditions in quantum theory', *PR*, **95**: 1612–1627.

Gell-Mann, M. and Goldberger, M. L. (1954). 'The scattering of low energy photons by particles of spin 1/2', *PR*, **96**: 1433–1438.

Gell-Mann, M. (1956). 'Dispersion relations in pion–pion and photon–nucleon scattering', in *High Energy Nuclear Physics, Proceedings of the Sixth Annual Rochester Conference* (Interscience, New York), sect. III: 30–36.

Gell-Mann, M. (1958). Remarks after Heisenberg's paper, in *1958 Annual International Conference on High Energy Physics at CERN* (CERN, Geneva), 126.

Gell-Mann, M. and Levy, M. (1960). 'The axial vector current in beta decay', *NC*, **14**: 705–725.

Gell-Mann, M. (1960). Remarks in *Proceedings of the 1960 Annual International Conference on High Energy Physics at Rochester* (Interscience, New York), 508.

Gell-Mann, M. (1962a). 'Symmetries of baryons and mesons', *PR*, **125**: 1067–1084.

Gell-Mann, M. (1962b). 'Applications of Regge poles', in *1962 International Conference on High Energy Physics at CERN* (CERN, Geneva), 533–542.

Gell-Mann, M. (1962c). 'Factorization of coupling to Regge poles', *PRL*, **81**: 263–264.

Gell-Mann, M. and Ne'eman, Y. (1964). *The Eightfold Way* (Benjamin, New York).

Gell-Mann, M. (1964a). 'A schematic model of baryons and mesons', *PL*, **8**: 214–215.

Gell-Mann, M. (1964b). 'The symmetry group of vector and axial vector currents', *Physics*, **1**: 63–75.

Gell-Mann, M., Ramond, P. and Slansky, R. (1978). 'Color embeddings, charge assignments, and proton stability in unified gauge theories', *RMP*, **50**: 721–744.

Gell-Mann. M. (1987). 'Particle theory from *S*-matrix to quarks', in *Symmetries in Physics (1600–1980)*, eds. Doncel, M. G., Hermann, A., Michael, L. and Pais, A. (Bellaterra, Barcelona), 474–497.

Gell-Mann, M. (1989). 'Progress in elementary particle theory, 1950–1964', in *Pions to Quarks*, eds. Brown, L. M., Dresden, M. and Hoddeson, L. (Cambridge University Press, Cambridge), 694–709.

Georgi, H. and Glashow, S. L. (1974). 'Unity of all elementary particles', *PRL*, **32**: 438–441.

Georgi, H., Quinn, H. R. and Weinberg, S. (1974). 'Hierarchy of interactions in unified gauge theories', *PRL*, **33**: 451–454.

Georgi, H. (1989a). 'Grand unified field theories', in *The New Physics*, ed. Davies, P. (Cambridge University Press, Cambridge), 425–445.

Georgi, H. (1989b). 'Effective quantum field theories', in *The New Physics*, ed. Davies, P. (Cambridge University Press, Cambridge), 446–457.

Gershstein, S. and Zel'dovich, J. (1955). 'On corrections from mesons to the theory of β-decay', *Zh. Eksp. Teor. Fiz.* **29**: 698–699 [English trans.: *JETP*, **2**: 576].

Gerstein I. and Jackiw, R. (1969). 'Anomalies in Ward identities for three-point functions', *PR*, **181**: 1955–1963.

Gervais, J. L. and Lee, B. W. (1969). 'Renormalization of the σ-model (II) Fermion fields and regularization', *NP*, **B12**: 627–646.

Giedymin, J. (1982). *Science and Convention* (Pergamon, Oxford).

Ginzburg, V. L. and Landau, L. D. (1950). 'On the theory of superconductivity', *Zh. Eksp. Teor. Fiz.*, **20**: 1064.

Glashow, S. L. (1959). 'The renormalizability of vector meson interactions', *NP*, **10**: 107–117.

Glashow, S. L. (1961). 'Partial symmetries of weak interactions', *NP*, **22**: 579–588.

Glashow, S. L., Iliopoulos, J. and Maiani, L. (1970). 'Weak interactions with lepton–hadron symmetry', *PR*, **D2**: 1285–1292.

Glashow, S. L. (1980). 'Toward a unified theory: threads in a tapestry', *RMP*, **52**: 539–543.

Glauser, R. J. (1953). 'On the gauge invariance of the neutral vector meson theory', *PTP*, **9**: 295–298.

Gödel, K. (1949). 'An example of a new type of cosmological solution of Einstein's field equations of gravitation', *RMP*, **21**: 447–450.

Gödel, K. (1952). 'Rotating universes in general relativity theory', *Proc. Int. Congr. of Mathematicians, Cambridge, Mass. 1950* (American Mathematical Society, Providence), vol. I: 175–181.

Goldberger, M. L. (1955a). 'Use of causality conditions in quantum theory', *PR*, **97**: 508–510.

Goldberger, M. L. (1955b). 'Causality conditions and dispersion relations I. Boson fields', *PR*, **99**: 979–985.

Goldberger, M. L. and Treiman, S. B. (1958a). 'Decay of the pi meson', *PR*, **110**: 1178–1184.

Goldberger, M. L. and Treiman, S. B. (1958b). 'Form factors in β decay and ν capture', *PR*, **111**: 354–361.

Goldhaber, G. *et al.* (1976). 'Observation in e^+e^- annihilation of a narrow state at 1865 Mev/c^2 decaying to Kπ and K$\pi\pi\pi$', *PRL*, **37**: 255–259.

Goldstone, J. (1961). 'Field theories with "superconductor" solutions', *NC*, **19**, 154–164.

Goldstone, J., Salam, A. and Weinberg, S. (1962) 'Broken symmetries', *PR*, **127**, 965–970.

Goto, T. and Imamura, T. (1955). 'Note on the non-perturbation-approach to quantum field theory', *PTP*, **14**: 396–397.

Green, M. B. (1985). 'Unification of forces and particles in superstring theories', *Nature*, **314**: 409–414.

Green, M. B. and Schwarz, J. H. (1985). 'Infinity cancellations in SO(32) superstring theory', *PL*, **151B**: 21–25.

Green, M. B., Schwarz, J. H. and Witten, E. (1987). *Superstring Theory* (Cambridge University Press, Cambridge).

Gross, D. and Llewellyn Smith, C. H. (1969). 'High-energy neutrino–nucleon scattering, current algebra and partons', *NP*, **B14**: 337–347.

Gross, D. and Treiman, S. (1971). 'Light cone structure of current commutators in the gluon–quark model', *PR*, **D4**: 1059–1072.

Gross, D. and Jackiw, R. (1972). 'Effect of anomalies on quasi-renormalizable theories', *PR*, **D6**: 477–493.

Gross, D. and Wilczek, F. (1973a). 'Ultraviolet behavior of non-Abelian gauge theories', *PRL*, **30**: 1343–1346.

Gross, D. and Wilczek, F. (1973b). 'Asymptotic free gauge theories: I', *PR*, **D8**: 3633–3652.

Gross, D. (1985). 'Beyond quantum field theory', in *Recent Developments in Quantum Field Theory*, eds. Ambjorn, J. Durhuus, B. J. and Petersen, J. L. (Elsevier, New York), 151–168.

Gross, D. (1992). 'Asymptotic freedom and the emergence of QCD' (a talk given at the Third International Symposium on the History of Particle Physics, June 26, 1992; Princeton Preprint PUPT 1329).

Gürsey, F. (1960a). 'On the symmetries of strong and weak interactions', *NC*, **16**: 230–240.

Gürsey, F. (1960b). 'On the structure and parity of weak interaction currents', in *Proceedings of 10th International High Energy Physics Conference*, 570–577.

Gürsey, F. and Radicati, L. A. (1964). 'Spin and unitary spin independence', *PRL*, **13**: 173–175.

Gürsey, F. and Sikivie, P. (1976). 'E$_7$ as a unitary group', *PRL*, **36**: 775–778.

Guralnik, G. S., Hagen, C. R. and Kibble, T. W. B. (1964). 'Global conservation laws and massless particles', *PRL*, **13**: 585–587.

Haag, R. (1958). 'Quantum field theories with composite particles and asymptotic conditions', *PR*, **112**: 669–673.

Haas, A. (1910a). 'Über die elektrodynamische Bedeutung des Planck'schen strahlungsgesetzes und über eine neue Bestimmung des elektrischen Elementarquantums und der Dimensionen des Wasserstoffatoms', *Wiener Ber. II*, **119**:119–144.

Haas, A. (1910b). 'Über eine neue theoretische Methode zur Bestmmung des elektrischen Elementarquantums und des Halbmessers des Wasserstoffatoms', *Phys. Z.*, **11**: 537–538.

Haas, A. (1910c). 'Der zusammenhang des Planckschen elementaren Wirkungsquantums mit dem Grundgrössen der Elektronentheorie', *J. Radioakt.*, **7**: 261–268.

Hacking, I. (1983). *Representing and Intervening* (Cambridge University Press, Cambridge).

Hamprecht, B. (1967). 'Schwinger terms in perturbation theory', *NC*, **50A**: 449–457.

Han, M. Y. and Nambu, Y. (1965). 'Three-triplet model with double SU(3) symmetry', *PR*, **B139**: 1006–1010.

Hanson, N. R. (1958). *Patterns of Discovery* (Cambridge University Press, Cambridge).

Hara, Y. (1964). 'Unitary triplets and the eightfold way', *PR*, **B134**: 701–704.

Harman, P. M. (1982a). *Energy, Force, and Matter. The Conceptual Development of Nineteenth-Century Physics* (Cambridge University Press, Cambridge).

Harman, P. M. (1982b). *Metaphysics and Natural Philosophy. The Problem of Substance in Classical Physics* (Barnes and Noble, Totowa, New Jersey).

Harrington, B. J., Park, S. Y. and Yildiz, A. (1975). 'Spectrum of heavy mesons in e^+e^- annihilation', *PRL*, **34**: 168–171.

Hartle, J. B. and Hawking, S. W. (1983). 'Wave function of the universe', *PR*, **D28**: 2960–2975.

Hasert, F. J. *et al.* (1973a). 'Search for elastic muon-neutrino electron scattering', *PL*, **46B**: 121–124.

Hasert, F. J. *et al.* (1973b). 'Observation of neutrino-like interactions without muon or electron in the Gargamelle neutrino experiment', *PL*, **46B**: 138–140.

Hawking, S. and Ellis, G. (1968). 'The cosmic black body radiation and the existence of singularities in our universe', *Astrophys. J.*, **152**: 25–36.

Hawking, S. and Penrose, R. (1970). 'The singularities of gravitational collapse and cosmology', *PRS*, **A314**: 529–548.

Hawking, S. (1975). 'Particle creation by black holes', *CMP*, **43**: 199–220.

Hawking, S. (1979). 'The path-integral approach to quantum gravity', chap. 15 of *General Relativity: An Einstein Centenary Survey*, eds. Hawking, S. and Israel, W. (Cambridge University Press, Cambridge).

Hawking, S. (1980). *Is the End in Sight for Theoretical Physics?* (Cambridge University Press, Cambridge).

Hawking, S. (1987). 'Quantum cosmology', in *Three Hundred Years of Gravitation*, eds. Hawking, S. W. and Israel, W. (Cambridge, University Press, Cambridge), 631.

Hayakawa, S. (1983). 'The development of meson physics in Japan', in *The Birth of Particle Physics*, eds. Brown, L. M. and Hoddeson, L. (Cambridge University Press, Cambridge), 82–107.

Hayashi, K. and Bregman, A. (1973). 'Poincaré gauge invariance and the dynamical role of spin in gravitational theory', *AP*, **75**: 562–600.

Hehl, F. W. *et al.* (1976). 'General relativity with spin and torsion: foundations and prospects', *RMP*, **48**: 393–416.

Heilbron, J. L. (1981). 'The electric field before Faraday', in *Conceptions of Ether*, eds. Cantor, G. N. and Hodge, M. J. S. (Cambridge University Press, Cambridge), 187–213.

Heisenberg, W. (1925). 'Über quantentheoretische Umdeutung kinematischer und mechanischer Beziehung', *ZP*, **33**: 879–883.

Heisenberg, W. (1926a). 'Über die spektra von Atomsystemen mit zwei Elektronen', *ZP*, **39**: 499–518.

Heisenberg, W. (1926b). 'Schwankungserscheinungen und Quantenmechanik', *ZP*, **40**: 501.

Heisenberg, W. (1926c). 'Quantenmechanik', *Naturwissenschaften* **14**: 989–994.

Heisenberg, W. (1926d). Letters to Pauli, W.: 28 Oct. 1926, 4 Nov. 1926, 23 Nov. 1926, in Pauli, W., *Wissenschaftlicher Briefwechsel mit Bohr, Einstein, Heisenberg, u. A., Band I: 1919–1929*, eds. Hermann, A., Meyenn, K. V. and Weisskopf, V. E. (Springer-Verlag, Berlin, 1979).

Heisenberg, W. (1927). 'Über den anschaulichen Inhalt der quantentheoretischen kinematik u. mechanik', *ZP*, **43**: 172–198.

Heisenberg, W. (1928). 'Zur Theorie des Ferromagnetismus', *ZP*, **49**: 619–636.

Heisenberg, W. and Pauli, W. (1929). 'Zur Quantenelektrodynamik der Wellenfelder. I', *ZP*, **56**: 1–61.

Heisenberg, W. and Pauli, W. (1930). 'Zur Quantenelektrodynamik der Wellenfelder. II', *ZP*, **59**: 168–190.

Heisenberg, W. (1932a). 'Über den Bau der Atomkerne', *ZP*, **77**: 1–11.

Heisenberg, W. (1932b). 'Über den Bau der Atomkerne', *ZP*, **78**: 156–164.

Heisenberg, W. (1933). 'Über den Bau der Atomkerne', *ZP*, **80**: 587–596.

Heisenberg, W. (1934). 'Remerkung zur Diracschen Theorie des Positrons', *ZP*, **90**: 209–231.

Heisenberg, W. (1943a). 'Die "beobachtbaren Grössen" in der Theorie der Elementarteilchen', *ZP*, **120**: 513–538.

Heisenberg, W. (1943b). 'Die "beobachtbaren Grössen" in der Theorie der Elementarteilchen. II', *ZP*, **120**: 673–702.

Heisenberg, W. (1944). 'Die "beobachtbaren Grössen" in der Theorie der Elementarteilchen. III', *ZP*, **123**: 93–112.

Heisenberg, W. (1955). 'The development of the interpretation of the quantum theory', in *Niels Bohr and the Development of Physics*, ed. Pauli, W. (McGraw-Hill, New York), 12–29.

Heisenberg, W. (1957). 'Quantum theory of fields and elementary particles', *RMP*, **29**: 269–278.

Heisenberg, W. (1958). 'Research on the non-linear spinor theory with indefinite metric in Hilbert space', in *1958 Annual International Conference on High Energy Physics at CERN* (CERN, Geneva), 119–126.

Heisenberg, W. (1960). 'Recent research on the nonlinear spinor theory of elementary particles', in *Proceedings of the 1960 Annual International Conference on High Energy Physics at Rochester* (Interscience, New York), 851–857.

Heisenberg, W. (1961). 'Planck's discovery and the philosophical problems of atomic physics', in *On Modern Physics* (by Heisenberg and others), trans. Goodman, M. and Binns, J. W. (C. N. Potter, New York).

Heisenberg, W. (1971). *Physics and Beyond: Encounters and Conversations* (Harper and Row, New York).

Heitler, W. and Herzberg, G. (1929). *Naturwissenschaften*, **17**: 673.

Heitler, W. (1936). *The Quantum Theory of Radiation* (Clarendon, Oxford).

Heitler, W. (1961). 'Physical aspects in quantum-field theory', in *The Quantum Theory of Fields*, ed. Stoops, R. (Interscience, New York), 37–60.

Hendry, J. (1984). *The Creation of Quantum Mechanics and the Bohr–Pauli Dialogue* (Reidel, Dordrecht/Boston/Lancaster).

Hertz, H. (1894). *Die Prinzipien der Mechanik in neuem Zusammenhang Dargestellt* (Barth, Leipzig).

Hesse, M. B. (1961). *Forces and Fields* (Nelson, London).

Hesse, M. B. (1974). *The Structure of Scientific Inference*, (Macmillan, London).

Hesse, M. B. (1980). *Revolutions and Reconstructions in the Philosophy of Science* (Harvester, Brighton, Sussex).

Hesse, M. B. (1981). 'The hunt for scientific reason', in *PSA 1980* (the Philosophy of Science Association, 1981) **2**: 3–22.

Hesse, M. B. (1985). 'Science beyond realism and relativism', (unpublished manuscript).

Hessenberg, G. (1917). 'Vektorielle Begründung der Differentialgeometrie', *Math. Ann.*, **78**: 187.

Higgs, P. W. (1964a). 'Broken symmetries, massless particles and gauge fields', *PL*, **12**: 132–133.

Higgs, P. W. (1964b). 'Broken symmetries and the masses of gauge bosons', *PRL*, **13**: 508–509.

Hilbert, D. (1915). 'Die Grundlagen der Physik', *Nachr. Ges. Wiss. Göttingen Math.-phys. Kl.*, 395–407.

Hilbert, D. (1917). 'Die Grundlagen der Physik: zweite Mitteilung', *Nachr. Ges. Wiss. Göttingen Math.-phys. Kl.*, 55–76, 201, 477–480.

Holton, G. (1973). *Thematic Origins of Scientific Thought: Kepler to Einstein* (Harvard University Press, Cambridge, MA).

Hosotani, Y. (1983). 'Dynamical gauge symmetry breaking as the Casimir effect', *PL*, **129B**: 193–197.

Houard, J. C. and Jouvet, B. (1960). 'Etude d'un modèle de champ à constante de renormalisation nulle', *NC*, **18**: 466–481.

Houtappel, R. M. F., Van Dam, H. and Wigner, E. P. (1965). 'The conceptual basis and use of the geometric invariance principles', *RMP*, **37**: 595–632.

Hubble, E. P. (1929). 'A relation between distance and radial velocity among extragalactic nebulae', *Proc. Nat. Acad. of Sci. US*, **15**: 169–173.

Hurst, C. A. (1952). 'The enumeration of graphs in the Feynman–Dyson technique', *PRS*, **A214**: 44–61.

Iagolnitzer, D. and Stapp, H. P. (1969). 'Macroscopic causality and physical region analyticity in the S-matrix theory', *CMP*, **14**: 15–55.

Isham, C. (1990). 'An introduction to general topology and quantum topology', in *Physics, Geometry, and Topology*, ed. Lee, H. C. (Plenum, New York and London), 129–189.

Ito, D., Koba, Z. and Tomonaga, S. (1948). 'Corrections due to the reaction of "cohesive force field" to the elastic scattering of an electron. I', *PTP*, **3**: 276–289.

Iwanenko, D. (1932a). 'The neutron hypothesis', *Nature*, **129**: 798.

Iwanenko, D. (1932b). 'Sur la constitution des noyaux atomiques', *Compt. Rend. Acad. Sci. Paris*, **195**: 439–441.

Iwanenko, D. (1934). 'Interaction of neutrons and protons', *Nature*, **133**: 981–982.

Jackiw, R. and Johnson, K. (1969). 'Anomalies of the axial-vector current', *PR*, **182**: 1459–1469.

Jackiw, R. and Preparata, G. (1969). 'Probe for the constituents of the electromagnetic current and anomalous commutators', *PRL*, **22**: 975–977.

Jackiw, R., Van Royen, R., and West, G. (1970). 'Measuring light-cone singularities', *PR*, **D2**: 2473–2485.

Jackiw, R. (1972). 'Field investigations in current algebra', in *Lectures on Current Algebra and Its Applications*, by Treiman, S. B., Jackiw, R. and Gross, D. J. (Princeton University Press, Princeton), 97–254.

Jackiw, R. and Rebbi, C. (1976). 'Vacuum periodicity in a Yang–Mills quantum theory', *PRL*, **37**: 172–175.

Jackiw, R. (1985). 'Topological investigations of quantified gauge theories', in *Current Algebra and Anomalies*, by Treiman, S. B., Jackiw, R., Zumino, B. and Witten, E. (Princeton University Press, Princeton), 211–359.

Jackiw, R. (1991). 'Breaking of classical symmetries by quantum effects', MIT preprint CTP #1971 (May 1991).

Jacob, M. (1974). *Dual Theory* (Amsterdam, North-Holland).

Jaffe, A. (1965). 'Divergence of perturbation theory for bosons', *CMP*, **1**: 127–149.

Jaffe, A. (1995). Conversation with Schweber, S. S. and Cao, T. Y. on 9 February, 1995.

Jammer, M. (1974). *The Philosophy of Quantum Mechanics* (McGraw-Hill, New York).

Johnson, K. (1961). 'Solution of the equations for the Green's functions of a two dimensional relativistic field theory', *NC*, **20**: 773–790.

Johnson, K. (1963). 'γ_5 Invariance', *PL*, **5**: 253–254.

Johnson, K. and Low, F. (1966). 'Current algebra in a simple model', *PTP*, **37–38**: 74–93.

Jordan, P. (1925). 'Über das thermische Gleichgewicht zwischen quantenatomen und Hohlraumstrahlung', *ZP*, **33**: 649–655.

Jordan, P. (1927a). 'Zur quantenmechanik der gasentartung', *ZP*, **44**: 473–480.

Jordan, P. (1927b). 'Über Wellen und Korpuskeln in der quantenmechanik', *ZP*, **45**: 766–775.

Jordan, P. and Klein, O. (1927). 'Zum Mehrkörperproblem der Quantentheorie', *ZP*, **45**: 751–765.

Jordan, P. (1928). 'Der charakter der quantenphysik', *Naturwissenschaften*, **41**: 765–772. (English translation taken from J. Bromberg, 1976).

Jordan, P. and Wigner, E. (1928). 'Über das Paulische Äquivalenzverbot', *ZP*, **47**: 631–651.

Jordan, P. (1973). 'Early years of quantum mechanics: Some reminiscences', in *The Physicist's Conception of Nature*, ed. Mehra, (Reidel, Dordrecht), 294–300.

Jost, R. (1947). 'Über die falschen Nullstellen der Eigenwerte der S-matrix', *Helv. Phys. Acta.*, **20**: 256–266.

Jost, R. (1965). *The General Theory of Quantum Fields* (American Mathematical Society, Providence).

Jost, R. (1972). 'Foundation of quantum field theory', in *Aspects of Quantum Theory*, eds. Salam, A. and Wigner, E. P. (Cambridge University Press, Cambridge), 61–77.

Kadanoff, L. P. (1966). 'Scaling laws for Ising models near T$_c$', *Physics*, **2**: 263–272.

Kadanoff, L. P., Götze, W., Hamblen, D., Hecht, R., Lewis, E., Palciauskas, V., Rayl, M., Swift, J., Aspnes, D. and Kane, J. (1967). 'Static phenomena near critical points: theory and experiment', *RMP*, **39**: 395–431.

Källen, G. (1953). 'On the magnitude of the renormalization constants in quantum electrodynamics', *Dan. Mat.-Fys. Medd.*, **27**: 1–18.

Källen, G. (1966). 'Review of consistency problems in quantum electrodynamics', *Acta Phys. Austr. Suppl.* **2**: 133–161.

Kaluza, T. (1921). 'Zum Unitätsproblem der Physik', *Preussische Akad. Wiss. Sitzungsber.*, **54**: 966–972.

Kamefuchi, S. (1951). 'Note on the direct interaction between spinor fields', *PTP*, **6**: 175–181.

Kamefuchi, S. (1960). 'On Salam's equivalence theorem in vector meson theory', *NC*, **18**: 691–696.

Kant, I. (1783). *Prolegomena zu einer jeden künftigen Metaphysik die als Wissenschaft wird auftreten können* (J. F. Hartknoch, Riga), English transl. revised by Ellington, J. W. (Hackett, Indianapolis, 1977).

Kastrup, H. A. (1966). 'Conformal group in space-time', *PR*, **142**: 1060–1071.

Kemmer, N. (1938). 'The charge-dependence of nuclear forces', *Proc. Cambridge Philos. Soc.*, **34**: 354–364.

Kerr, R. P. (1963). 'Gravitational field of a spinning mass as an example of algebraically special metrics', *PRL*, **11**: 237–238.

Kerszberg, P. (1989). 'The Einstein–de Sitter controversy of 1916–1917 and the rise of relativistic cosmology', in *Einstein and the History of General Relativity*, eds. Howard, D. and Stachel, J. (Birkhäuser, Boston/Basel/Berlin), 325–366.

Kibble, T. W. B. (1961). 'Lorentz invariance and the gravitational field', *JMP*, **2**: 212.

Kibble, T. W. B. (1967). 'Symmetry breaking in non-Abelian gauge theories', *PR*, **155**: 1554–1561.

Kinoshita, T. (1950). 'A note on the C meson hypothesis', *PTP*, **5**: 535–536.

Klein, F. (1872). 'A comparative review of recent researches in geometry', English trans. in *N. Y. Math. Soc. Bull.* **2** (1893): 215–249.

Klein, O. (1926). 'Quantentheorie und fündimentionale Relativitätstheorie', *ZP*, **37**: 895–906.

Klein, O. (1927). 'Zur fundimentionalen Darstellung der Relativitästheorie', *ZP*, **46**: 188.

Klein, O. (1938). *Entretiens sur les idées fondamentalles de la physique moderne* (Hermann, Paris).

Kline, M. (1972). *Mathematical Thought from Ancient to Modern Times* (Oxford University Press, Oxford and New York).

Koba, Z. and Tomonaga, S. (1947). 'Application of the "self-consistent" subtraction method to the elastic scattering of an electron', *PTP*, **2**: 218.

Koester, D., Sullivan, D. and White, D. H. (1982). 'Theory selection in particle physics: a quantitative case study of the evolution of weak-electromagnetic unification theory', *Social Studies of Science*, **12**: 73–100.

Kogut, J. and Susskind, L. (1975). 'Hamiltonian Formulation of Wilson's lattice gauge theories', *PR*, **D11**: 395–408.

Komar, A. and Salam, A. (1960). 'Renormalization problem for vector meson theories', *NP*, **21**: 624–630.

Komar, A. and Salam, A. (1962). 'Renormalization of gauge theories', *PR*, **127**: 331–334.

Koyré, A. (1965). *Newtonian Studies* (Harvard University Press, Cambridge, MA).

Krajewski, W. (1977). *Correspondence Principle and Growth of Science* (Reidel, Dordrecht).

Kramers, H. (1924). 'The law of dispersion and Bohr's theory of spectra', *Nature*, **113**: 673–674.

Kramers, H. and Heisenberg, W. (1925). 'Über die streuung von Strahlen durch Atome', *ZP*, **31**: 681–708.

Kramers, H. (1938a). *Quantentheorie des Elektrons und der Strahlung*, part 2 of *Hand- und Jahrbuch der Chemische Physik. I* (Akad. Verlag, Leipzig).

Kramers, H. (1938b). 'Die Wechselwirkung zwischen geladenen Teilchen und Strahlungsfeld', *NC*, **15**: 108–114.

Kramers, H. (1944). 'Fundamental difficulties of a theory of particles', *Ned. Tijdschr. Natuurk.*, **11**: 134–147.

Kramers, H. (1947). A review talk at the Shelter Island conference, June 1947 (unpublished). For its content and significance in the development of renormalization theory, see Schweber, S. S.: 'A short history of Shelter Island I', in *Shelter Island II*, eds. Jackiw, R., Khuri, N. N., Weinberg, S. and Witten, E. (MIT Press, Cambridge, MA, 1985).

Kretschmann, E. (1917). 'Über den Physikalischen Sinn der Relativitats-postulaten', *Ann. Phys.*, **53**: 575–614.

Kripke, (1972). 'Naming and necessity', in *Semantics of Natural Language*, eds. Davidson, D. and Harman, G. (Reidel, Dordrecht).

Kronig, R. (1946). 'A supplementary condition in Heisenberg's theory of elementary particles', *Physica*, **12**: 543–544.

Kruskal, M. D. (1960). 'Minimal extension of the Schwarzschild metric', *PR*, **119**: 1743–1745.

Kuhn, T. S. (1962). *The Structure of Scientific Revolutions* (University of Chicago Press, Chicago).

Kuhn, T. S. (1970). *The Structure of Scientific Revolutions*, second enlarged edition (University of Chicago Press, Chicago).

Kuhn, T. S. (1976). 'Theory-change as structure-change: comments on the Sneed formalism', *Erkenntnis*, **10**: 179–199.

Kuhn, T. S. (1990). 'The road since structure', in *PSA 1990* (Philosophy of Science Association, East Lansing), vol. 2, 3–13.

Kuhn, T. S. (1991). 'The trouble with the historical philosophy of science', Robert and Maurine Rothschild Distinguished Lecture delivered at Harvard University on 19 November 1991.

Kuhn, T. S. (1993). Afterwords to *World Changes – Thomas Kuhn and the Nature of Science*, ed. Horwich, P. (MIT Press, Cambridge, MA).

Lamb, W. E. and Retherford, R. C. (1947). 'Fine structure of the hydrogen atom by a microwave method', *PR*, **72**: 241–243.

Lagrange, J. L. (1788). *Mécanique analytique* (République, Paris).

Lagrange, J. L. (1797). *Théorie des fonctions analytiques* (République, Paris).

Landau, L. D. (1937). 'On the theory of phase transitions', *Phys. Z. Sowjetunion*, **11**: 26–47, 545–555; also in *Collected Papers* (of Landau, L. D.), ed. ter Haar, D. (New York, 1965), 193–216.

Landau, L. D., Abrikosov, A. A. and Khalatnikov, I. M. (1954a). 'The removal of infinities in quantum electrodynamics', *DAN*, **95**: 497–499.

Landau, L. D., Abrikosov, A. A. and Khalatnikov, I. M. (1954b). 'An asymptotic expression for the electro Green function in quantum electrodynamics', *DAN*, **95**: 773–776.

Landau, L. D., Abrikosov, A. A. and Khalatnikov, I. M. (1954c). 'An asymptotic expression for the photon Green function in quantum electrodynamics', *DAN*, **95**: 1117–1120.

Landau, L. D., Abrikosov, A. A. and Khalatnikov, I. M. (1954d). 'The electron mass in quantum electrodynamics', *DAN*, **96**: 261–263.

Landau, L. D. (1955). 'On the quantum theory of fields', in *Niels Bohr and the Development of Physics*, ed. Pauli, W. (Pergamon, London), 52–69.

Landau, L. D. and Pomeranchuck, I. (1955). 'On point interactions in quantum electrodynamics', *DAN*, **102**: 489–491.

Landau, L. D., Abrikosov, A. A. and Khalatnikov, I. M. (1956). 'On the quantum theory of fields', *NC* (Suppl.), **3**: 80–104.

Landau, L. D. (1958). Letter to Heisenberg, Feb. 1958, quoted by Brown, L. M. and Rechenberg, H. (1988): 'Landau's work on quantum field theory and high energy physics (1930–1961)', Max Planck Institute Preprint MPI-PAE/Pth 42/88 (July 1988), 30.

Landau, L. D. (1959). 'On analytic properties of vertex parts in quantum field theory', *NP*, **13**: 181–192.

Landau, L. D. (1960a). 'On analytic properties of vertex parts in quantum field theory', in *Proceedings of the Ninth International Annual Conference on High Energy Physics (Moscow)*, II, 95–101.

Landau, L. D. (1960b). 'Fundamental problems', in *Theoretical Physics in the Twentieth Century*, eds. Fierz, M. and Weisskopf, V. F. (Interscience, New York).

Landau, L. D. (1965). *Collected Papers* (Pergamon, Oxford).

Larmor, J. (1894). 'A dynamical theory of the electric and luminiferous medium', *Philos. Trans. Roy. Soc.*, **185**: 719–822.

Laudan, (1981). 'A confutation of convergent realism', *Philos. Sci.*, **48**: 19–49.

Lee, B. W. (1969). 'Renormalization of the σ-model', *NP*, **B9**: 649–672.

Lee, B. W. (1972). In *Proceedings of High Energy Physics Conference (1972)*, ed. Jacob, M. (National Accelerator Laboratory, Batavia, IL), vol. 4, 251.

Lee, T. D. (1954). 'Some special examples in renormalizable field theory', *PR*, **95**: 1329–1334.

Lee, T. D. and Wick, G. C. (1974). 'Vacuum stability and vacuum excitation in a spin-0 field theory', *PR*, **D9**: 2291–2316.

Lee, T. D., and Yang, C. N. (1962). 'Theory of charge vector mesons interacting with the electromagnetic field', *PR*, **128**: 885–898.

Lee, Y. K., Mo, L. W. and Wu, C. S. (1963). 'Experimental test of the conserved vector current theory on the beta spectra of B^{12} and N^{12}', *PRL*, **10**: 253–258.

Lehmann, H., Symanzik, K. and Zimmerman, W. (1955). 'Zur Formalisierung quantisierter Feldtheorie', *NC*, (10)**1**: 205–225.

Lehmann, H., Symanzik, K. and Zimmerman, W. (1957). 'On the formulation of quantized field theories', *NC*, (10)**6**: 319–333.

Lemaître, G. (1925). 'Note on de Sitter's universe', *J. Math. Phys.* (Cambridge, Mass), **4**: 189–192.

Lemaître, G. (1927). 'Un univers homogène de masse constante et de rayon croissant, rendant compte de la vitesse radiale des nebuleues extragalactiques', *Ann. Soc. Sci. Bruxelles*, **A47**: 49–59. [English translation: *Monthly Notices of the Royal Astronomical Society*, **91** (1931): 483–490.]

Lemaître, G. (1932). 'La expansion de l'espace', *Rev. Quest. Sci.* **20**: 391–410.

Lemaître, G. (1934). 'Evolution of the expanding universe', *Proc. Nat. Acad. Sci.*, **20**: 12–17.

Lense, J. and Thirring, H. (1918). 'Ueber den Einfluss der Eigenrotation der Zentralkörper auf die Bewegung der Planeten und Monde nach der Einsteinschen Gravitationstheorie', *Phys. Z.*, **19**: 156–163.

Lepage, G.P. (1989). 'What is renormalization?', Cornell University preprint CLNS

89/970, also in *From Action to Answers*, eds. DeGrand, T. and Toussaint, T. (World Scientific, Singapore, 1990), 446–457.

Leutwyler, H. and Stern, J. (1970). 'Singularities of current commutators on the light cone', *NP*, **B20**: 77–101.

Levi-Civita, T. (1917). 'Nozione di parallelismo in una varietà qualunque', *Rend. Circ. Mat. Palermo*, **42**: 173–205.

Lewis, G. N. (1926). 'The conservation of photons', *Nature*, **118**: 874–875.

Lewis, H. W. (1948). 'On the reactive terms in quantum electrodynamics', *PR*, **73**: 173–176.

Lie, S. and Engel, F. (1893). *Theorie der Transformationsgruppen*, (Teubner, Leipzig).

Lodge, O. (1883). 'The ether and its functions', *Nature*, **27**: 304–306, 328–330.

London, F. (1927). 'Quantenmechanische Deutung der Theorie von Weyl', *ZP*, **42**: 375–389.

Lorentz, H. A. (1892). 'La théorie électromagnetique de Maxwell et son application aux corps mouvants', *Arch. Néerlandaises*, **25**: 363–552.

Lorentz, H. A. (1895). *Versuch einer Theorie der electrischen und optischen Erscheinungen in bewegten Körpen* (Brill, Leiden).

Lorentz, H. A. (1899). 'Théorie simplifiée des phénomènes électriques et optiques dans les corps en mouvement', *Versl. K. Akad. Wet. Amst.*, **7**: 507–522; reprinted in *Collected Papers* (Nijhoff, The Hague, 1934–39), **5**: 139–155.

Lorentz, H. A. (1904). 'Electromagnetic phenomena in a system moving with any velocity less than that of light', *Proc. K. Akad. Amsterdam*, **6**: 809–830.

Lorentz, H. A. (1915). *The Theory of Electrons and its Applications to the Phenomena of Light and Radiant Heat*, second edition (Dover, New York).

Lorentz, H. A. (1916) *Les Théories Statistiques et Thermodynamique* (Teubner, Leipzig and Berlin).

Low, F. E. (1954). 'Scattering of light of very low frequency by system of spin 1/2', *PR*, **96**: 1428–1432.

Low, F. E. (1962). 'Bound states and elementary particles', *NC*, **25**: 678–684.

Low, F. E. (1967). 'Consistency of current algebra', in *Proceedings of Fifth Annual Eastern Theoretical Physics Conference*, ed. Feldman, D. (Benjamin, New York), 75.

Low, F. E. (1988). An interview at MIT, 26 July 1988.

Ludwig, G. (1968). *Wave Mechanics* (Pergamon, Oxford).

Mach, E. (1872). *Die Geschichte und die Wurzel des Satzes von der Erhaltung der Arbeit* (Calve, Prague).

Mach, E. (1883). *Die Mechanik in ihrer Entwicklung. Historisch-Kritisch dargestellt* (Brockhaus, Leipzig).

Mack, G. (1968). 'Partially conserved dilatation current', *NP*, **B5**: 499–507.

Majorana, E. (1933). 'Über die Kerntheorie', *ZP*, **82**: 137–145.

Mandelstam, S. (1958). 'Determination of the pion-nucleon scattering amplitude from dispersion relations and unitarity. General theory', *PR*, **112**: 1344–1360.

Mandelstam, S. (1968a). 'Feynman rules for electromagnetic and Yang-Mills fields from the gauge-independent field-theoretic formalism', *PR*, **175**: 1580–1603.

Mandelstam, S. (1968b). 'Feynman rules for the gravitational field from the coordinate-independent field-theoretic formalism', *PR*, **175**: 1604–1623.

Marciano, W. and Pagels, H. (1978). 'Quantum chromodynamics', *Phys. Rep.*, **36**: 137–276.

Marshak, R. E. and Bethe, H. A. (1947). 'On the two-meson hypothesis', *PR*, **72**: 506–509.

Matthews, P. T. (1949). 'The S-matrix for meson-nucleon interactions', *PR*, **76**: 1254–1255.

Maxwell, G. (1971). 'Structural realism and the meaning of theoretical terms', in *Minnesota Studies in the Philosophy of Science*, vol. 4 (University of Minnesota Press, Minneapolis).

Maxwell, J. C. (1861/62). 'On physical lines of force', *Philos. Mag.* (4) **21**: 162–175, 281–291, 338–348; **23**: 12–24, 85–95.

Maxwell, J. C. (1864). 'A dynamical theory of the electromagnetic field', in *Scientific Papers*, vol. I: 526–597.

Maxwell, J. C. (1873). *A Treatise on Electricity and Magnetism* (Clarendon, Oxford).

McGlinn, W. D. (1964). 'Problem of combining interaction symmetries and relativistic invariance', *PRL*, **12**: 467–469.

McGuire, J. E. (1974). 'Forces, powers, aethers and fields', in *Methodological and Historical Essays in the Natural and Social Sciences*, eds. Cohen, R. S. and Wartofsky, M. W. (Reidel, Dordrecht), 119–159.

McMullin, E. (1982). 'The motive for metaphor', *Proceedings of American Catholic Philosophical Association*, **55**: 27.

McMullin, E. (1984). 'A case for scientific realism', in *Scientific Realism*, ed. Leplin, J. (University of California Press, Berkeley/Los Angeles/London).

Merton, R. (1938). 'Science, technology and society in seventeenth century England', *Osiris*, **4**: 360–632.

Merz, J. T. (1904). *A History of European Thought in the Nineteenth Century* (W. Blackwood, Edinburgh, 1904–1912).

Meyerson, É. (1908). *Identité et Reálité* (Labrairie Félix Alcan, Paris).

Michell, J. (1784). 'On the means of discovering the distance, magnitude, etc., of the fixed stars', *Philos. Trans. R. Soc. London*, **74**: 35–57.

Mie, G. (1912a, b; 1913). 'Grundlagen einer Theorie der Materie', *Ann. Phys.*, (4)**37**: 511–534; **39**: 1–40; **40**: 1–66.

Miller, A. (1975). 'Albert Einstein and Max Wertheimer: a gestalt psychologist's view of the genesis of special relativity theory', *History of Science*, **13**: 75–103.

Mills, R. L. and Yang, C. N. (1966). 'Treatment of overlapping divergences in the photon self-energy function', *PTP* (Suppl.), **37/38**: 507–511.

Minkowski, H. (1907). 'Das Relativitätsprinzip', *Ann. Phys.*, **47**: 927.

Minkowski, H. (1908). 'Die grundgleichungen für die elektromagnetischen vorguge in bewegten körper', *Gött. Nachr.*, 53–111.

Minkowski, H. (1909). 'Raum und Zeit', *Phys. Z.* **10**: 104–111.

Misner, C. W., Thorne, K. S. and Wheeler, J. A. (1973). *Gravitation* (Freeman, San Francisco).

Moulines, C. U. (1984). 'Ontological reduction in the natural sciences', in *Reduction in Science*, eds. Balzer *et al.* (Reidel, Dordrecht), 51–70.

Moyer, D. F. (1978). 'Continuum mechanics and field theory: Thomson and Maxwell', *Stud. Hist. Philos. Sci.*, **9**: 35–50.

Nagel, E. (1961). *The Structure of Science* (Harcourt, New York).

Nambu, Y. (1955). 'Structure of Green's functions in quantum field theory', *PR*, **100**: 394–411.

Nambu, Y. (1956). 'Structure of Green's functions in quantum field theory. II', *PR*, **101**: 459–467.

Nambu, Y. (1957). 'Parametric representations of general Green's functions. II', *NC*, **6**: 1064–1083.

Nambu, Y. (1957a). 'Possible existence of a heavy neutral meson', *PR*, **106**: 1366–1367.

Nambu, Y. (1959). Discussion remarks, *Proceedings of the International Conference on High Energy Physics, IX (1959)* (Academy of Science, Moscow, 1960), **2**: 121–122.

Nambu, Y. (1960a). 'Dynamical theory of elementary particles suggested by super-conductivity', in *Proceedings of the 1960 Annual International Conference on High Energy Physics at Rochester* (Interscience, New York), 858–866.

Nambu, Y. (1960b). 'Quasi-particles and gauge invariance in the theory of super-conductivity', *PR*, **117**: 648–663.

Nambu, Y. (1960c). 'A "superconductor" model of elementary particles and its consequences', in *Proceedings of the Midwest Conference on Theoretical Physics* (Purdue University, West Lafayette).

Nambu, Y. (1960d). 'Axial vector current conservation in weak interactions', *PRL*, **4**: 380–382.

Nambu, Y. and Jona-Lasinio, G. (1961a). 'A dynamical model of elementary particles based on an analogy with superconductivity. I', *PR*, **122**: 345–358.

Nambu, Y. and Jona-Lasinio, G. (1961b). 'A dynamical model of elementary particles based on an analogy with superconductivity. II', *PR*, **124**: 246–254.

Nambu, Y. and Lurie, D. (1962). 'Chirality conservation and soft pion production', *PR*, **125**: 1429–1436.

Nambu, Y. (1965). 'Dynamical symmetries and fundamental fields', in *Symmetry Principles at High Energies*, eds. Kursunoglu, B., Perlmutter, A. and Sakmar, A. (Freeman, San Francisco and London), 274–285.

Nambu, Y. (1989). 'Gauge invariance, vector-meson dominance, and spontaneous symmetry breaking', in Brown, Dresden, and Hoddeson (1989), 639–642.

Navier, C. L. (1821). 'Sur les lois de l'equilibre et du mouvement des corps solides elastiques', *Mém. Acad. (Paris)*, **7**: 375.

Nersessian, N. J. (1984). *Faraday to Einstein: Constructing Meaning in Scientific Theories* (Martinus Nijhoff, Dordrecht).

Neumann, C. (1868). 'Resultate einer Untersuchung über die Prinzipien der elektro-dynamik', *Nachr. Ges. Wiss. Göttingen. Math.-phys. Kl.*, **20**: 223–235.

Neveu, A., and Scherk, J. (1972). 'Connection between Yang–Mills fields and dual models', *NP*, **B36**: 155–161.

Neveu, A. and Schwarz, J. H. (1971a). 'Factorizable dual model of pions', *NP*, **B31**: 86–112.

Neveu, A. and Schwarz, J. H. (1971b). 'Quark model of dual pions', *PR*, **D4**: 1109–1111.

Newman, J. R. (ed., 1946). *The Common Sense of the Exact Sciences (by W. K. Clifford)* (Simon and Schuster).

Newton, I. (1934). *Sir Isaac Newton's Mathematical Principle of Natural Philosophy and His System of the World*, ed. Cajori, F. (University of California Press, Berkeley).

Newton, I. (1978). *Unpublished Scientific Papers of Isaac Newton*, eds. Hall, A. R. and Hall, M. B. (Cambridge, University Press, Cambridge).

Nishijima, K. (1957). 'On the asymptotic conditions in quantum field theory', *PTP*, **17**: 765–802.

Nishijima, K. (1958). 'Formulation of field theories of composite particles', *PR*, **111**: 995–1011.

Nissani, N. (1984). 'SL(2, C) gauge theory of gravitation: Conservation laws', *Phys. Rep.*, **109**: 95–130.

Norton, J. (1984). 'How Einstein found his field equations, 1912–1915', *Hist. Stud. Phys. Sci.*, **14**: 253–316.

Norton, J. (1985). 'What was Einstein's principle of equivalence?', *Stud. Hist. Philos. Sci.*, **16**: 203–246.

Okubo, S. (1966). 'Impossibility of having the exact U_6 group based upon algebra of currents', *NC*, **42A**: 1029–1034.

Olive, D. I. (1964). 'Exploration of S-matrix theory', *PR*, **135B**: 745–760.

Oppenheimer, J. R. (1930a). 'Note on the theory of the interaction of field and matter', *PR*, **35**: 461–477.

Oppenheimer, J. R. (1930b). 'On the theory of electrons and protons', *PR*, **35**: 562–563.

Oppenheimer, J. R. and Snyder, H. (1939). 'On continued gravitational contraction', *PR*, **56**: 455.

Oppenheimer, J. R. and Volkoff, G. (1939). 'On massive neutron cores', *PR*, **54**: 540.

O'Raifeartaigh, L. (1965). 'Mass difference and Lie algebras of finite order', *PRL*, **14**: 575–577.

Pais, A. (1945). 'On the theory of the electron and of the nucleon', *PR*, **68**: 227–228.

Pais, A. (1966). 'Dynamical symmetry in particle physics', *RMP*, **368**: 215–255.

Pais, A. and Treiman, S. (1975). 'How many charm quantum numbers are there?', *PRL*, **35**: 1556–1559.

Pais, A. (1986). *Inward Bound* (Oxford University Press, Oxford).

Papapetrou, A. (1949). 'Non-symmetric stress-energy-momentum tensor and spin-density', *Philos. Mag.*, **40**: 937.

Parisi, G. (1973). 'Deep inelastic scattering in a field theory with computable large-momenta behavior', *NC*, **7**: 84–87.

Pasternack, S. (1938). 'Note on the fine structure of H_α and D_α ', *PR*, **54**: 1113–1115.

Pati, J. C. and Salam, A. (1973a). 'Unified lepton–hadron symmetry and a gauge theory of the basic interactions', *PR*, **D8**: 1240–1251.

Pati, J. C. and Salam, A. (1973b). 'Is baryon number conserved?', *PRL*, **31**: 661–664.

Pauli, W. (1926). Letter to Heisenberg, 19 Oct. 1926, in *Wissenschaftlicher Briefwechsel mit Bohr, Einstein, Heisenberg, U. A., Band I: 1919–1929*, eds. Hermann, A., Meyenn, K. V. and Weisskopf, V. E. (Springer-Verlag, Berlin, 1979).

Pauli, W. (1930). A letter of 4 December 1930, in *Collected Scientific Papers*, eds. Kronig, R. and Weisskopf, V. F. (Interscience, New York, 1964), II: 1313.

Pauli, W. (1933). Pauli's remarks at the seventh Solvay conference, Oct. 1933, in *Rapports du Septième Conseil de Physique Solvay*, 1933 (Gauthier–Villarg, Paris, 1934).

Pauli, W. and Weisskopf, V. (1934). 'Über die quantisierung der skalaren relativistischen wellengleichung', *Helv. Phys. Acta* **7**: 709–731.

Pauli, W. and Fierz, M. (1938). 'Zur Theorie der Emission langwelliger Lichtquanten', *NC*, **15**: 167–188

Pauli, W. (1941). 'Relativistic field theories of elementary particles', *RMP*, **13**: 203–232.

Pauli, W. and Villars, F. (1949). 'On the invariant regularization in relativistic quantum theory', *RMP*, **21**: 434– 444.

Pavelle, R. (1975): 'Unphysical solutions of Yang's gravitational-field equations', *PRL*, **34**, 1114.

Peierls, R. (1934). 'The vacuum in Dirac's theory of the positive electron', *PRS*, **A146**: 420–441.

Penrose, R. (1967a). 'Twistor algebra', *JMP*, **8**: 345.

Penrose, R. (1967b). 'Twistor quantization and curved space-time', *Int. J. Theor. Phys.*, **1**: 61–99.

Penrose, R. (1968). 'Structure of space-time', in *Lectures in Mathematics and Physics* (Battele Rencontres), eds. DeWitt, C. M. and Wheeler, J. A. (Benjamin, New York), 121–235.

Penrose, R. (1969). 'Gravitational collapse: the role of general relativity', *NC*, **1**: 252–276.

Penrose, R. (1972). 'Black holes and gravitational theory', *Nature*, **236**: 377–380.

Penrose, R. (1975). 'Twistor theory', in *Quantum Gravity*, eds. Isham, C. J., Penrose, R. and Sciama, J. W. (Claredon, Oxford).

Penrose, R. (1987). In *Three Hundred Years of Gravitation*, eds. Hawking, S. W. and Israel, W. (Cambridge, University Press, Cambridge).

Perring, F. (1933). 'Neutral particles of intrinsic mass 0', *Compt. Rend. Acad. Sci. Paris*, **197**: 1625–1627.

Peruzzi, I. *et al.* (1976). 'Observation of a narrow charged state at 1875 Mev/c^2 decaying to an exotic combination of K$\pi\pi$', *PRL*, **37**: 569–571.

Peterman, A. and Stueckelberg, E. C. G. (1951). 'Restriction of possible interactions in quantum electrodynamics', *PR*, **82**: 548–549.

Peterman, A. (1953a). 'Divergence of perturbation expression', *PR*, **89**: 1160–1161.

Peterman, A. (1953b). 'Renormalisation dans les séries divergentes', *Helv. Phys. Acta*, **26**: 291–299.

Pickering, A. (1984). *Constructing Quarks: A Sociological History of Particle Physics* (Edinburgh University Press, Edinburgh).

Pines, D. and Schrieffer, L. R. (1958). 'Gauge invariance in the theory of superconductivity', *NC*, **10**: 407–408.

Planck, M. (1900). 'Zur Theorie des Gesetzes der Energieverteilung im Normalspectrum', *Verh. Deutsch. Phys. Ges.*, **2**: 237–245.

Poincaré, H. (1890). *Electricité et optique: les théories de Maxwell et la théorie électromagnétique de la lumière*, ed. Blondin (G. Carré, Paris).

Poincaré, H. (1895). 'A propos de la theorie de M. Larmor', in *Oeuvres* (Gauthier-Villars, Paris), **9**: 369–426.

Poincaré, H. (1897). 'Les idées de Hertz sur la méchanique', in *Oeuvres* (Gauthier-Villars, Paris), **9**: 231–250.

Poincaré, H. (1898). 'De la mesure du temps', *Rev. Métaphys. Morale*, **6**: 1–13.

Poincaré, H. (1899). 'Des fondements de la géométrie, a propos d'un livre de M. Russell', *Rev. Métaphys. Morale*, **7**: 251–279.

Poincaré, H. (1900). 'La théorie de Lorentz et le principe de la réaction', *Arch. Néerlandaises*, **5**: 252–278.

Poincaré, H. (1902). *La science et l'hypothèse* (Flammarion, Paris).

Poincaré, H. (1904). 'The principles of mathematical physics', in *Philosophy and Mathematics*, volume I of *Congress of Arts and Sciences: Universal Exposition, St. Louis, 1904*, ed. Rogers, H. (H. Mifflin, Boston, 1905), 604–622.

Poincaré, H. (1905). 'Sur la dynamique de l'électron', *Comptes Rendus Acad. Sci.* **140**: 1504–1508.

Poincaré, H. (1906). 'Sur la dynamique de l'électron', *Rendiconti Circ. Mat. Palermo*, **21**: 129–175.

Polchinski, J. (1984). 'Renormalization and effective Lagrangians', *NP*, **B231**: 269–295.

Politzer, H. D. (1973). 'Reliable perturbative results for strong interactions?', *PRL*, **30**: 1346–1349.

Polkinghorne, J. C. (1958). 'Renormalization of axial vector coupling', *NC*, **8**: 179–180, 781.

Polkinghorne, J. C. (1967). 'Schwinger terms and the Johnson–Low model', *NC*, **52A**: 351–358.

Polkinghorne, J. C. (1989). Private correspondence, 24 Oct. 1989.

Polyakov, A. M. (1974). 'Particle spectrum in quantum field theory', *JETP (Lett.)*, **20**: 194–195.

Popper, K. (1970). 'A realist view of logic, physics, and history' in *Physics, Logic, and History*, eds. Yourgrau, W. and Breck, A. D. (Plenum, New York), 1–39.

Popper, K. (1974). 'Scientific reduction and the essential incompleteness of all science', in *Studies in the Philosophy of Biology, Reduction and Related Problems*, eds. Ayala, F. J. and Dobzhansky, T. (Macmillan, London), 259–284.

Post, H. R. (1971). 'Correspondence, invariance and heuristics', *Stud. Hist. Philos. Sci.*, **2**: 213–255.

Putnam, H. (1975). *Mind, Language and Reality* (Cambridge University Press, Cambridge).

Putnam, H. (1978). Meaning and the Moral Sciences (Routledge and Kegan Paul, London).

Putnam, H. (1981). *Reason, Truth and History* (Cambridge University Press, Cambridge).

Radicati, L. A. (1987). 'Remarks on the early development of the notion of symmetry breaking', in *Symmetries in Physics* (1600– 1980), eds. Doncel, M. G., Hermann, A., Michael, L. and Pais, A. (Bellaterra, Barcelona), 197–207.

Raine, D. J. (1975). 'Mach's principle in general relativity', *Mon. Not. R. Astron. Soc.*, **171**: 507–528.

Raine, D. J. (1981). 'Mach's principle and space-time structure', *Rep. Prog. Phys.*, **44**: 1151–1195.

Ramond, P. (1971a). 'Dual theory for free fermions', *PR*, **D3**: 2415–2418.

Ramond, P. (1971b). 'An interpretation of dual theories', *NC*, **4A**: 544–548.

Rasetti, F. (1929). 'On the Raman effect in diatomic gases. II', *Proc. Nat. Acad. Sci.*, **15**: 515–519.

Rayski, J. (1948). 'On simultaneous interaction of several fields and the self-energy problem', *Acta Phys. Polonica*, **9**: 129–140.

Redhead, M. L. G. (1983). 'Quantum field theory for philosophers', in *PSA 1982* (Philosophy of Science Association, East Lansing, MI, 1983), 57–99.

Regge, T. (1958a). 'Analytic properties of the scattering matrix', *NC*, **8**: 671–679.

Regge, T. (1958b). 'On the analytic behavior of the eigenvalue of the S-matrix in the complex plane of the energy', *NC*, **9**: 295– 302.

Regge, T. (1959). 'Introduction to complex orbital momenta', *NC*, **14**: 951–976.

Regge, T. (1960). 'Bound states, shadow states and Mandelstam representation', *NC*, **18**: 947–956.

Reiff, J. and Veltman, M. (1969). 'Massive Yang–Mills fields', *NP*, **B13**: 545–564.

Reinhardt, M. (1973). 'Mach's principle – A critical review', *Z. Naturf.* **28A**: 529–537.

Resnik, M. (1981). 'Mathematics as a science of patterns: ontology and reference', *Nous*, **15**: 529–550.

Resnik, M. (1982). 'Mathematics as a science of patterns: epistemology', *Nous*, **16**: 95–105.

Riazuddin, A. and Sarker, A. Q. (1968). 'Some radiative meson decay processes in current algebra', *PRL*, **20**: 1455–1458.

Ricci, G. and Levi-Civita, T. (1901). 'Méthodes de calcul differentiel absolu et leurs applications', *Math. Ann.*, **54**: 125–201.

Rickaysen, J. G. (1958). 'Meissner effect and gauge invariance', *PR*, **111**: 817–821.

Riemann, B. (1853, 1858, 1867). In his *Gesammelte Mathematische Werke und Wissenschaftlicher Nachlass.*, ed. Weber, H. (Teubner, Leipzig, 1892).

Riemann, B. (1854). 'Über die Hypothesen, welche der Geometrie zu Grunde liegen', *Ges. Wiss. Göttingen. Abhandl.*, **13** (1867): 133–152.

Riemann, B. (1861a). In *Schwere, Elektricität, und Magnetismus: nach den Vorlesungen von Bernhard Riemann*, ed. K. Hattendoff (Hannover, 1876).

Riemann, B. (1861b). Quoted by Kline (1972), 894–896.

Rindler, W. (1956). 'Visual horizons in world models', *Mon. Not. R. Astron. Soc.*, **116**: 662–677.

Rindler, W. (1977). *Essential Relativity* (Springer-Verlag, New York).

Rivier, D. and Stueckelberg, E. C. G. (1948). 'A convergent expression for the magnetic moment of the neutron', *PR*, **74**: 218.

Robertson, H. P. (1939). Lecture notes, printed posthumously in *Relativity and Cosmology*, by Robertson, H. P. and Noonan, T. W. (Saunders, Philadelphia, 1968).

Rodichev, V. I. (1961). 'Twisted space and non-linear field equations', *Zh. Eksper. Ther. Fiz.*, **40**: 1469.

Rohrlich, F. (1973). 'The electron: development of the first elementary particle theory', in *The Physicist's Conception of Nature*, ed. J. Mehra (Reidel, Dordrecht), 331–369.

Rohrlich, F. and Hardin, L. (1983). 'Established theories', *Philos. Sci.*, **50**: 603–617.

Rosenberg, L. (1963). 'Electromagnetic interactions of neutrinos', *PR*, **129**: 2786–2788.

Rosenfeld, L. (1963). Interview with Rosenfeld on 1 July 1963; Archive for the History of Quantum Physics.

Rosenfeld, L. (1968). 'The structure of quantum theory', in *Selected Papers*, eds. Cohen, R. S. and Stachel, J. J. (Reidel, Dordrecht, 1979).

Rosenfeld, L. (1973). 'The wave-particle dilemma', in *The Physicist's Conception of Nature*, ed. Mehra, J. (Reidel, Dordrecht), 251–263.

Russell, B. (1927). *The Analysis of Matter* (Allen & Unwin, London).

Sakata, S. and Tanikawa, Y. (1940). 'The spontaneous disintegration of the neutral mesotron (neutretto)', *PR*, **57**: 548.

Sakata, S. and Inoue, T. (1943). First published in Japanese in *Report of Symposium on Meson Theory*, then in English, 'On the correlations between mesons and Yukawa particles', *PTP*, **1**: 143– 150.

Sakata, S. and Hara, O. (1947). 'The self-energy of the electron and the mass difference of nucleons', *PTP*, **2**: 30–31.

Sakata, S. (1947). 'The theory of the interaction of elementary particles', *PTP*, **2**: 145–147.

Sakata, S. (1950). 'On the direction of the theory of elementary particles', *Iwanami*, **II**: 100-103 (English trans. *PTP (Suppl.)*, **50** (1971): 155–158.

Sakata, S., Umezawa, H. and Kamefuchi, S. (1952). 'On the structure of the interaction of the elementary particles', *PTP*, **7**: 377–390.

Sakata, S. (1956). 'On the composite model for the new particles', *PTP*, **16**: 686–688.

Sakita, B. (1964). 'Supermultiplets of elementay particles', *PR*, **B136**: 1756–1760.

Sakurai, J. J. (1960). 'Theory of strong interactions', *AP*, **11**: 1–48.

Salam, A. (1951a). 'Overlapping divergences and the S-matrix', *PR*, **82**: 217–227.

Salam, A. (1951b). 'Divergent integrals in renormalizable field theories', *PR*, **84**: 426–431.

Salam, A. and Ward, J. C. (1959). 'Weak and electromagnetic interactions', *NC*, **11**: 568–577.

Salam, A. (1960). 'An equivalence theorem for partially gauge-invariant vector meson interactions', *NP*, **18**: 681–690.

Salam, A. and Ward, J. C. (1961). 'On a gauge theory of elementary interactions', *NC*, **19**: 165–170.

Salam, A. (1962a). 'Lagrangian theory of composite particles', *NC*, **25**: 224–227.

Salam, A. (1962b). 'Renormalizability of gauge theories', *PR*, **127**: 331–334.

Salam, A. and Ward, J. C. (1964). 'Electromagnetic and weak interactions', *PL*, **13**: 168–171.

Salam, A. (1968). 'Weak and electromagnetic interactions', in *Elementary Particle Theory: Relativistic Group and Analyticity. Proceedings of Nobel Conference VIII*, ed. Svartholm, N. (Almqvist and Wiksell, Stockholm). 367–377.

Salam, A. and Wigner, E. P. (eds.) (1972). *Aspects of Quantum Theory* (Cambridge University Press, Cambridge).

Scadron, M. and Weinberg, S. (1964b). 'Potential theory calculations by the quasi-particle method', *PR*, **B133**: 1589–1596.

Schafroth, M. R. (1951). 'Bemerkungen zur Frohlichsen theorie der supraleitung', *Helv. Phys. Acta*, **24**: 645–662.

Schafroth, M. R. (1958). 'Remark on the Meissner effect', *PR*, **111**: 72–74,

Scherk, J. (1975). 'An introduction to the theory of dual models and strings', *RMP*, **47**: 123–164.

Schlick, M. (1918). *General Theory of Knowledge*, trans. Blumberg, A. E. and Feigl, H. (Springer-Verlag, New York).

Schmidt, B. G. (1971). 'A new definition of singular points in general relativity', *General Relativity and Gravitation*, **1**: 269–280.

Schrödinger, E. (1922). 'Über eine bemerkenswerte Eigenschaft der Quantenbahnen eine einzelnen Elektrons', *ZP*, **12**: 13–23.

Schrödinger, E. (1926a). 'Zur Einsteinschen Gastheorie', *Phys. Z.*, **27**: 95–101.

Schrödinger, E. (1926b). 'Quantisierung als Eigenwertproblem, Erste Mitteilung', *Ann. Phys.* **79**: 361–376.

Schrödinger, E. (1926c). 'Quantisierung, Zweite Mitteilung', *Ann. Phys.* **79**: 489–527.

Schrödinger, E. (1926d). 'Über das Verhältnis der Heisenberg–Born–Jordanschen Quantenmechanik zu der meinen', *Ann. Phys.* **79**: 734–756.

Schrödinger, E. (1926e). 'Quantisierung, Dritte Mitteilung', *Ann. Phys.* **80**: 437–490.

Schrödinger, E. (1926f). 'Quantisierung, Vierte Mitteilung', *Ann. Phys.* **81**: 109–139.

Schrödinger, E. (1926g). Letter to Max Planck, 31 May 1926, in *Letters On Wave Mechanics*, ed. Przibram, K., trans. Klein, M. (Philosophical Library, New York, 1967), 8–11.

Schrödinger, E. (1950). *Spacetime Structure* (Cambridge University Press, Cambridge).

Schrödinger, E. (1961). 'Wave field and particle: their theoretical relationship', 'Quantum steps and identity of particles', and 'Wave identity', in *On Modern Physics* (by Heisenberg and others), trans. Goodman, M. and Binns, J. W. (C. N. Potter, New York), 48–54.

Schwarzschild, K. (1916a). 'Über das Gravitationsfeld eines Massenpunktes nach der Einsteinschen Theorie', *Preuss. Akad. Wiss. Sitzungsber.*, 189–196.

Schwarzschild, K. (1916b). 'Über das Gravitationsfeld eines Kugel aus inkompressibler Flüssigkeit nach der Einsteinschen Theorie', *Preuss. Akad. Wiss. Sitzungsber.*, 424–434.

Schweber, S. S., Bethe, H. A. and Hoffmann, F. de. (1955). *Mesons and Fields*, vol. I (Row, Peterson and Co., New York).

Schweber, S. S. (1985). 'A short history of Shelter Island I', in *Shelter Island I*, eds. Jackiw, R.; Khuri, N, N.; Weinberg, S. and Witten, E. (MIT Press, Cambridge, MA), 301–343.

Schweber, S. S. (1986). 'Shelter Island, Pocono, and Oldstone: the emergence of American quantum electrodynamics after World War II', *Osiris*, second series **2**: 65–302.

Schwinger, J. (1948a). 'On quantum electrodynamics and the magnetic moment of the electron', *PR*, **73**: 416–417.

Schwinger, J. (1948b). 'Quantum electrodynamics. I. A covariant formulation', *PR*, **74**: 1439–1461.

Schwinger, J. (1949a). 'Quantum electrodynamics. II. Vacuum polarization and self energy', *PR*, **75**: 651–679.

Schwinger, J. (1949b). 'Quantum electrodynamics. III. The electromagnetic properties of the electro-radiative corrections to scattering' *PR*, **76**: 790–817.

Schwinger, J. (1951). 'On gauge invariance and vacuum polarization', *PR*, **82**: 664–679.

Schwinger, J. (1957). 'A theory of the fundamental interactions', *AP*, **2**: 407–434.

Schwinger, J. (1959). 'Field theory commutators', *PRL*, **3**: 296–297.

Schwinger, J. (1962a). 'Gauge invariance and mass', *PR*, **125**: 397–398.

Schwinger, J. (1962b). 'Gauge invariance and mass. II', *PR*, **128**: 2425–2429.

Schwinger, J. (1966). 'Relativistic quantum field theory. Nobel lecture', *Science*, **153**: 949–953.

Schwinger, J. (1967). 'Chiral dynamics', PL, **24B**: 473–476.

Schwinger, J. (1970). *Particles, Sources and Fields* (Addison-Wesley, Reading, MA).

Schwinger, J. (1973a). *Particles, Sources and Fields*, vol. 2 (Addison-Wesley, Reading, MA).

Schwinger, J. (1973b). 'A report on quantum electrodynamics', in *The Physicist's Conception of Nature*, ed. Mehra, J. (Reidel, Dordrecht), 413–429.

Schwinger, J. (1983). 'Renormalization theory of quantum electrodynamics: an individual view', in *The Birth of Particle Physics* (eds. Brown, L. M. and Hoddeson, L. (Cambridge University Press, Cambridge), 329–375.

Sciama, D. W. (1958). 'On a non-symmetric theory of the pure gravitational field', *Proc. Cam. Philos. Soc.* **54**: 72.

Seelig, C. (1954). *Albert Einstein: Eine dokumentarische Biographie* (Europa-Verlag, Zürich).

Sellars, W. (1961). 'The language of theories', in *Current Issues in the Philosophy of Science*, eds. Feigl, H. and Maxwell, G. (Holt, Rinehart, and Winston, New York), 57.

Sellars, W. (1963). 'Theoretical explanation', *Philosophy of Science*, **2**: 61.

Sellars, W. (1965). 'Scientific realism or irenic instrumentalism', in *Boston Studies in the Philosophy of Science*, eds. Cohen, R. S. and Wortofsky, M. W. (Humanities Press, New York), vol. 2: 171.

Shimony, A. (1978). 'Metaphysical problems in the foundations of quantum mechanics', *International Philosophical Quarterly*, **18**: 3–17.

Slater, J. C. (1924). 'Radiation and Atoms', *Nature*, 113: 307–308.

Slavnov, A. A. (1972). 'Ward Identities in gauge theories', *Theor. Math. Phys.*, **19**: 99–104.

Sneed, J. D. (1971). *The Logical Structure of Mathematical Physics* (Reidel, Dordrecht).

Sneed, J. D. (1979). 'Theoritization and invariance principles', in *The Logic and Epistemology of Scientific Change* (North-Holland, Amsterdam), 130–178.

Sommerfeld, A. (1911a). Letter to Planck, M. 24 April 1911.

Sommerfeld, A. (1911b). In *Die Theorie der Strahlung und der Quanten*, ed. Eucken, A. (Halle, 1913), 303.

Spector, M. (1978). *Concept of Reduction in Physical Science* (Temple University Press, Philadelphia).

Speziali, P. (ed.) (1972). *Albert Einstein–Michele Besso. Correspondance 1903–1955* (Hermann, Paris).

Stachel, J. (1980a). 'Einstein and the rigidly rotating disk', in *General Relativity and Gravitation One Hundred Years after the Birth of Albert Einstein*, ed. Held, A. (Plenum, New York), vol. 1: 1–15.

Stachel, J. (1980b). 'Einstein's search for general covariance, 1912–1915', paper presented to the 9th International Conference on General Relativity and Gravitation, Jena, 1980; later printed in *Einstein and the History of General Relativity*, eds. Howard, D. and Stachel, J. (Birkhäuser, Boston), 63–100.

Stachel, J. (1993). 'The meaning of general covariance: the hole story', in *Philosophical Problems of the Internal and External Worlds: Essays Concerning the Philosophy of Adolf Grübaum*, eds. Janis, A. I., Rescher, N. and Massey, G. J. (University of Pittsburgh Press, Pittsburgh), 129–160.

Stachel, J. (1994). 'Changes in the concepts of space and time brought about by relativity', in *Artifacts, Representations, and Social Practice*, eds. Gould, C. C. and Cohen, R. S. (Kluwer, Dordrecht/Boston/London), 141–162.

Stapp, H. P. (1962a). 'Derivation of the CPT theorem and the connection between spin and statistics from postulated of the S-matrix theory', *PR*, **125**: 2139–2162.

Stapp, H. P. (1962b). 'Axiomatic S-matrix theory', *RMP*, **34**: 390–394.

Stapp, H. P. (1965). 'Space and time in S-matrix theory', *PR*, **B139**: 257–270.

Stapp, H. P. (1968). 'Crossing, hermitian analyticity, and the connection between spin and statistics', *JMP*, **9**: 1548–1592.

Stark, J. (1909). 'Zur experimentellen Entscheidung zwischen Ätherwellen- und Lichtquantenhypothese. I. Röntgenstrahlen', *Phys. Z.*, **10**: 902–913.

Stegmüller, W. (1976). *The Structure and Dynamics of Theories* (Springer-Verlag, New York).

Stegmüller, W. (1979). *The Structuralist View of Theories* (Springer-Verlag, New York).

Stein, H. (1981). '"Subtle forms of matter" in the period following Maxwell', in *Conceptions of Ether*, eds. Cantor, G. N. and Hodge, M. J. S. (Cambridge University Press, Cambridge), 309–340.

Steinberger, J. (1949). 'On the use of subtraction fields and the lifetimes of some types of meson decay', *PR*, **70**: 1180–1186.

Strawson, P. T. (1950). *Mind*, **54**: 320.

Streater, R. F. and Wightman, A. S. (1964). *PTC, Spin and Statistics, and All That* (Benjamin, New York).

Stueckelberg, E. C. G. (1938). 'Die Wechselwirkungskräfte in der Elektrodynamik und in der Feldtheorie der Kern kräfte', *Helv. Phys. Acta*, **11**: 225–244 and 299–329.

Stueckelberg, E. C. G. and Peterman, A. (1953). 'La normalisation des constances dans la théorie des quanta', *Helv. Phys. Acta.*, **26**: 499–520.

Sudarshan, E. C. G. and Marshak, R. E. (1958).'Chirality invariance and the universal Fermi interaction', *PR*, **109**: 1860– 1862.

Sullivan, D., Koester, D., White, D. H. and Kern, K. (1980). 'Understanding rapid theoretical change in particle physics: a month-by-month co-citation analysis', *Scientometrics*, **2**: 309–319.

Sutherland, D. G. (1967). 'Current algebra and some non-strong mesonic decays', *NP*, **B2**: 473–440.

Symanzik, K. (1969). 'Renormalization of certain models with PCAC', *(Lett.) NC*, **2**: 10–12.

Symanzik, K. (1970). 'Small distance behavior in field theory and power counting', *CMP*, **18**: 227–246.

Symanzik, K. (1971). 'Small-distance behavior analysis and Wilson expansions', *CMP*, **23**: 49–86.

Symanzik, K. (1972). In *Renormalization of Yang–Mills Fields and Applications to Particle Physics* (the Proceedings of the Marseille Conference, 19–23 June 1972), ed. Korthals-Altes, C. P. (Centre de Physique Théorique, CNRS, Marseille).

Symanzik, K. (1973). 'Infrared singularities and small-distance-behavior analysis', *CMP*, **34**: 7–36.

Symanzik, K. (1983). 'Continuum limit and improved action in lattice theories', *NP*, **B226**: 187–227.

Takabayasi, T. (1983). 'Some characteristic aspects of early elementary particle theory in Japan', in *The Birth of Particle Physics*, eds. Brown, L. M. and Hoddeson, L. (Cambridge University Press, Cambridge), 294–303.

Takahashi, Y. (1957). 'On the generalized Ward Identity', *NC*, **6**: 371–375.

Tamm, I. (1934). 'Exchange forces between neutrons and protons, and Fermi's theory', *Nature*, **133**: 981.

Tanikawa, Y. (1943). First published in Japanese in *Report of the Symposium on Meson Theory*, then in English, 'On the cosmic-ray meson and the nuclear meson', *PTP*, **2**: 220.

Taylor, J. C. (1958). 'Beta decay of the pion', *PR*, **110**: 1216.

Taylor, J. C. (1971). 'Ward identities and the Yang–Mills field', *NP*, **B33**: 436–444.

Taylor, J. C. (1976). *Gauge Theories of Weak Interactions* (Cambridge University Press, Cambridge).

Thirring, H. (1918). 'Über die Wirkung rotierender ferner Massen in der Einsteinschen Gravitationstheorie', *Phys. Z.*, **19**: 33–39.

Thirring, H. (1921). 'Berichtigung zu meiner Arbeit: "Über die Wirkung rotierender ferner Massen in der Einsteinschen Gravitationstheorie" ', *Phys. Z.*, **22**: 29–30.

Thirring, W. (1953). 'On the divergence of perturbation theory for quantum fields', *Helv. Phys. Acta.* **26**: 33–52.

Thomson, J. J. (1881). 'On the electric and magnetic effects produced by the motion of electrified bodies', *Philos. Mag.*, **11**: 227–249.

Thomson, J. J. (1904). *Electricity and Matter* (Charles Scribner's Sons, New York).

Thomson, W. (1884). *Notes of Lectures on Molecular Dynamics and the Wave Theory of Light* (Johns Hopkins University, Baltimore).

't Hooft, G. (1971a). 'Renormalization of massless Yang–Mills fields', *NP*, **B33**: 173–99.

't Hooft, G. (1971b). 'Renormalizable Lagrangians for massive Yang–Mills fields', *NP*, **B35**: 167–188.

't Hooft, G. (1971c). 'Prediction for neutrino-electron cross-sections in Weinberg's model of weak interactions', *PL*, **37B**: 195–196.

't Hooft and Veltman, M. (1972a). 'Renormalization and regularization of gauge fields', *NP*, **B44**: 189–213.

't Hooft and Veltman, M. (1972b). 'Combinatorics of gauge fields', *NP*, **B50**: 318–353.

't Hooft, G. (1972c). Remarks after Symansik's presentation, in *Renormalization of Yang–Mills Fields and Applications to Particle Physics* (the Proceedings of the Marseille Conference, 19–23 June 1972), ed. Korthals-Altes, C. P. (Centre de Physique Théorique, CNRS, Marseille).

't Hooft, G. (1973). 'Dimensional regularization and the renormalization group', CERN Preprint Th.1666, May 2, 1973, and *NP*, **B61**: 455, in which his unpublished remarks at the Marseille Conference (1972c) were publicized.

't Hooft, G. (1974). 'Magnetic monopoles in unified gauge theories', *NP*, **B79**: 276–284.

't Hooft, G. (1976a). 'Symmetry breaking through Bell–Jackiw anomalies', *PRL*, **37**: 8–11.

't Hooft, G. (1976b). 'Computation of the quantum effects due to a four-dimensional pseudoparticle', *PR*, **D14**: 3432–3450.

Toll, J. (1952). 'The dispersion relation for light and its application to problems involving electron pairs', unpublished Ph. D. dissertation, Princeton University.

Toll, J. (1956). 'Causality and the dispersion relation: logical foundations', *PR*, **104**: 1760–1770.

Tomonaga, S. (1943). 'On a relativistic reformulation of quantum field theory', *Bull. IPCR (Rikeniko)*, **22**: 545–557. (In Japanese; the English translation appeared in *PTP*, **1**: 1–13).

Tomonaga, S. (1946). 'On a relativistically invariant formulation of the quantum theory of wave fields', *PTP*, **1**: 27–42.

Tomonaga, S. (1948). 'On infinite reactions in quantum field theory', *PR*, **74**: 224–225.

Tomonaga, S. and Koba, Z. (1948). 'On radiation reactions in collision processes. I', *PTP*, **3**: 290–303.

Tomonaga, S. (1966). 'Development of quantum electrodynamics. Personal recollections', *Noble Lectures (Physics): 1963–1970* (Elsevier, Amsterdam/London/New York), 126–136.

Torretti, R. (1983). *Relativity and Geometry* (Pergamon, Oxford).

Toulmin, S. (1961). *Foresight and Understanding* (Hutchinson, London).

Touschek, B. F. (1957). 'The mass of the neutrino and the nonconservation of parity', *NC*, **5**: 1281–1291.

Umezawa, H., Yukawa, J. and Yamada, E. (1948). 'The problem of vacuum polarization', *PTP*, **3**: 317–318.

Umezawa, H. and Kawabe, R. (1949a). 'Some general formulae relating to vacuum polarization', *PTP*, **4**: 423–442.

Umezawa, H. and Kawabe, R. (1949b). 'Vacuum polarization due to various charged particles', *PTP*, **4**: 443–460.

Umezawa, H. and Kamefuchi, R. (1951). 'The vacuum in quantum field theory', *PTP*, **6**: 543–558.

Umezawa, H. (1952). 'On the structure of the interactions of the elementary particles, II', *PTP*, **7**: 551–562.

Umezawa, H. and Kamefuchi, S. (1961). 'Equivalence theorems and renormalization problem in vector field theory (the Yang–Mills field with non-vanishing masses)', *NP*, **23**: 399–429.

Utiyama, R. (1956). 'Invariant theoretical interpretation of interaction', *PR*, **101**: 1597–1607.

Valatin, J. G. (1954a). 'Singularities of electron kernel functions in an external electromagnetic field', *PRS*, **A222**: 93–108.

Valatin, J. G. (1954b). 'On the Dirac–Heisenberg theory of vacuum polarization', *PRS*, **A222**: 228–39.

Valatin, J. G. (1954c). 'On the propagation functions of quantum electrodynamics', *PRS*, **A225**, 535–548.

Valatin, J. G. (1954d). 'On the definition of finite operator quantities in quantum electrodynamics', *PRS*, **A226**: 254–265.

van Dam, H. and Veltman, M. (1970). 'Massive and massless Yang–Mills and gravitational fields', *NP*, **B22**: 397–411.

van der Waerden, B. L. (1967). *Sources of Quantum Mechanics* (North-Holland, Amsterdam).

van Nieuwenhuizen, P. (1981). 'Supergravity', *PL (Rep.)*, **68**: 189–398.

Vaughn, M. T., Aaron, R. and Amado, R. D. (1961). 'Elementary and composite particles', *PR*, **124**: 1258–1268.

Veblen, O. and Hoffmann, D. (1930). 'Projective relativity', *PR*, **36**: 810–822.

Velo, G. and Wightman, A. (eds.) (1973). *Constructive Quantum Field Theory* (Springer-Verlag, Berlin and New York).

Veltman, M. (1963a). 'Higher order corrections to the coherent production of vector bosons in the Coulomb field of a nucleus', *Physica*, **29**: 161–185.

Veltman, M. (1963b). 'Unitarity and causality in a renormalizable field theory with unstable particles', *Physica*, **29**: 186–207.

Veltman, M. (1966). 'Divergence conditions and sum rules', *PRL*, **17**: 553–556.

Veltman, M. (1967). 'Theoretical aspects of high energy neutrino interactions', *PRS*, **A301**: 107–112.

Veltman, M. (1968a). 'Relations between the practical results of current algebra techniques and the originating quark model', Copenhagen Lectures, July 1968, reprinted in *Gauge Theory – Past and Future*, eds. Akhoury, R., DeWitt, B., Van Nieuwenhuizen, P. and Veltman, H. (World Scientific, Singapore, 1992).

Veltman, M. (1968b). 'Perturbation theory of massive Yang–Mills fields', *NP*, **B7**: 637–650.

Veltman, M. (1969). 'Massive Yang-Mills fields', in *Proceedings of Topical Conference on Weak Interactions* (CERN Yellow Report 69-7), 391–393.

Veltman, M. (1970). 'Generalized Ward identities and Yang–Mills fields', *NP*, **B21**: 288–02.

Veltman, M. (1973). 'Gauge field theories (with an appendix "Historical review and bibliography"', in the *Proceedings of the 6th International Symposium on Electron and Photon Interactions at High Energies*, eds. Rollnik, H. and Pheil, W. (North-Holland, Amsterdam).

Veltman, M. (1977). 'Large Higgs mass and μ-e universality', *PL*, **70B**: 253–254.

Veltman, M. (1991). Interview with the author, 9–11 Dec. 1991.

Veltman, M. (1992). 'The path to renormalizability', a talk given at the *Third International Symposium on the History of Particle Physics, SLAC*, 24–27 June 1992.

Waller, I. (1930). 'Bemerküngen über die Rolle der Eigenenergie des Elektrons in der Quantentheorie der Strahlung', *ZP*, **62**: 673–676.

Ward, J. C. (1950). 'An identity in quantum electrodynamics', *PR*, **78**: 182.

Ward, J. C. (1951). 'On the renormalization of quantum electrodynamics', *Proc. Phys. Soc. London*, **A64**: 54–6.

Ward, J. C. and Salam. A. (1959). 'Weak and electromagnetic interaction', *NC*, **11**: 568–577.

Weinberg, S. (1960). 'High energy behavior in quantum field theory', *PR*, **118**: 838–849.

Weinberg, S. (1964a). 'Systematic solution of multiparticle scattering problems', *PR*, **B133**: 232–256.

Weinberg, S. (1964b). 'Derivation of gauge invariance and the equivalence principle from Lorentz invariance of the S-matrix', *PL*, **9**: 357–359.

Weinberg, S. (1964c). 'Photons and gravitons in S-matrix theory: derivation of charge conservation and equality of gravitational and inertial mass', *PR*, **B135**: 1049–1056.

Weinberg, S. (1965a). In *Lectures on Particles and Field Theories*, eds. Deser, S. and Ford, K. W. (Prentice/Englewood, NJ).

Weinberg, S. (1965b). 'Photons and gravitons in perturbation theory: derivation of Maxwell's and Einstein's equations', *PR*, **B138**: 988–1002.

Weinberg, S. (1967a). 'Dynamical approach to current algebra', *PRL*, **18**: 188–191.

Weinberg, S. (1967b). 'A model of leptons', *PRL*, **19**: 1264–66.

Weinberg, S. (1968). 'Nonlinear realizations of chiral symmetry', *PR*, **166**: 1568–1577.

Weinberg, S. (1977). 'The search for unity: notes for a history of quantum field theory', *Daedalus*, **106**: 17–35.

Weinberg, S. (1978). 'Critical phenomena for field theorists', in *Understanding the Fundamental Constituents of Matter*, ed. Zichichi, A. (Plenum, New York), 1–52.

Weinberg, S. (1979). 'Phenomenological Lagrangian', *Physica*, **96A**: 327–340.

Weinberg, S. (1980a). 'Conceptual foundations of the unified theory of weak and electromagnetic interactions', *RMP*, **52**: 515–523.

Weinberg, S. (1980b). 'Effective gauge theories', *PL*, **91B**: 51–55.

Weinberg, S. (1983). 'Why the renormalization group is a good thing', in *Asymptotic Realms of Physics: Essays in Honor of Francis. E. Low*, eds. Guth, A. H.; Huang, K. and Jaffe, R. L. (MIT Press, Cambridge, MA), 1–19.

Weinberg, S. (1985). 'The ultimate structure of matter', in *A Passion for Physics: Essays in Honor of Jeffrey Chew*, eds. De Tar, C., Finkelstein, J. and Tan, C. I. (Taylor & Francis, Philadelphia), 114–127.

Weinberg, S. (1986a). 'Particle physics: past and future', *Int. J. Mod. Phys.* **A1/1**: 135–145.

Weinberg, S. (1986b). 'Particles, fields, and now strings', in *The Lesson of Quantum Theory*, eds. De Boer, J., Dal, E. and Ulfbeck, O. (Elsevier, Amsterdam), 227–239.

Weinberg, S. (1987). 'Newtonianism, reductionism and the art of congressional testimony', in *Three Hundred Years of Gravitation*, eds. Hawking, S. W. and Israel, W. (Cambridge, University Press, Cambridge).

Weinberg, S. (1992). *Dreams of a Final Theory* (Pantheon).

Weinberg, S. (1995). *The Quantum Theory of Fields* (Cambridge University Press, Cambridge).

Weisberger, W. I. (1965) 'Renormalization of the weak axial-vector coupling constant', *PRL*, **14**: 1047–1055.

Weisberger, W. I. (1966). 'Unsubstracted dispersion relations and the renormalization of the weak axial-vector coupling constants', *PR*, **143**: 1302–1309.

Weisskopf, V. F. (1934). 'Über die Selbstenergie des Elektrons', *ZP*, **89**: 27–39.

Weisskopf, V. F. (1936). 'Über die Elektrodynamic des Vakuums auf Grund der Quantentheorie des Elektrons', *K. Danske Vidensk. Selsk., Mat.-Fys. Medd.*, **14**: 1–39.

Weisskopf, V. F. (1939). 'On the self-energy and the electromagnetic field of the electron', *PR*, **56**: 72–85.

Weisskopf, V. F. (1949). 'Recent developments in the theory of the electron', *RMP*, **21**: 305–328.

Weisskopf, V. F. (1972). *Physics in the Twentieth Century* (MIT Press, Cambridge, MA).

Weisskopf, V. F. (1983). 'Growing up with field theory: the development of quantum electrodynamics', in *The Birth of Particle Physics*, eds. Brown, L. M. and Hoddeson, L. (Cambridge University Press, Cambridge), 56–81.

Welton, T. A. (1948). 'Some observable effects of the quantum-mechanical fluctuations of the electromagnetic field', *PR*, **74**: 1157–1167.

Wentzel, G. (1933a, b; 1934). 'Über die Eigenkräfte der Elementarteilchen. I, II and III', *ZP*, **86**: 479–494, 635–645; **87**: 726–733.

Wentzel, G. (1943). *Einführung in der Quantentheorie der Wellenfelder* (Franz Deuticke, Vienna); English translation: *Quantum Theory of Fields*, (Interscience, New York, 1949).

Wentzel, G. (1956). 'Discussion remark', in *Proceedings of 6th Rochester Conference on High Energy Nuclear Physics* (New York), VIII-15 to VIII-17.

Wentzel, G. (1958). 'Meissner effect', *PR*, **112**: 1488–1492.

Wentzel, G. (1959). 'Problem of gauge invariance in the theory of the Meissner effect', *PRL*, **2**: 33–34.

Wentzel, G. (1960). 'Quantum theory of fields (until 1947)', in *Theoretical Physics in the Twentieth Century: A Memorial Volume to Wolfgang Pauli* (Interscience, New York), 44–67.

Wess, J. (1960). 'The conformal invariance in quantum field theory', *NC*, **13**: 1086.

Wess, J. and Zumino, B. (1971). 'Consequences of anomalous Ward identities', *PL*, **37B**: 95–97.

Wess, J. and Zumino, B. (1974). 'Supergauge transformations in four dimensions', *NP*, **B70**: 39–50.

Weyl, H. (1918a). *Raum-Zeit-Materie* (Springer, Berlin).

Weyl, H. (1918b). 'Gravitation und Elektrizität', *Preussische Akad. Wiss. Sitzungsber.*, 465–478.

Weyl, H. (1921). 'Electricity and gravitation', *Nature*, **106**: 800–802.

Weyl, H. (1922). 'Gravitation and electricity', in *The Principle of Relativity* (Methuen, 1923), 201–216.

Weyl, H. (1929). 'Elektron und Gravitation. I', *ZP*, **56**: 330–352.

Weyl, H. (1930). *Gruppentheorie und Quantenmechanik* (translated by Robertson as *The Theory of Groups and Quantum Mechanics* (E. P. Dutton, New York, 1931).

Wheeler, J. (1962). *Geometrodynamics* (Academic, New York).

Wheeler, J. (1964a). 'Mach's principle as a boundary condition for Einstein's equations', in *Gravitation and Relativity*, eds. Chiu, H.-Y. and Hoffman, W. F. (Benjamin, New York), 303–349.

Wheeler, J. (1964b). 'Geometrodynamics and the issue of the final state', in *Relativity, Group and Topology*, eds. DeWitt, C. and DeWitt, B. (Gordon and Breach, New York), 315.

Wheeler, J. (1973). 'From relativity to mutability', in *The Physicist's Conception of Nature*, ed. Mehra, J. (Reidel, Dordrecht), 202–247.

Whewell, W. (1847). *The Philosophy of the Inductive Sciences*, second edition (Parker & Son, London).

Whittaker, E. (1951). *A History of the Theories of Aether and Electricity* (Nelson, London).

Widom, B. (1965a). 'Surface tension and molecular correlations near the critical point', *J. Chem. Phys.*, **43**: 3892–3897.

Widom, B. (1965b). 'Equation of state in the neighborhood of the critical point', *J. Chem. Phys.*, **43**: 3898–3905.

Wien, W. (1909). *Encykl. der Math. Wiss* (Leipzig), vol. 3: 356.

Wightman, A. S. (1956). 'Quantum field theories in terms of expectation values', *PR*, **101**: 860–866.

Wightman, A. S. (1976). 'Hilbert's sixth problem: mathematical treatment of the axioms of physics', in *Mathematical Developments Arising from Hilbert Problems*, ed. Browder, F. E. (American Mathematical Society, Providence), 147–240.

Wightman, A. S. (1978). 'Field theory, Axiomatic', in the *Encyclopedia of Physics* (McGraw-Hill, New York), 318–321.

Wightman, A. S. (1986). 'Some lessons of renormalization theory', in *The Lesson of Quantum Theory*, eds. de Boer, J., Dal, E. and Ulfbeck, D. (Elsevier, Amsterdam), 201–225.

Wightman, A. S. (1989). 'The general theory of quantized fields in the 1950s', in *Pions to Quarks*, eds. Brown, L. M., Dresden, M. and Hoddeson, L. (Cambridge University Press, Cambridge), 608–629.

Wightman, A. S. (1992). Interview.

Wigner, E. P. and Witmer, E. E. (1928). 'Über die Struktur der zweiatomigen Molekulspektren nach der Quantenmechanik', *ZP*, **51**: 859–886.

Wigner, E. P. (1931). *Gruppentheorie und ihre Anwendung auf die Quantenmechanik der Atomspektren* (Braunschweig).

Wigner, E. (1937). 'On the consequences of the symmetry of the nuclear Hamiltonian on the spectroscopy of nuclei', *PR*, **51**: 106–119.

Wigner, E. (1949). 'Invariance in physical theory', in *Proc. Amer. Philos. Soc.*, **93**: 521–526.

Wigner, E. (1960). 'The unreasonable effectiveness of mathematics in the natural sciences', *Commun. Pure. Appl. Math.* **13**: 1–14.

Wigner, E. (1963). Interview with Wigner on 4 Dec 1963; Archive for the History of Quantum Physics.

Wigner, E. (1964a). 'Symmetry and conservation laws', in *Proc. Nat. Acad. Sci.*, **5**: 956–965.

Wigner, E. (1964b). 'The role of invariance principles in natural philosophy', in *Proceedings of the Enrico Fermi International School of Physics* (XXIX), ed. Wigner, E. (New York), ix-xvi.

Wilson, K. G. (1965). 'Model Hamiltonians for local quantum field theory', *PR*, **B140**: 445–457.

Wilson, K. G. (1969). 'Non-Lagrangian models of current algebra', *PR*, **179**: 1499–1512.

Wilson, K. G. (1970a). 'Model of coupling-constant renormalization', *PR*, **D2**: 1438–1472.

Wilson, K. G. (1970b). 'Operator-product expansions and anomalous dimensions in the Thirring model', *PR*, **D2**: 1473–1477.

Wilson, K. G. (1970c). 'Anomalous dimensions and the breakdown of scale invariance in perturbation theory', *PR*, **D2**: 1478–1493.

Wilson, K. G. (1971a). 'The renormalization group and strong interactions', *PR*, **D3**: 1818–1846.

Wilson, K. G. (1971b). 'Renormalization group and critical phenomena. I. Renormalization group and the Kadanof scaling picture', *PR*, **B4**: 3174–3183.

Wilson, K. G. (1971c). 'Renormalization group and critical phenomena. II. Phase-space cell analysis of critical behavior', *PR*, **B4**: 3184–3205.

Wilson, K. G. (1972). 'Renormalization of a scalar field in strong coupling', *PR*, **D6**: 419–426.

Wilson, K. G. and Fisher, M. E. (1972). 'Critical exponents in 3.99 dimensions', *PRL*, **28**: 240–243.

Wilson, K. G. (1974). 'Confinement of quarks', *PR*, **D10**: 2445–2459.

Wilson, K. G. and Kogut, J. (1974). 'The renormalization group and the ϵ expansion', *PL (Rep.)*, **12C**: 75–199.

Wilson, K. G. (1975). 'The renormalization group: critical phenomena and the Kondo problem', *RMP*, **47**: 773–840.

Wilson, K. G. (1979). 'Problems in physics with many scales of length', *Scientific American*, **241**(2): 140–157.

Wilson, K. G. (1983). 'The renormalization group and critical phenomena', *RMP*, **55**: 583–600.

Wise, N. (1981). 'German concepts of force, energy and the electromagnetic ether: 1845–1880', in Contor and Hodge (1981), 269–308.

Wu, T. T. and Yang, C. N. (1967). 'Some solutions of the classical isotopic gauge field equations', in *Properties of Matter under Unusual Conditions*, eds. Mark, H. and Fernbach, S. (Wiley, New York), 349–354.

Wu, T. T. and Yang, C. N. (1975). 'Concept of nonintegrable phase factors and global formulation of gauge fields', *PR*, **D12**: 3845–3857.

Yang, C. N. (1970). 'Charge quantization of the gauge group, and flux quantization', *PR*, **D1**: 8.

Yang, C. N. (1974). 'Integral formalism for gauge fields', *PRL*, **33**: 445–447.

Yang, C. N. (1977). 'Magnetic monopoles, fiber bundles, and gauge fields', *Ann. N. Y. Acad. Sci.*, **294**: 86–97.

Yang, C. N. (1980). 'Einstein's impact on theoretical physics', *Physics Today*, **33**(6): 42–49.

Yang, C. N. (1983). *Selected Papers (1945–1980)* (Freeman, New York).

Yang, C. N. and Mills, R. L. (1954a). 'Isotopic spin conservation and a generalized gauge invariance', *PR*, **95**: 631.

Yang, C. N. and Mills, R. L. (1954b). 'Conservation of isotopic spin and isotopic gauge invariance', *PR*, **96**: 191–195.

Yates, F. (1964). *Giordano Bruno and the Hermetic Tradition* (Routledge and Kegan Paul, London).

Yoshida, R. (1977). *Reduction in the Physical Sciences*. (Dalhousie University Press, Halifax).

Yoshida, R. (1981). 'Reduction as replacement', *Brit. J. Philos. Sci.*, **32**: 400.

Yukawa, H. (1935). 'On the interaction of elementary particles', *Proc. Phys.-Math. Soc. Japan*, **17**: 48–57.

Yukawa, H. and Sakata, S. (1935a). 'On the theory of internal pair production', *Proc. Phys.-Math. Soc. Japan*, **17**: 397–407.

Yukawa, H. and Sakata, S. (1935b). 'On the theory of the β-disintegration and the allied phenomenon', *Proc. Phys.-Math. Soc. Japan*, **17**: 467–479.

Zachariasen, F. and Zemach, C. (1962). 'Pion resonances', *PR*, **128**: 849–858.

Zahar, E. (1982). 'Poincaré et la découverte du Principe de Relativité', LSE-Preprint, 6 Feb. 1982.

Zilsel, E. (1942). 'The sociological roots of science', *Amer. J. Sociology*, **47**: 544–562.

Zimmermann, W. (1958). 'On the bound state problem in quantum field theory', *NC*, **10**: 597–614.

Zimmermann, W. (1967). 'Local field equation for A^4-coupling in renormalized perturbation theory', *CMP*, **6**: 161–188.

Zumino, B. (1975). 'Supersymmetry and the vacuum', *NP*, **B89**: 535–546.

Zweig, G. (1964a). 'An SU_3 model for strong interaction symmetry and its breaking: I and II', CERN preprint TH401 (17 Jan. 1964) and TH 412 (21 Feb. 1964).

Zweig, G. (1964b). 'Fractionally charged particles and SU(6)', in *Symmetries in Elementary Particle Physics*, ed. Zichichi, A. (Academic Press, New York), 192–243.

Zwicky, F. (1935). 'Stellar guests', *Scientific Monthly*, **40**: 461.

Zwicky, F. (1939). 'On the theory and observation of highly collapsed stars', *PR*, **55**: 726–743.

Name index

420

Subject index

426